Methods in Molecular Biology

Series Editor
John M. Walker
School of Life and Medical Sciences
University of Hertfordshire
Hatfield, Hertfordshire, UK

For further volumes:
http://www.springer.com/series/7651

For over 35 years, biological scientists have come to rely on the research protocols and methodologies in the critically acclaimed *Methods in Molecular Biology* series. The series was the first to introduce the step-by-step protocols approach that has become the standard in all biomedical protocol publishing. Each protocol is provided in readily-reproducible step-by-step fashion, opening with an introductory overview, a list of the materials and reagents needed to complete . the experiment, and followed by a detailed procedure that is supported with a helpful notes section offering tips and tricks of the trade as well as troubleshooting advice. These hallmark features were introduced by series editor Dr. John Walker and constitute the key ingredient in ,each and every volume of the *Methods in Molecular Biology* series. Tested and trusted, comprehensive and reliable, all protocols from the series are indexed in PubMed.

Phytochromes

Methods and Protocols

Edited by

Andreas Winkler

Institute of Biochemistry, Graz University of Technology, Graz, Austria; BioTechMed Graz, Graz, Austria

Humana Press

Editor
Andreas Winkler
Institute of Biochemistry
Graz University of Technology
Graz, Austria

BioTechMed Graz
Graz, Austria

ISSN 1064-3745 ISSN 1940-6029 (electronic)
Methods in Molecular Biology
ISBN 978-1-0716-4790-5 ISBN 978-1-0716-4791-2 (eBook)
https://doi.org/10.1007/978-1-0716-4791-2

This Humana imprint is published by the registered company Springer Science+Business Media, LLC, part of Springer Nature.
The registered company address is: 1 New York Plaza, New York, NY 10004, U.S.A.

Preface

Phytochrome photoreceptors have intrigued scientists ever since their initial discovery in plants and their identification as key players in the control of processes like seed germination or flowering. More recently, related red light photoreceptors have also been identified in other kingdoms of life—bacteria, fungi, and algae—highlighting the importance of light sensing for many organisms. While impressive progress has been made in understanding phytochrome structure and function over the past decades, different systems appear to have reused the core concept of red-light perception for different physiological outputs. To this end, understanding the subtleties of signal integration and allosteric regulation of various effector modules or interaction partners will be key to appreciating common aspects of phytochrome function and differentiating them from system-specific tuning mechanisms. Scientists interested in this field of research will benefit from the collection of chapters in this edition on phytochromes that combines valuable information ranging from phytochrome identification and analysis, via their expression and characterization, all the way to the appreciation of their structure-function relationship and eventually also their application as optogenetic tools.

This volume is divided into four parts: the first six chapters revolve around phytochrome discovery, the analysis of sequence diversity, and phytochrome expression. Due to the need for specific cofactors involved in light perception in all phytochrome systems, the latter part alone is covered in four chapters that outline different approaches to obtain the holoproteins of interest.

In the next part on biophysical and computational characterization of phytochromes, five chapters revolve around different spectroscopic techniques to address characteristic cofactor properties and also on computational approaches to study phytochrome properties on different timescales. The chapters describing the latter approach also provide a nice transition to the structural analysis of the photoreceptors of interest in the next part of this volume.

In total, five chapters cover the breadth of structural analyses describing mostly complementary techniques to the classical approach using x-ray crystallography. Due to size and conformational heterogeneity of phytochromes, techniques such as cryo-EM, solution scattering, solid-state NMR, and hydrogen deuterium exchange coupled to mass spectrometry (HDX-MS) have recently provided valuable insights. Protocols on how to perform these experiments effectively in the context of phytochrome research are provided.

In the last part, six chapters demonstrate how to analyze phytochrome function in vitro and in vivo, but also how to engineer phytochromes to be used in the recently developed field of optogenetics. In summary, the current volume on phytochromes provides the reader with a useful and versatile laboratory bench and computational resource that is an essential reference for those interested in going in the direction of phytochrome research, but also more established researches that are keen to broaden their horizon and to integrate different methodologies to better understand the intricate molecular details of their systems of interest.

Graz, Austria *Andreas Winkler*

Contents

Part III Structural Analysis Approaches

Part IV Addressing Phytochrome Functions and Applications

Contributors

ELMAR BEHRMANN • *Institute of Biochemistry, Department of Chemistry and Biochemistry, University of Cologne, Cologne, Germany*

CORNELIA BÖHM • *Institute of Biochemistry, Graz University of Technology, Graz, Austria*

HERNÁN R. BONOMI • *Generate Biomedicines, Somerville, MA, USA*

CHENG CHI • *Peking University Institute of Advanced Agricultural Sciences, Shandong Laboratory of Advanced Agriculture Sciences at Weifang, Weifang, Shandong, China*

ROLF DILLER • *Department of Physics, Biophysics, RPTU University Kaiserslautern-Landau, Kaiserslautern, Germany*

OLIVER MAXIMILIAN EDER • *Institute of Biochemistry, Graz University of Technology, Graz, Austria*

ANNA K. EHRET • *Signaling Research Centers BIOSS and CIBSS, Freiburg, Germany; Department of Immunology, Faculty of Biology, University of Freiburg, Freiburg, Germany; Centre for Chronic Immunodeficiency (CCI), Faculty of Medicine, University of Freiburg, Freiburg, Germany; Spemann Graduate School of Biology and Medicine (SGBM), University of Freiburg, Freiburg, Germany*

LISA ENTRINGER • *Signaling Research Centers BIOSS and CIBSS, Freiburg, Germany; Department of Immunology, Faculty of Biology, University of Freiburg, Freiburg, Germany*

ZIYI FENG • *State Key Laboratory of Plant Environmental Resilience, College of Biological Sciences, China Agricultural University, Beijing, China*

ANDREA FLORES-IBARRA • *Institute of Biochemistry, Department of Chemistry and Biochemistry, University of Cologne, Cologne, Germany*

NICOLE FRANKENBERG-DINKEL • *Department of Biology, Microbiology, RPTU University Kaiserslautern-Landau, Kaiserslautern, Germany*

ANNE SMEDEGAARD FREDERIKSEN • *Signaling Research Centers BIOSS and CIBSS, Freiburg, Germany; Department of Immunology, Faculty of Biology, University of Freiburg, Freiburg, Germany*

MAXIMILIAN FUCHS • *Institute of Biochemistry, Graz University of Technology, Graz, Austria*

KEIJI FUSHIMI • *Department of Biological Sciences, Graduate School of Science, Tokyo Metropolitan University, Hachioji, Tokyo, Japan; Department of Biological Chemistry, Chubu University, Kasugai, Aichi, Japan*

WOLFGANG GÄRTNER • *Institut für Analytische Chemie, Universität Leipzig, Leipzig, Germany*

KAOLING GUAN • *Department of Plant Physiology, Justus Liebig University Giessen, Giessen, Germany; Department of Molecular Biosciences, University of Texas at Austin, Austin, TX, USA*

MONIKA GUNKEL • *Institute of Biochemistry, Department of Chemistry and Biochemistry, University of Cologne, Cologne, Germany*

SARA HARTMANN • *Signaling Research Centers BIOSS and CIBSS, Freiburg, Germany; Department of Immunology, Faculty of Biology, University of Freiburg, Freiburg, Germany; Centre for Chronic Immunodeficiency (CCI), Faculty of Medicine, University of Freiburg, Freiburg, Germany; Spemann Graduate School of Biology and Medicine (SGBM), University of Freiburg, Freiburg, Germany*

HIROKI HOSHINO • *Department of Biological Sciences, Graduate School of Science, Tokyo Metropolitan University, Hachioji, Tokyo, Japan*
CHRISTINA HUBER • *Department of Biology, Microbiology, RPTU University Kaiserslautern-Landau, Kaiserslautern, Germany*
JON HUGHES • *Department of Plant Physiology, Justus Liebig University, Giessen, Germany*
VINCENT IDSTEIN • *Signaling Research Centers BIOSS and CIBSS, Freiburg, Germany; Department of Immunology, Faculty of Biology, University of Freiburg, Freiburg, Germany; Centre for Chronic Immunodeficiency (CCI), Faculty of Medicine, University of Freiburg, Freiburg, Germany; Spemann Graduate School of Biology and Medicine (SGBM), University of Freiburg, Freiburg, Germany*
JANNE A. IHALAINEN • *Nanoscience Centre, Department of Biological and Environmental Sciences, University of Jyväskylä, Jyväskylä, Finland*
MARTA JANČIĆ • *Institute of Biochemistry, Graz University of Technology, Graz, Austria*
MARIA KEHR • *Department of Biochemistry, University of Bayreuth, Bayreuth, Germany*
MOONA KURTTILA • *Nanoscience Centre, Department of Biological and Environmental Sciences, University of Jyväskylä, Jyväskylä, Finland*
CHRISTINA LANG • *Department of Plant Physiology, Justus Liebig University, Giessen, Germany*
JIGANG LI • *State Key Laboratory of Plant Environmental Resilience, College of Biological Sciences, China Agricultural University, Beijing, China*
MEGAN J. MACKINTOSH • *Institute of Chemistry, The Hebrew University of Jerusalem, Jerusalem, Israel*
TEK NARSINGH MALLA • *Department of Physics, University of Wisconsin-Milwaukee, Milwaukee, WI, USA; Vaccine Research Center Electron Microscopy Unit, Frederick National Laboratory for Cancer Research, Frederick, MD, USA*
JÖRG MATYSIK • *Institut für Analytische Chemie, Universität Leipzig, Leipzig, Germany*
STEFANIE S. M. MEIER • *Department of Biochemistry, University of Bayreuth, Bayreuth, Germany*
KEITA MIYAKE • *Graduate School of Arts and Sciences, University of Tokyo, Meguro, Tokyo, Japan*
ANDREAS MÖGLICH • *Department of Biochemistry, University of Bayreuth, Bayreuth, Germany; Bayreuth Center for Biochemistry & Molecular Biology, Universität Bayreuth, Bayreuth, Germany; North-Bavarian NMR Center, Universität Bayreuth, Bayreuth, Germany*
PROBAL NAG • *Department of Physics, Technical University Dortmund, Dortmund, Germany; Institute of Chemistry, The Hebrew University of Jerusalem, Jerusalem, Israel*
AKIRA NAGATANI • *Graduate School of Science, Kyoto University, Kyoto, Japan*
MASAYOSHI NAKASAKO • *Department of Physics, Faculty of Science and Technology, Keio University, Yokohama, Kanagawa, Japan; RIKEN SPring-8 Center, Hyogo, Japan*
REI NARIKAWA • *Department of Biological Sciences, Graduate School of Science, Tokyo Metropolitan University, Hachioji, Tokyo, Japan*
MAO OIDE • *Institute for Protein Research, Osaka University, Osaka, Japan; PRESTO, Japan Science and Technology Agency, Tokyo, Japan*
LISANDRO H. OTERO • *Departamento de Biología Molecular, Facultad de Ciencias Exactas, Físico-Químicas y Naturales, Instituto de Biotecnología Ambiental y Salud, CONICET, Universidad Nacional de Río Cuarto, Córdoba, Argentina; Centro de Re-diseño e Ingeniería de Proteínas, Universidad Nacional de San Martín, Campus Miguelete, Buenos Aires, Argentina*

JIMENA RINALDI • *Instituto de Nanosistemas, Universidad Nacional de San Martín, Campus Miguelete, Buenos Aires, Argentina*

VALENTIN ROHR • *Institut für Analytische Chemie, Universität Leipzig, Leipzig, Germany*

MARISSA RUSS • *Signaling Research Centers BIOSS and CIBSS, Freiburg, Germany; Department of Immunology, Faculty of Biology, University of Freiburg, Freiburg, Germany*

GIACOMO SALVADORI • *Institute for Computational Biomedicine (INM-9), Forschungszentrum Jülich, Jülich, Germany*

WOLFGANG W. SCHAMEL • *Signaling Research Centers BIOSS and CIBSS, Freiburg, Germany; Department of Immunology, Faculty of Biology, University of Freiburg, Freiburg, Germany; Centre for Chronic Immunodeficiency (CCI), Faculty of Medicine, University of Freiburg, Freiburg, Germany; Spemann Graduate School of Biology and Medicine (SGBM), University of Freiburg, Freiburg, Germany*

IGOR SCHAPIRO • *Department of Physics, Technical University Dortmund, Dortmund, Germany; Institute of Chemistry, The Hebrew University of Jerusalem, Jerusalem, Israel; Research Center Chemical Sciences and Sustainability, University Alliance Ruhr, Bochum, Germany*

MARIUS SCHMIDT • *Department of Physics, University of Wisconsin-Milwaukee, Milwaukee, WI, USA*

SINTAYEHU MANAYE SHENKUTIE • *Department of Plant Physiology, Justus Liebig University Giessen, Giessen, Germany; Department of Chemistry, Hawassa University, Hawassa, Ethiopia*

CHEN SONG • *Institut für Analytische Chemie, Universität Leipzig, Leipzig, Germany*

EMINA A. STOJKOVIĆ • *Department of Biology, Northeastern Illinois University, Chicago, IL, USA*

MERLE STRACK • *Department of Physics, Biophysics, RPTU University Kaiserslautern-Landau, Kaiserslautern, Germany*

TAKAHISA SUZUKI • *Department of Biological Sciences, Graduate School of Science, Tokyo Metropolitan University, Hachioji, Tokyo, Japan; School of Life Science and Technology, Institute of Science Tokyo, Yokohama, Kanagawa, Japan*

HEIKKI TAKALA • *Department of Biological and Environmental Science, Nanoscience Center, University of Jyvaskyla, Jyvaskyla, Finland*

MASSIMO TOTARO • *Institute of Biochemistry, Graz University of Technology, Graz, Austria*

IIDA TUURE • *Department of Biological and Environmental Science, Nanoscience Center, University of Jyvaskyla, Jyvaskyla, Finland*

ILSE VILLMANN • *Justus-Liebig-University Giessen, Institute for Plant Physiology, Giessen, Germany*

JIZONG WANG • *Peking University Institute of Advanced Agricultural Sciences, Shandong Laboratory of Advanced Agriculture Sciences at Weifang, Weifang, Shandong, China; State Key Laboratory of Gene Function and Modulation Research, School of Advanced Agricultural Sciences, Peking University, Beijing, China*

RUIHAN WANG • *Peking University Institute of Advanced Agricultural Sciences, Shandong Laboratory of Advanced Agriculture Sciences at Weifang, Weifang, Shandong, China; State Key Laboratory of Gene Function and Modulation Research, School of Advanced Agricultural Sciences, Peking University, Beijing, China*

ZHENGDONG WANG • *Peking University Institute of Advanced Agricultural Sciences, Shandong Laboratory of Advanced Agriculture Sciences at Weifang, Weifang, Shandong, China; State Key Laboratory of Gene Function and Modulation Research, School of Advanced Agricultural Sciences, Peking University, Beijing, China; Peking-Tsinghua Joint Center for Life Sciences, Peking University, Beijing, China*

CHONGYAO WEI • *Department of Physics, Technical University Dortmund, Dortmund, Germany; Institute of Chemistry, The Hebrew University of Jerusalem, Jerusalem, Israel*
ANDREAS WINKLER • *Institute of Biochemistry, Graz University of Technology, Graz, Austria; BioTechMed Graz, Graz, Austria*
CHENGWEI YI • *Department of Biochemistry, University of Bayreuth, Bayreuth, Germany*
TOMO YOKOTA • *Department of Biological Sciences, Graduate School of Science, Tokyo Metropolitan University, Hachioji, Tokyo, Japan*
AYANO YONEDA • *Department of Biological Sciences, Graduate School of Science, Tokyo Metropolitan University, Hachioji, Tokyo, Japan*
O. SASCHA YOUSEFI • *Signaling Research Centers BIOSS and CIBSS, Freiburg, Germany; Department of Immunology, Faculty of Biology, University of Freiburg, Freiburg, Germany; Centre for Chronic Immunodeficiency (CCI), Faculty of Medicine, University of Freiburg, Freiburg, Germany*
MATHIAS ZEIDLER • *Justus-Liebig-University Giessen, Institute for Plant Physiology, Giessen, Germany*

Part I

Phytochrome Discovery, Analysis and Expression

Chapter 1

Discovery and Engineering of Unique Cyanobacteriochrome GAF Domains

Takahisa Suzuki, Ayano Yoneda, Tomo Yokota, Hiroki Hoshino, Keita Miyake, Keiji Fushimi, and Rei Narikawa

Abstract

The linear tetrapyrrole-binding cyanobacteriochrome (CBCR) photoreceptors have been extensively analyzed so far and found to be highly diverse in spectral properties. Here, we describe the methods to search novel CBCR molecules from the database, perform site-saturation random mutagenesis and colony-based protein expression, and express the CBCR molecules with various chromophore synthetic systems. These methods will contribute to further discovery and engineering of the CBCR molecules.

Key words Bilin, PHI-BLAST, Inverse PCR, Mixed bases, Phytoviolobilin

1 Introduction

Cyanobacterial photoreceptors, cyanobacteriochromes (CBCRs), are linear tetrapyrrole-binding photoreceptors distantly related to the canonical phytochrome photoreceptors [1, 2]. Only the cGMP phosphodiesterase/adenylate cyclase/FhlA (GAF) domain is needed for chromophore incorporation and photoreception. The CBCR GAF domains covalently bind the linear tetrapyrrole chromophore and show reversible photoconversion between the two light-absorbing states triggered by the primary photoreaction of the C15=C16 *Z/E* isomerization of the chromophore. Extensive characterization of CBCR GAF domains has demonstrated a remarkable diversity in their spectral sensitivities, encompassing a broad range of wavelengths from the ultraviolet to the far-red region. Spectral and structural studies identified that the binding chromophore species and various spectral tuning mechanisms established such diversity.

Andreas Winkler (ed.), *Phytochromes: Methods and Protocols*, Methods in Molecular Biology, vol. 2970,
https://doi.org/10.1007/978-1-0716-4791-2_1,

In the early phase of the CBCR study, two approaches identifying novel CBCR GAF domains were dominant: One is to analyze all the CBCR GAF domains detected from a particular cyanobacterium [3–10], and another is to analyze the CBCR GAF domains belonging to the specific clade of the phylogenetic tree [11–13]. Nowadays, extremely abundant genomic and metagenomic sequences provide too many candidate molecules to identify novel CBCR GAF domains. Since too many molecules with almost identical sequence were hit by simple homology search, efficient screening methods to exclude such molecules and select target molecules are needed. In Subheading 3.1, we introduce a pattern-hit initiated (PHI)-basic local algorithm search tool (BLAST)-based method for identification of interesting targets and showcase data from our previous study [14].

Based on the multiple alignment and structural information, site-directed mutagenesis analyses have largely contributed to revealing and modifying the spectral tuning mechanisms of the CBCR GAF domains [15–17]. Furthermore, we have recently established site-saturation random mutagenesis and colony-based protein expression [18, 19]. In Subheading 3.2, we introduce this method with a model case using AM1_1499g1 [18–21].

Since the normal *Escherichia coli* cells do not produce the linear tetrapyrrole chromophores, we need to introduce chromophore synthetic enzymes into the cells to obtain chromophore-binding holo-CBCR GAF domains. In this context, various coexpression systems are available [20–23]. Most CBCR GAF domains covalently bind phycocyanobilin (PCB) [24, 25], and some of them exhibit isomerization activity from PCB to phycoviolobilin (PVB) having no double bond between Rings A and B and absorbing shorter wavelength green light [5, 14, 26] (Fig. 1). Phytochromobilin (PΦB), chromophore for the plant phytochromes, has the same structure as PCB at Ring A where the conserved canonical Cys residue within the GAF domain covalently binds (Fig. 1). The PCB-binding molecules can also bind PΦB and vice versa. The binding chromophore species depends on the chromophore synthetic enzymes that the corresponding organisms possess. PCB is synthesized from biliverdin (BV) by the two-step reduction catalyzed by the cyanobacterial PcyA enzyme (Fig. 1). In contrast, PΦB is synthesized from BV by the one-step reduction catalyzed by the plant HY2 enzyme (Fig. 1). Furthermore, some CBCR GAF domains have been reported to bind BV [8, 27] (*see* **Note 1**). In Subheading 3.3, we introduce heterologous expression of the CBCR GAF domains with various chromophore synthetic systems using AnPixJg2_BV4 and AM1_6305g1 as the model cases [21].

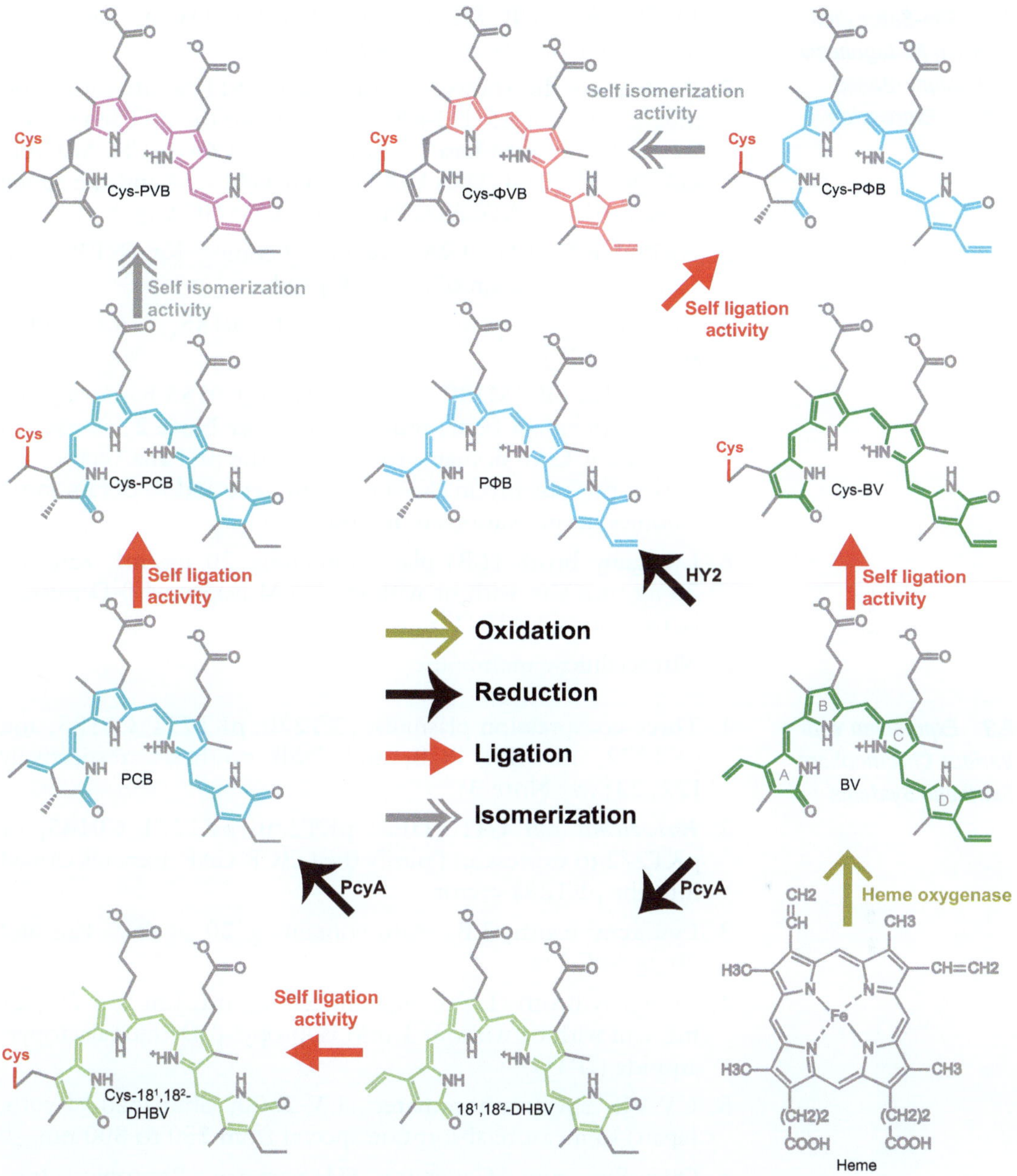

Fig. 1 Synthetic pathway and chemical structures of the linear tetrapyrrole chromophores

2 Materials

2.1 Identification of Novel CBCR GAF Domains

1. URL for the BLASTP search: https://blast.ncbi.nlm.nih.gov/Blast.cgi?PAGE=Proteins.

2.2 Site-Saturation Random Mutagenesis and Colony-Based Protein Expression

1. KOD One PCR Master Mix (Toyobo, Osaka, Japan) for inverse polymerase chain reaction (PCR).
2. Primers for the showcase to introduce S118X site-saturation mutagenesis on AM1_1499g1 cloned into the pET28a vector: forward mixed-base primers (CGGGTTCAAGAT-GATNNSTTTGGTGAAAATTATGCGACCA) and the single reverse primer (ATCATCTTGAACCCGTGCA).
3. NEBuilder® HiFi DNA Assembly Cloning Kit (NEB, MA, USA) for cyclization of the PCR product.
4. Coexpression plasmid, pKT271_C0185, for PCB synthesis [23].
5. *Escherichia coli* JM109 (DE3) pKT271_C0185 harboring the chloramphenicol (Cm)-resistant gene (*see* **Note 2**) to express the CBCR GAF domains cloned into the pET28a vector harboring the kanamycin (Km)-resistant gene and to obtain more colonies having various mutations.
6. Lysogeny broth (LB) plate containing 20 μg/mL Km and 20 μg/mL Cm with or without 1 mM isopropyl-β-D-thiogalactopyranoside (IPTG).
7. Nitrocellulose membrane.

2.3 Expression with Various Chromophore Synthetic Systems

1. Three coexpression plasmids, pKT270, pKT271_C0185, and pKT272, for BV, PCB, and PΦB synthesis, respectively [22, 23] (*see* **Note 3**).
2. *Escherichia coli* C41 (DE3) pKT270, pKT271_C0185, or pKT272 to express and purify the CBCR GAF domains cloned into the pET28a vector.
3. Lysogeny broth (LB) plate containing 20 μg/mL Km and 20 μg/mL Cm.
4. Lysogeny broth (LB) containing 20 μg/mL Km and 20 μg/mL Cm with or without 1 mM isopropyl-β-D-thiogalactopyranoside (IPTG).
5. UV-VIS spectrophotometer (UV-2600; Shimadzu, Kyoto, Japan) to measure absorption spectra from 250 to 800 nm.
6. Opto-Spectrum Generator (Hamamatsu Photonics Inc., Hamamatsu, Japan) for monochromic light sources (*see* **Note 4**).
7. 8.75 M guanidine/1.25% (v/v) HCl for acid denaturation of the CBCR GAF domains (*see* **Note 5**).

3 Methods

3.1 Identification of Novel CBCR GAF Domains

The PHI-BLAST, which is one option of the BLASTP search, can search sequences having a pattern we specify and homologous to the query sequence. Our previous study has speculated that a Cys residue near the chromophore would have a unique role in spectral tuning and searched target molecules using the typical extended red/green (XRG) CBCR GAF domain, AnPixJg2, with a modification as a query sequence [14]. We have focused on AnPixJg2 Trp289, which interacts with Rings D and A of the chromophore. The XRG CBCR GAF domains have the conserved WXDXXL motif containing the corresponding Trp residue.

1. Prepare a modified sequence that you consider to be present and enter this sequence in the query box (Fig. 2a). In the model case, we prepare the modified AnPixJg2 sequence in which Trp289 is replaced with Cys having the CXDXXL pattern (*see* **Note 6**) (Fig. 2a).
2. Select the PHI-BLAST algorithm and enter a pattern in the PHI pattern box. In the model case, we enter the CXDXXL pattern (Fig. 2b).
3. Press the BLAST button for execution (Fig. 2b). In the model case, we obtain many candidate sequences. Among them, the search hit sequence MCC3606390.1, which has been analyzed in the previous study, exhibits PCB-to-PVB isomerization activity (Fig. 2c) [16]. The sequence corresponding to the CXDXXL pattern is highlighted by the asterisks.

3.2 Site-Saturation Random Mutagenesis and Colony-Based Protein Expression

In this section, we introduce site-saturation random mutagenesis and colony-based protein expression with a model case focusing on S118X site-saturation mutagenesis of AM1_1499g1, whose detailed analysis will be presented elsewhere (Yoneda et al. in preparation).

1. Perform inverse PCR with the forward mixed-base primers and the single reverse primer using KOD One DNA polymerase (Fig. 3a) (*see* **Note 7**). The mixed bases are NNS (N: A/T/G/C, S: G/C) corresponding to 32 codons that cover all the amino acids and include only one stop codon (amber TAG codon) (*see* **Note 8**). The forward primer has 15-bp and 20- to 25-bp sequences before and after the NNS sequence, respectively (Fig. 3a). In contrast, the reverse primer is composed of ~20-bp sequence complementary to the upstream region of the NNS sequence (Fig. 3a). In the model case, we use the primers described in Subheading 2.2, **item 2** for the S118X site-saturation mutagenesis of AM1_1499g1, which is cloned into the pET28a vector.

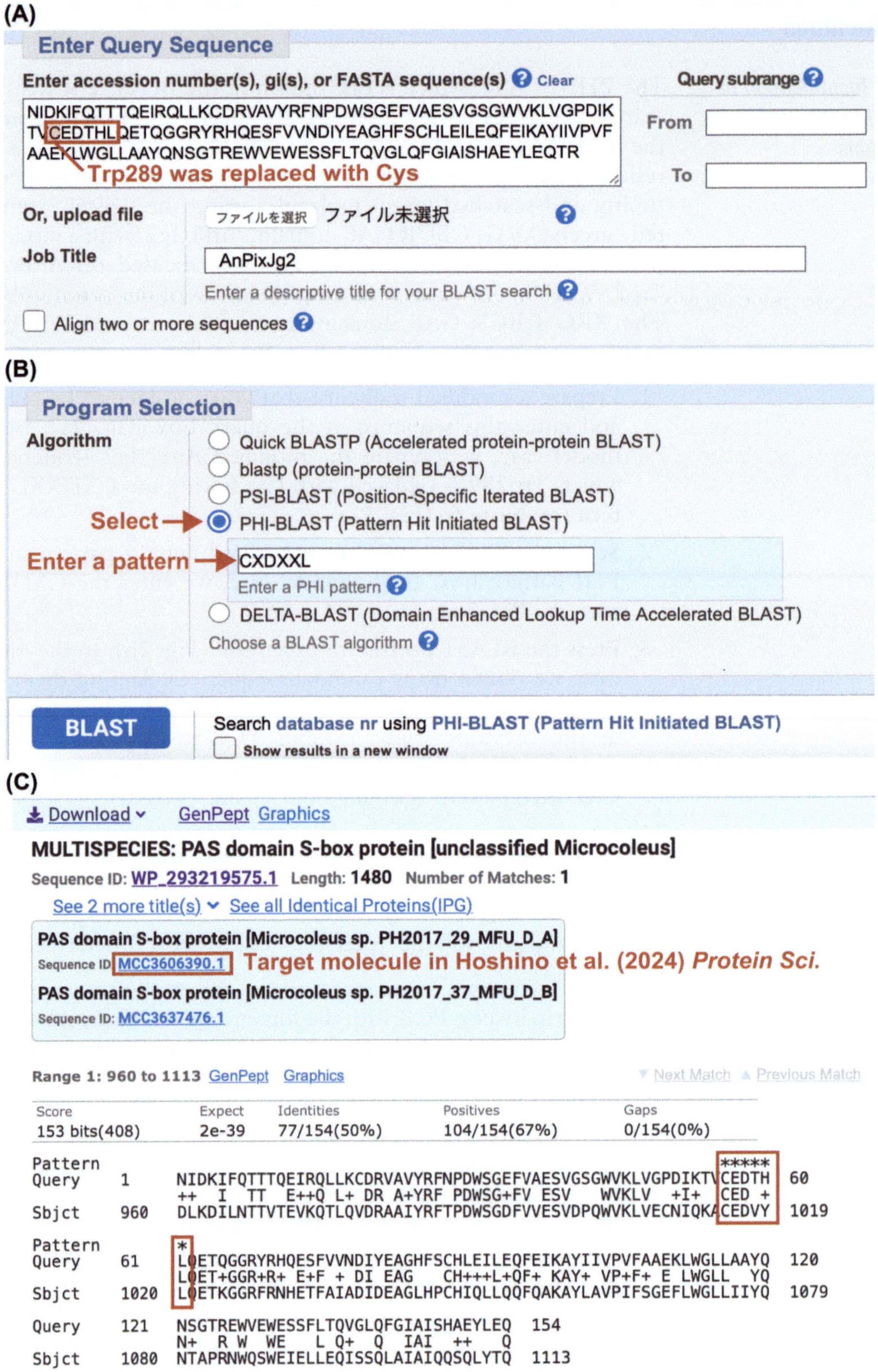

Fig. 2 Identification of novel CBCR GAF domains using the PHI-BLAST. (**a**–**c**) Screenshots of "Enter Query Sequence" (**a**), "Program Selection" (**b**), and an alignment of the query sequence and the target sequence (**c**)

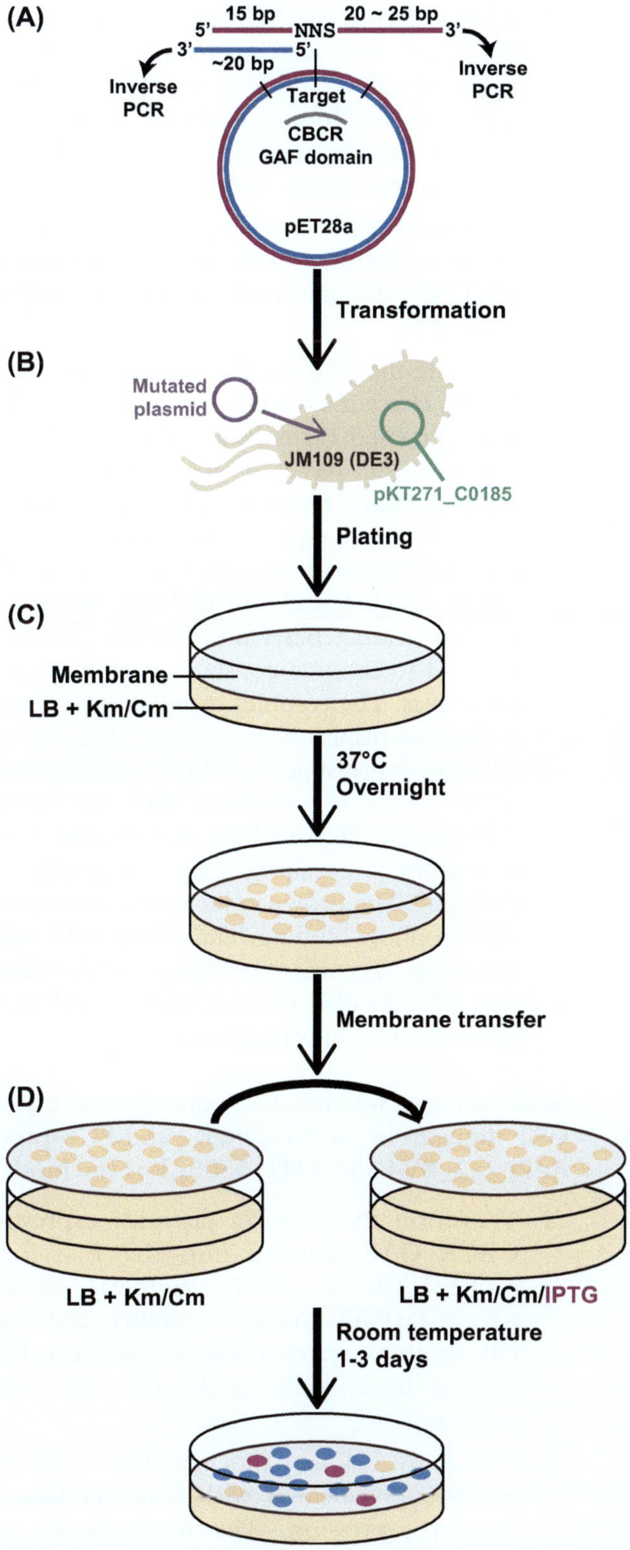

Fig. 3 Flowchart of site-saturation random mutagenesis and colony-based protein expression. (**a**) Inverse PCR for the site-saturation random mutagenesis. (**b**) Transformation to *E. coli* C41 (DE3) pKT271_C0185. (**c**) Colony formation. (**d**) Membrane transfer and induction of protein expression

2. After cyclizing the PCR product using the NEBuilder® HiFi DNA Assembly Kit (*see* **Note 9**), transform the mutated plasmids into the *E. coli* strain JM109 (DE3) pKT271_C0185 (Fig. 3b) (*see* **Notes 2** and **10**). Screen the transformed *E. coli* cells on the LB plate supplemented by Km and Cm and covered by the nitrocellulose membrane (Fig. 3c).
3. Incubate the plate overnight at 37 °C and transfer the membrane to the other LB plate supplemented by Km, Cm, and IPTG for induction of the protein expression (Fig. 3d) (*see* **Note 11**).
4. Incubate the plate for 2–3 days at room temperature (20–25 °C) and check the colony colors (Fig. 3d). In the model case, we have obtained many blue colonies and a few brown and red colonies (Fig. 4a–e). The wild-type AM1_1499g1 covalently binds PCB and exhibits orange/green reversible photoconversion [16]. Comprehensive Sanger sequence analysis has revealed that we have obtained all the S118X variant molecules except S118H and S118P (Fig. 4b–e). This site-saturation random mutagenesis method has proven to be a powerful method to simultaneously construct the site-saturated variant molecules. The colonies expressing wild-type and most variant molecules display a blue color (Fig. 4b–e). In contrast, the colonies expressing the S118C variant molecule display a red color, which is consistent with the fact that this molecule restores the PCB-to-PVB isomerization activity and exhibits blue-shifted yellow/teal reversible photoconversion [15, 16]. The brown color of the colonies indicates no or low chromophore binding. The colony color selection is effective in screening the variant molecules with unique spectral properties. The detailed characterization will be reported elsewhere (Yoneda et al. in preparation).

3.3 Expression with Various Chromophore Synthetic Systems

In this section, we introduce heterologous expression of the CBCR GAF domains in the BV-, PΦB-, and PCB-producing *E. coli* using AnPixJg2_BV4 and AM1_6305g1 as the model cases [20].

1. Transform the pET28a plasmids expressing the His-tagged CBCR GAF domains into the *E. coli* strain C41 (DE3) pKT271_C0185 (PCB synthetic system), C41 (DE3) pKT272 (PΦB synthetic system), and C41 (DE3) pKT270 (BV synthetic system) (Fig. 1) (*see* **Note 12**). Screen the transformed *E. coli* cells on the LB plate supplemented by Km and Cm.
2. Pick up the colonies, inoculate them with 1- to 2.5-L LB liquid media supplemented with Km and Cm and add IPTG for protein expression. The methods for overexpression and

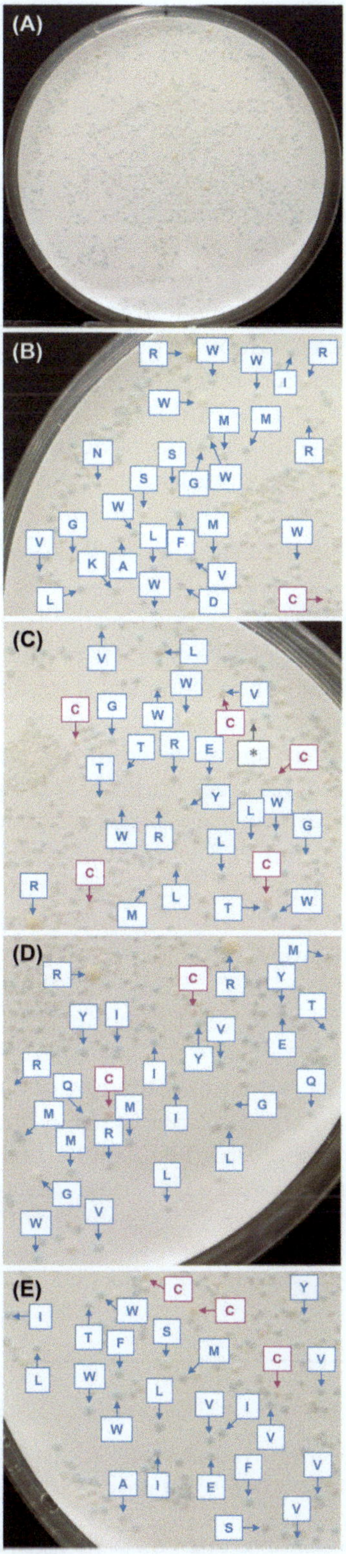

Fig. 4 Colonies expressing AM1_1499g1 S118X variant molecules. (**a**) The whole image of the LB plate. (**b**–**e**) Enlarged image of upper left (**b**), upper right (**c**), lower right (**d**), and lower left (**e**) regions of the LB plate. Plasmids of the selected colonies were sequenced, and mutation details were identified. The identified mutations are shown in the figures

purification using nickel-affinity chromatography of the CBCR GAF domains are described in detail in the previous studies [8, 14, 15, 19, 24].

3. Measure the absorption spectra of the purified CBCR GAF domains and characterize the photoconversion property of the purified CBCR GAF domains using monochromic light sources (*see* **Note 4**). In the model cases, AnPixJg2_BV4 holoproteins purified from the BV-, PΦB-, and PCB-producing *E. coli* exhibit far-red/orange, red/green, and blue-shifted red/green reversible photoconversion, respectively (Fig. 5a–c), whereas AM1_6305g1 holoproteins purified from the PΦB- and PCB-producing *E. coli* exhibit yellow/teal and blue-shifted yellow/teal reversible photoconversion, respectively (Fig. 5d–e). (*Z*-isomer)—(*E*-isomer) difference spectra of the native preparations are summarized in Fig. 6a.
4. Denature both light-absorbing states of the purified CBCR GAF domains with fivefold dilution into 7 M guanidinium chloride/1% (v/v) HCl to identify binding chromophore species and photochemical state. In the model cases, (*Z*-isomer)—(*E*-isomer) difference spectra of the acid-denatured preparations are summarized in Fig. 6b. The difference spectra of the acid-denatured AnPixJg2_BV4 from the BV-, PΦB-, and PCB-producing *E. coli* are assigned to the Cys-adducted BV, PΦB, and PCB chromophores (Figs. 1 and 6b). The difference spectrum of the acid-denatured AM1_6305g1 from the PCB-producing *E. coli* is assigned to the Cys-adducted PVB chromophore (Figs. 1 and 6b). That of the acid-denatured AM1_6305g1 from the PΦB-producing *E. coli* would be assigned to the Cys-adducted PΦB isomer whose structure is the same as PVB except that $C18^1{=}C18^2$ is a double bond, which we name phytoviolobilin (ΦVB) (Figs. 1 and 6b). In summary, the CBCR GAF domains incorporating five different chromophores (BV, PΦB, PCB, ΦVB, and PVB) have been characterized. The order of the conjugated system length is BV > PΦB > PCB > ΦVB > PVB (Fig. 1), and the absorbing wavelengths of both native and denatured preparations follow this order (Fig. 6a, b).

4 Notes

1. Based on the previous study, the canonical Cys residues of the CBCR GAF domains, JSC1_58120g3 and AnPixJg2_BV4, covalently bind to $C3^1$ and $C3^2$ of the BV chromophore, respectively [15, 27]. In both the cases, the double bond is present between C2 and C3, which is supported by the density maps of the structural analyses. We applied the chemical

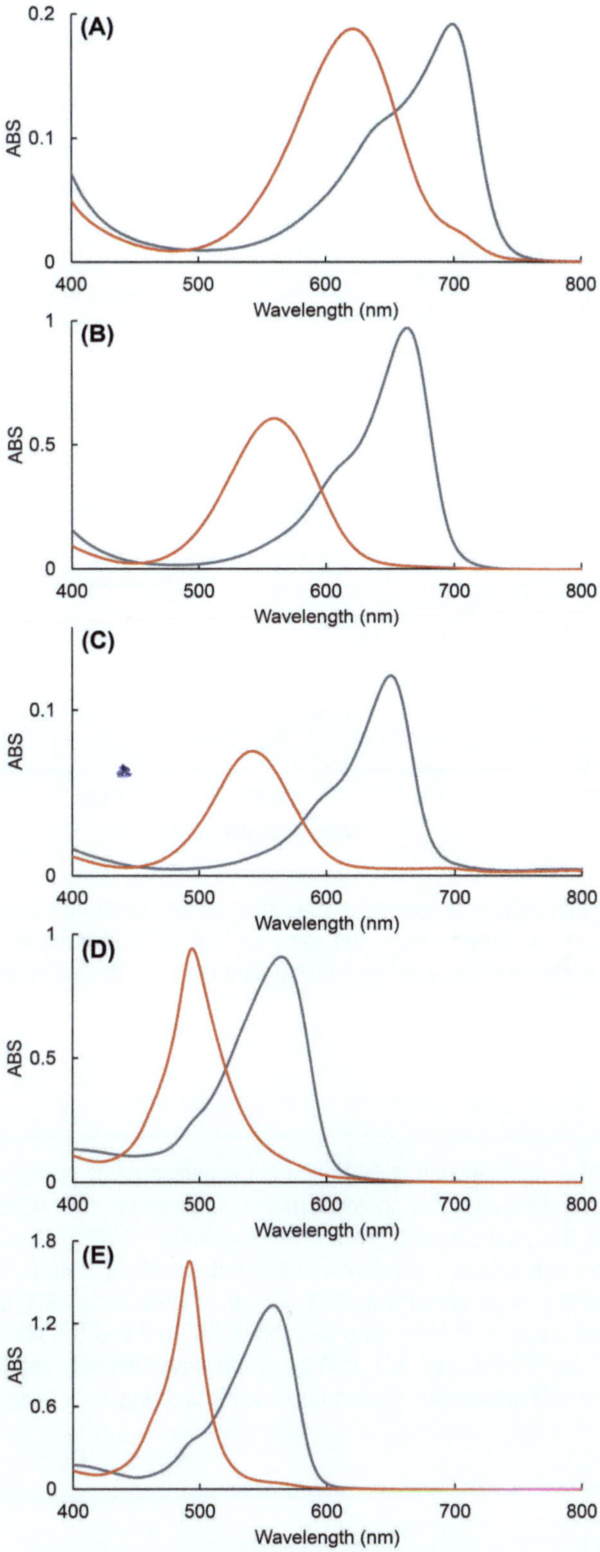

Fig. 5 Photoconversion of BV-binding, PΦB-binding, and PCB-binding AnPixJg2_BV4 and ΦVB-binding and PVB-binding AM1_6305g1. (**a**) BV-binding AnPixJg2_BV4 photoconversion between the dark-adapted far-red-

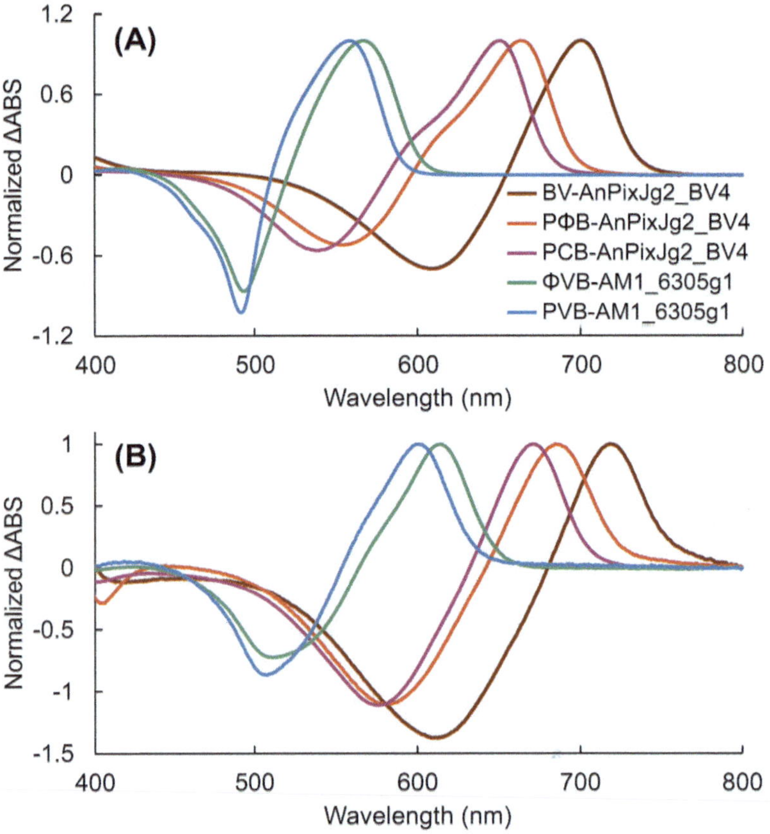

Fig. 6 (*Z*-isomer)—(*E*-isomer) difference spectra of the native (**a**) and denatured (**b**) preparations. The red-to-blue order of the spectra of the native ones is the same as that of the denatured ones, indicating that the chromophore structure largely contributes to the sensing light colors in the native molecules in this case

Fig. 5 (continued) absorbing state peaking at 699 nm and the photoproduct orange-absorbing state peaking at 621 nm. (**b**) PΦB-binding AnPixJg2_BV4 photoconversion between the dark-adapted red-absorbing state peaking at 663 nm and the photoproduct green-absorbing state peaking at 560 nm. (**c**) PCB-binding AnPixJg2_BV4 photoconversion between the dark-adapted red-absorbing state peaking at 650 nm and the photoproduct green-absorbing state peaking at 543 nm. (**d**) ΦVB-binding AM1_6305g1 photoconversion between the dark-adapted green-absorbing state peaking at 565 nm and the photoproduct teal-absorbing state peaking at 494 nm. (**e**) PVB-binding AM1_6305g1 photoconversion between the dark-adapted green-absorbing state peaking at 558 nm and the photoproduct teal-absorbing state peaking at 492 nm

structure of the BV chromophore binding to AnPixJg2_BV4, in which the canonical Cys residue covalently binds to C3^2 of the chromophore (Fig. 1).

2. To efficiently obtain the *E. coli* transformants containing both a plasmid encoding the CBCR GAF domains and a plasmid encoding chromophore synthetic enzymes, we recommend preparing chemically competent cells harboring various plasmids encoding the chromophore synthetic enzymes in advance by yourself. Although we use pKT271_C0185 for the PCB synthesis, pKT271 is also applicable [23]. pKT271_C0185 encodes one of the two PcyA genes, AM1_C0185, from *Acaryochloris marina* MBIC 11017, which transiently accumulates 18^1,18^2-dihydrobiliverdin (18^1,18^2-DHBV) for a short period and specifically provide PCB to the apoproteins (Fig. 1) [23, 27]. In contrast, pKT271_5311 encodes another PcyA gene, AM1_5311, which highly accumulates 18^1,18^2-DHBV for a long period [28] and can provide 18^1,18^2-DHBV to the CBCR GAF domains accepting verdin chromophores (Fig. 1) [23, 27].
3. While pKT270 and pKT271_C0185 are stable during amplification in *E. coli* cells, we need to treat pKT272 with special care. *E. coli* colonies transformed with pKT272 are heterogenous in their size. Although the smaller colonies retain the canonical pKT272, pKT272 derived from the larger colonies has a deletion mutation, in which a GGGG sequence within the *hy2* gene is replaced with GGG. Pick up smaller colonies transformed with pKT272 for plasmid preparation.
4. We use the Opto-Spectrum Generator (Hamamatsu Photonics Inc., Hamamatsu, Japan) for monochromic light sources. Monochromic LED and a combination of white light and interference filters are alternatives for monochromic light sources.
5. Although acid urea (8 M urea in 1 N HCl [pH 2.0]) is also applicable, the chromophore in this solution is unstable in some cases [28].
6. If a pattern not present in the query sequence is entered, the PHI-BLAST does not work. Thus, if we try to find molecules with novel patterns, we should modify the known sequences to give such patterns and the model case.
7. We use the KOD One DNA polymerase for amplification. The other proofreading polymerases such as PrimeSTAR Max (TaKaRa, Kusatsu, Japan) and Q5 High-Fidelity DNA Polymerases are alternatives.
8. The appearance frequency of the stop codons in the NNS mixed bases (1/32) is lower than that in the NNN mixed bases (3/64).

9. Because the nicked plasmid should be cyclized within the *E. coli* cells via the intrinsic process, the cyclizing process would not be essential. However, we can obtain much larger amounts of colonies with this process than without this process in our system.
10. We use the JM109 (DE3) strain for this analysis. JM109 (DE3) is the competent cell optimized for gene cloning with high competency, carrying the λDE3 gene in the chromosome for pET system. In this context, *E. coli* strains such as DH5α (DE3) and XL1-Blue (DE3) are alternatives.
11. The membrane transfer protocol enables us to separate cell growth and induction of protein expression. Induction of protein expression just after plating did not result in strong color development of the *E. coli* colonies, indicating inefficient holo-protein expression.
12. We use the C41 (DE3) strain for this analysis. C41 (DE3) is the competent cell optimized for protein expression, carrying the λDE3 gene in the chromosome for pET system. In this context, *E. coli* strains such as BL21 (DE3) and C43 (DE3) are alternatives.

Acknowledgments

This work was partly supported by JSPS KAKENHI (Grant Number JP22H02587; to R.N.), by Nakatani Foundation, Japan (to R.N.), by JST CREST (Grant Number JPMJCR24B5; to R.N.), and by TMU research fund for young scientists of the Tokyo Metropolitan University, Japan (to R.N.).

References

1. Fushimi K, Narikawa R (2019) Cyanobacteriochromes: photoreceptors covering the entire UV-to-visible spectrum. Curr Opin Struct Biol 57:39–46
2. Rockwell NC, Lagarias JC (2024) Cyanobacteriochromes: a rainbow of photoreceptors. Ann Rev Microbiol 78:61–81
3. Rockwell NC, Martin SS, Feoktistova K, Lagarias JC (2011) Diverse two-cysteine photocycles in phytochromes and cyanobacteriochromes. Proc Natl Acad Sci USA 108: 11854–11859
4. Rockwell NC, Martin SS, Lagarias JC (2012) Red/green cyanobacteriochromes: sensors of color and power. Biochemistry 51:9667–9677
5. Rockwell NC, Martin SS, Gulevich AG, Lagarias JC (2012) Phycoviolobilin formation and spectral tuning in the DXCF cyanobacteriochrome subfamily. Biochemistry 51:1449–1463
6. Ma Q, Hua H-H, Chen Y et al (2012) A rising tide of blue-absorbing biliprotein photoreceptors: characterization of seven such bilin-binding GAF domains in *Nostoc* sp. PCC7120. FEBS J 279:4095–4108
7. Narikawa R, Enomoto G, Ni-Ni-Win et al (2014) A new type of dual-Cys cyanobacteriochrome GAF domain found in cyanobacterium *Acaryochloris marina*, which has an unusual red/blue reversible photoconversion cycle. Biochemistry 53:5051–5059
8. Narikawa R, Nakajima T, Aono Y et al (2015) A biliverdin-binding cyanobacteriochrome from the chlorophyll *d*-bearing cyanobacterium *Acaryochloris marina*. Sci Rep 5:7950

9. Cho SM, Jeoung SC, Song J-Y et al (2015) Genomic survey and biochemical analysis of recombinant candidate cyanobacteriochromes reveals enrichment for near UV/violet sensors in the halotolerant and alkaliphilic cyanobacterium *Microcoleus* IPPAS B353. J Biol Chem 290:28502–28514
10. Hasegawa M, Fushimi K, Miyake K et al (2018) Molecular characterization of DXCF cyanobacteriochromes from the cyanobacterium *Acaryochloris marina* identifies a blue-light power sensor. J Biol Chem 293:1713–1727
11. Rockwell NC, Martin SS, Lagarias JC (2015) Identification of DXCF cyanobacteriochrome lineages with predictable photocycles. Photochem Photobiol Sci 14:929–941
12. Fushimi K, Rockwell NC, Enomoto G et al (2016) Cyanobacteriochrome photoreceptors lacking the canonical Cys residue. Biochemistry 55:6981–6995
13. Rockwell NC, Martin SS, Lagarias JC (2016) Identification of cyanobacteriochromes detecting far-red light. Biochemistry 55:3907–3919
14. Hoshino H, Miyake K, Fushimi K, Narikawa R (2024) Red/green cyanobacteriochromes acquire isomerization from phycocyanobilin to phycoviolobilin. Protein Sci 33:e5132
15. Fushimi K, Miyazaki T, Kuwasaki Y et al (2019) Rational conversion of chromophore selectivity of cyanobacteriochromes to accept mammalian intrinsic biliverdin. Proc Natl Acad Sci USA 116:8301–8309
16. Fushimi K, Hasegawa M, Ito T et al (2020) Evolution-inspired design of multicolored photoswitches from a single cyanobacteriochrome scaffold. Proc Natl Acad Sci USA 117:15573–15580
17. Rockwell NC, Moreno MV, Martin SS, Lagarias JC (2022) Protein-chromophore interactions controlling photoisomerization in red/green cyanobacteriochromes. Photochem Photobiol Sci 21:471–491
18. Suzuki T, Yoshimura M, Hoshino H et al (2023) Introduction of reversible cysteine ligation ability to the biliverdin-binding cyanobacteriochrome photoreceptor. FEBS J 290: 4999–5015
19. Suzuki T, Yoshimura M, Arai M, Narikawa R (2024) Crucial residue for tuning thermal relaxation kinetics in the biliverdin-binding cyanobacteriochrome photoreceptor revealed by site-saturation mutagenesis. J Mol Biol 436:168451
20. Gambetta GA, Lagarias JC (2001) Genetic engineering of phytochrome biosynthesis in bacteria. Proc Natl Acad Sci USA 98:10566–10571
21. Bhoo SH, Davis SJ, Walker J et al (2001) Bacteriophytochromes are photochromic histidine kinases using a biliverdin chromophore. Nature 414:776–779
22. Mukougawa K, Kanamoto H, Kobayashi T et al (2006) Metabolic engineering to produce phytochromes with phytochromobilin, phycocyanobilin, or phycoerythrobilin chromophore in *Escherichia coli*. FEBS Lett 580:1333–1338
23. Miyake K, Fushimi K, Kashimoto T et al (2020) Functional diversification of two bilin reductases for light perception and harvesting in unique cyanobacterium *Acaryochloris marina* MBIC 11017. FEBS J 287:4016–4031
24. Narikawa R, Fukushima Y, Ishizuka T et al (2008) A novel photoactive GAF domain of cyanobacteriochrome AnPixJ that shows reversible green/red photoconversion. J Mol Biol 380:844–855
25. Hirose Y, Shimada T, Narikawa R et al (2008) Cyanobacteriochrome CcaS is the green light receptor that induces the expression of phycobilisome linker protein. Proc Natl Acad Sci USA 105:9528–9533
26. Ishizuka T, Narikawa R, Kohchi T et al (2007) Cyanobacteriochrome TePixJ of *Thermosynechococcus elongatus* harbors phycoviolobilin as a chromophore. Plant Cell Physiol 48:1385–1390
27. Moreno MV, Rockwell NC, Mora M et al (2020) A far-red cyanobacteriochrome lineage specific for verdins. Proc Natl Acad Sci USA 117:27962–27970
28. Fushimi K, Nakajima T, Aono Y et al (2016) Photoconversion and fluorescence properties of a red/green-type cyanobacteriochrome AM1_C0023g2 that binds not only phycocyanobilin but also biliverdin. Front Microbiol 7: 588

Chapter 2

Analysis of Sequence Diversity in Subfamilies of Phytochrome-Linked Effectors

Cornelia Böhm and Massimo Totaro

Abstract

Sequence Similarity Networks allow the integration of protein sequence relationships with complementary data such as functional or spectral information, enabling the visualization of potential functional connections. Here, we describe the use of basic local alignment search tool (BLAST)-generated E-values or Bit-scores within a network of bacteriophytochrome homologs to emphasize various characteristics of interest and their sequence-level conservation.

Key words Sequence similarity network, Sequence alignment, Visualization, Network analysis, Bacteriophytochromes, Effector families

1 Introduction

According to the sequence–structure–function paradigm, protein function is defined by structure, which is determined by the amino acid sequence [1]. Within individual subfamilies of protein classes, identifying conserved features, such as the length of structural elements, and correlating them with other conserved characteristics can be challenging. Sequence Similarity Networks (SSNs) integrate protein sequence relationships with additional orthogonal information, providing a deeper understanding of functional connections [2]. Unlike phylogenetic analysis based on distance or parsimony methods, SSNs are not constrained by sample size; in fact, larger datasets enhance the representation of existing sequence relationships. These relationships are determined using National Center for Biotechnology Information (NCBI) basic local alignment search tool (BLAST+)-generated E-values or Bit-scores [3]. Duplicate results and self-BLAST entries need to be removed. For small datasets, this can be done manually; for larger ones, a software solution is recommended. To assist with this, we provide analysis scripts available on GitHub (https://gitlab.tugraz.at/bioc/ssn).

Andreas Winkler (ed.), *Phytochromes: Methods and Protocols*, Methods in Molecular Biology, vol. 2970,
https://doi.org/10.1007/978-1-0716-4791-2_2,

Curated results are subsequently represented in network form using Cytoscape [4] and yWorks Organic Layout [5], with nodes corresponding to individual proteins and edges representing E-values/Bit-scores. Various characteristics of interest, such as spectral properties, different effector modules, or sensor–effector linker length, can be highlighted in the visualized output. Thus, SSNs are a powerful tool for investigating systems like bacteriophytochrome (BphP) effector subfamilies, providing insights into and conclusions about the complex signal integration mechanisms involved. This has been successfully used by us in characterizing bacteriophytochrome-regulated diguanylate cyclases [6]. Considering the complexity of phytochrome signaling in general [7, 8], conservation and correlation of individual characteristics among and across different effector (sub)families is expected to remain an issue of high interest.

2 Materials

The computational requirements for this bioinformatics method are as follows:

2.1 BLAST+

BLAST+ [3] can be run locally in Windows or Linux. NCBI provides installers and source codes for download (https://blast.ncbi.nlm.nih.gov/doc/blast-help/downloadblastdata.html) as well as user and help manuals.

2.2 Analysis Scripts

Different formats of the analysis scripts are provided, so that each user can choose the most convenient way to run their analysis. Our recommendation is to use the Google Colab notebook format, since it requires no local setup.

In general, the basic requirements are a shell to run commands and a specific programming language interpreter that depends on the chosen method.

The provided script files have been tested with GNU Bash 5.1.x, GNU Awk 5.3.0, Python 3.12.x, and pandas 2.2.2.

2.3 Cytoscape

Visualization requires Cytoscape [4] (downloadable from https://cytoscape.org/download.html). yFiles [5] for Organic Layout representation are also available for download (https://www.yworks.com/yfiles-download).

3 Methods

3.1 Database Generation and Curation

To generate the SSN data, starting from the unaligned FASTA (Fast-All) file (*see* **Note 1**), an automated script is provided as an online Google Colab notebook (https://colab.research.google.com/drive/1RQssmD8X7ZOGaxOYDUYA5kpxmDIGYmkx).

After loading the page and granting it the necessary permissions, follow the instructions contained therein.

Alternatively, the analysis can be performed manually with the following steps:

3.1.1 (Manual) Database Compilation and BLAST

1. Compile a list of unaligned protein sequences in FASTA format (*see* **Notes 2** and **3**) with unique identifier codes.
2. Create a table containing all required information about the individual entries, such as class/phylum/order, effector module, linker length, and spectral properties in comma-separated values (CSV) format (*see* **Note 2** and Fig. 1).
3. Build a BLAST database using the command "makeblastdb" in the NCBI BLAST+ environment. The command-line input is *makeblastdb -in FILENAME.fasta -dbtype prot -title TITLE -parse_seqids -out DATABASENAME_db*, with FILENAME = name of the FASTA file, TITLE = title of the database, DATABASENAME = name of the created database file (*see* Fig. 2 and **Notes 4** and **5**).
4. Perform an all-vs-all BLAST, where every entry of the database is pairwise BLASTed against every other entry, using the command *blastp -db DATABASENAME_db -query FILENAME.fasta -out OUTPUTNAME.tsv -outfmt "6 qseqid sseqid evalue bitscore"* with OUTPUTNAME = name of the tab-separated values (TSV) output file (*see* Fig. 3 and **Notes 6** and **7**).

3.1.2 (Manual) BLAST Result Curation

The Colab notebook can automatically process the TSV file generated by blastp. Alternatively:

1. Clone the analysis Git repository (https://gitlab.tugraz.at/D5B8E35025578B91/ssn).

	A	B	C	D	E	F	G	H
1	Source	Organism	Order	Class	feature 1	NCBI	UNIPROT	length
2	BphP1	species 1	order 1	Actinobacteria	A	ID_1	ID_A	681
3	BphP2	species 2	order 3	Betaproteobacteria	B	ID_2	ID_B	676
4	BphP3	species 3	order 4	Gammaproteobacteria	D	ID_3	ID_A	689
5	BphP4	species 4	order 4	Gammaproteobacteria	D	ID_4	ID_B	689
6	BphP5	species 5	order 5	Gammaproteobacteria	D	ID_5	ID_A	695
7	BphP6	species 6	order 2	Gammaproteobacteria	C	ID_6	ID_B	683
8	BphP7	species 7	order 3	Betaproteobacteria	B	ID_7	ID_A	659
9	BphP8	species 8	order 2	Gammaproteobacteria	D	ID_8	ID_B	689
10	BphP9	species 9	order 1	Actinobacteria	A	ID_9	ID_A	679

Fig. 1 Information file containing unique ID as well as all relevant features to be visualized in the SSN

```
C:\blast-2.16.0+\bin>makeblastdb -in FILENAME.fasta -dbtype prot -title TITLE -parse_seqids -out DATABASENAME_db


Building a new DB, current time: 12/13/2024 09:24:06
New DB name:   C:\blast-2.16.0+\bin\DATABASENAME_db
New DB title:  TITLE
Sequence type: Protein
Keep MBits: T
Maximum file size: 3000000000B
Adding sequences from FASTA; added 50 sequences in 0.194746 seconds.
```

Fig. 2 NCBI BLAST+ command-line output after database creation. If the database was created successfully, ten new files should have been added to the corresponding folder

```
OUTPUTNAME.tsv - Notepad
File  Edit  Format  View  Help
BphP1   BphP1   0.0     1339
BphP1   BphP15  0.0     1187
BphP1   BphP9   0.0     1066
BphP1   BphP16  0.0     1022
BphP1   BphP13  0.0     1010
BphP1   BphP13  1.6     23.5
BphP1   BphP10  0.0     915
BphP1   BphP27  5.75e-177       513
BphP1   BphP20  1.67e-164       481
BphP1   BphP25  1.41e-163       478
BphP1   BphP18  2.86e-163       478
BphP1   BphP12  1.97e-162       476
BphP1   BphP21  4.98e-156       460
BphP1   BphP19  9.14e-147       435
BphP1   BphP17  1.17e-92        295
BphP1   BphP22  1.35e-83        271
BphP1   BphP26  3.40e-83        270
```

Fig. 3 All-vs-all BLAST output file in TSV format containing both E-values and Bit-scores

```
OUTPUTNAME_out_evalues.csv - Notepad
File  Edit  Format  View  Help
Source;Target;evalue
BphP10;BphP1;0.0
BphP11;BphP1;1.1599999999999999e-67
BphP12;BphP1;4.68e-166
BphP13;BphP1;0.0
BphP14;BphP1;1.44e-72
BphP15;BphP1;0.0
BphP16;BphP1;0.0
BphP17;BphP1;1.17e-92
BphP18;BphP1;2.02e-168
```

```
OUTPUTNAME_out_bitscores.csv - Notepad
File  Edit  Format  View  Help
Source;Target;bitscore
BphP10;BphP1;931.0
BphP11;BphP1;227.0
BphP12;BphP1;485.0
BphP13;BphP1;1010.0
BphP14;BphP1;241.0
BphP15;BphP1;1187.0
BphP16;BphP1;1022.0
BphP17;BphP1;295.0
BphP18;BphP1;491.0
```

Fig. 4 Colab notebook E-value and Bit-score output files

2. Select the correct script to execute and follow the respective instructions.
3. OUTPUTNAME.tsv is the file generated in the previous steps (*see* Subheading 3.1.1 and Fig. 4).
4. Following execution of the script, two files will be available: bts.csv and evs.csv (*see* **Note 6**), already properly formatted.

3.2 SSN Visualization Using Cytoscape

1. Import bts.csv or evs.csv under "File"—"Import"—"Network from File." Under "Advanced Options," check semicolon as the delimiter (*see* Fig. 5 and **Note 8**).
2. Import the information CSV file under "File"—"Import"—"-Table from File." Again, choose the correct delimiter (*see* **Note 9**).
3. Under "Layout," choose "yFiles Organic Layout" from the yFiles family of diagramming software components.
4. Remove edges of higher E-values/lower Bit-scores corresponding to less closely related sequence relationships: Choose "Filter"—"Column Filter," select "evalue"/"bit-score," and create a filter (*see* Fig. 6). The range defined by the numbers corresponds to the selected edges to be deleted. Under "Edit"—"Remove Selected Nodes and Edges," proceed with deletion. Reorganize the organic layout to see the effect of the deletions (*see* Fig. 7 and **Notes 10** and **11**). Proceed until individual nodes or clusters begin to split off.

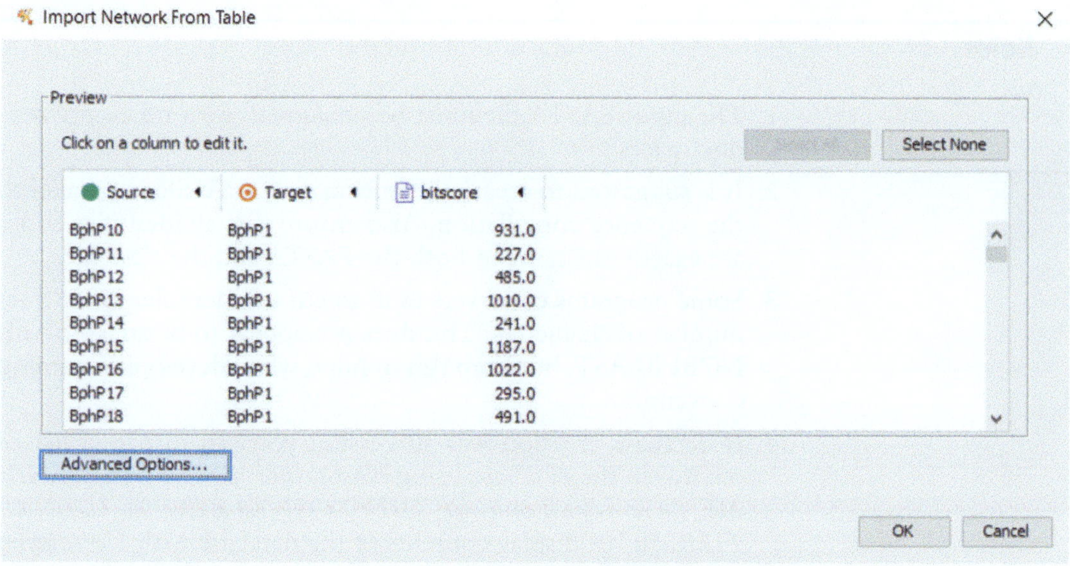

Fig. 5 Cytoscape file import. Correct assignment of source and target nodes is important

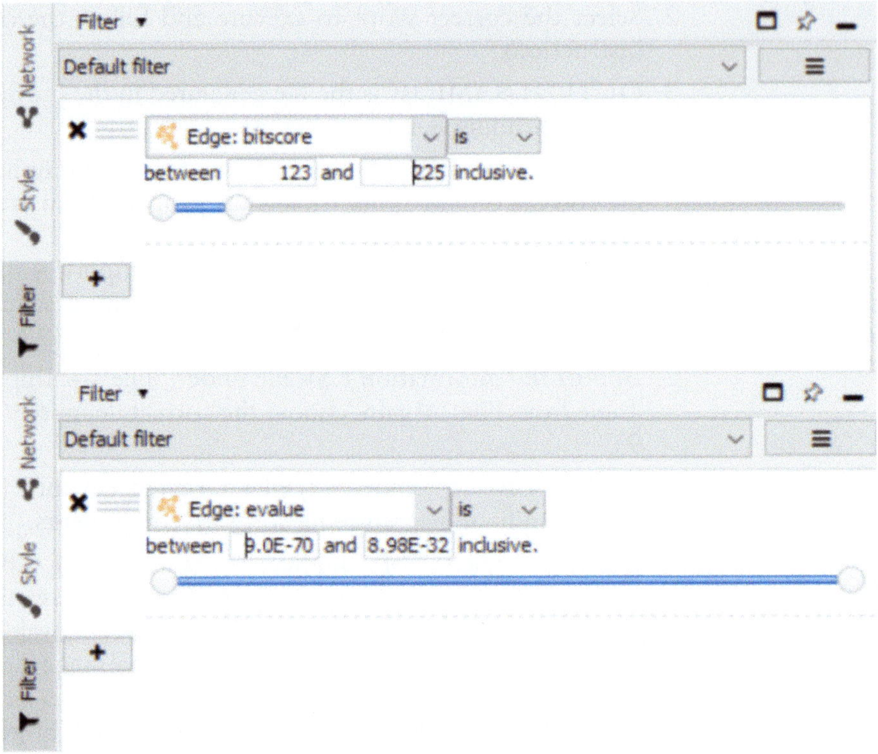

Fig. 6 Filter applied to delete Bit-scores/E-values corresponding to less sequence identity

5. Under "Style," assign different layout options such as node form, fill color, or border color for different features to be visualized (*see* Fig. 8 and **Note 12**).

4 Notes

1. The input FASTA file must be unaligned, with no empty residue spaces.
2. It is suggested to create the information CSV file in parallel to the sequence compilation. Also ensure that all identifier codes are exactly the same in both the FASTA and the CSV file.
3. Some programs or servers tend to cut off identifiers after a set number of characters. This does not appear to be an issue with NCBI BLAST, but keep this in mind when developing naming conventions.
4. If database creation was successful, you can display a list of records in the database using *blastdbcmd -db DATABASENAME_db -entry all -outfmt "OID: %o GI: %g ACC: %a IDENTIFIER: %i"* or display contents of the database with *blastdbcmd -db DATABASENAME_db -entry all.*

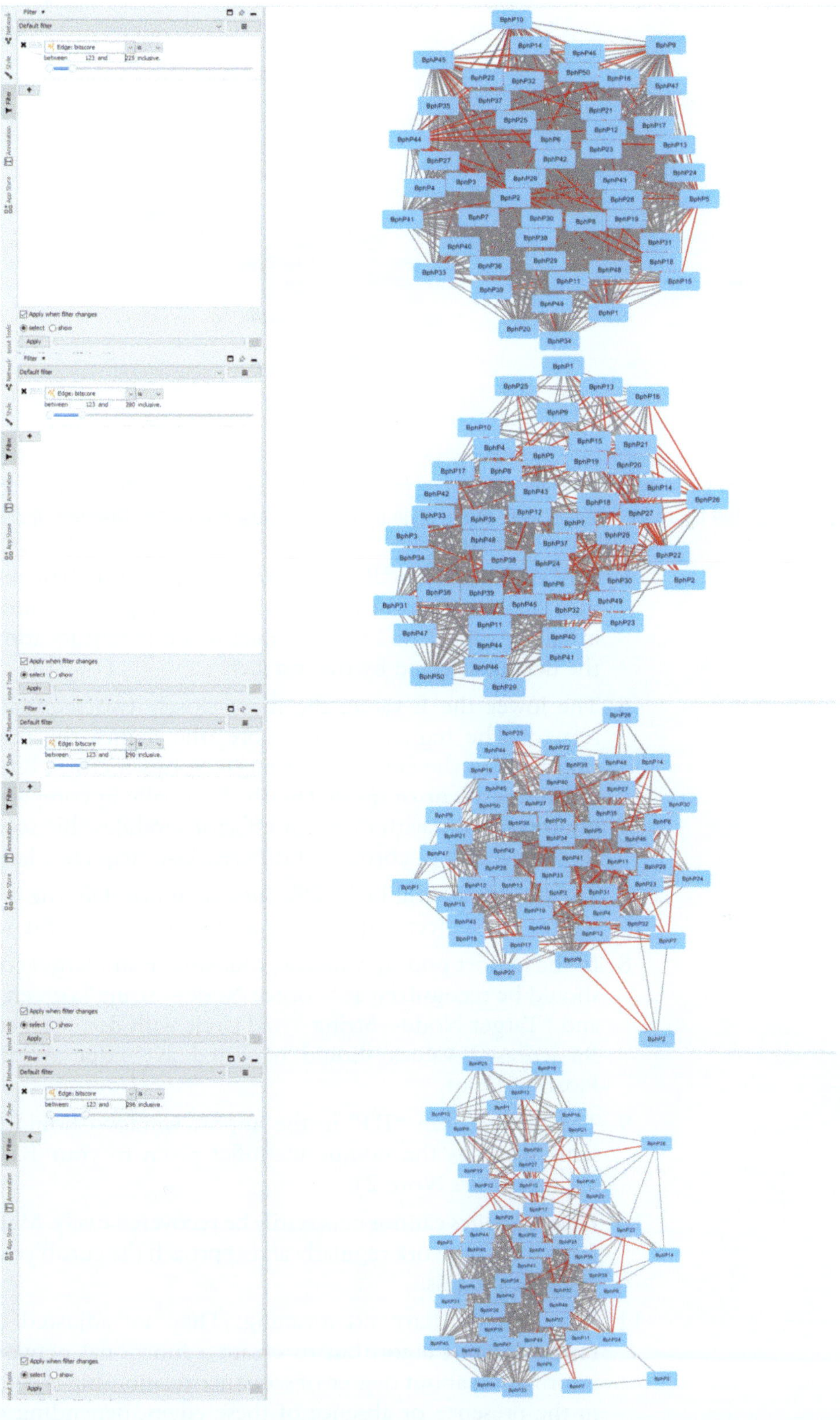

Fig. 7 Stepwise deletion of edges representing more distant relationships, here for a Bit-score-based network. Proceed until individual nodes begin to split off entirely. Zoom into the figure for detailed view of the Bit-score deletion process in Cytoscape

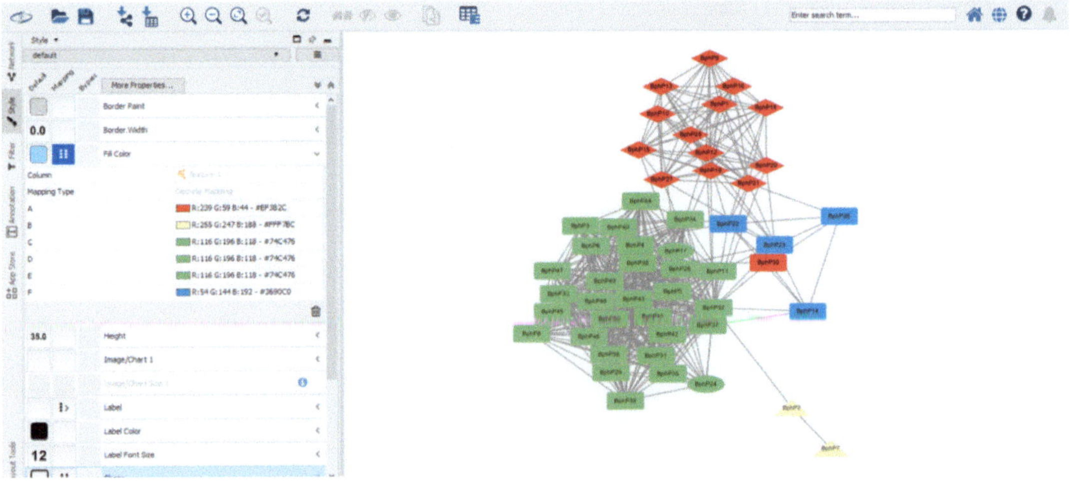

Fig. 8 Application of style options to highlight various features, represented for example by node color and shape. Zoom into the figure for detailed view of the style options in Cytoscape using "Discrete Mapping"

5. If BLAST+ returns "BLAST Database error: Database memory map file error" and there is sufficient computer memory, try installing the BLAST+ executables in a different location than the default selected by the installer.
6. The lower the E-value, the more closely two sequences are related. The higher the Bit-score, the higher the sequence similarity. In the choice between these two metrics, Bit-scores are independent of query length. Especially in comparison of photoreceptors with different effector modules, Bit-scores are recommended to correct for differences in sequence length.
7. The output file will be in TSV format and contain the query as well as the subject Seq-ID, in addition to E-value/Bit-score.
8. In the import pop-up window, your source and target columns should be recognized as "Source Node—String" (green circle) and "Target Node—String" (red circle with dot), respectively. Bit-scores/E-values should be listed as "Edge Attribute—Floating Point."
9. The first column "ID" in the import window should be the same name as the unique identifier given to your BLASTed sequences (*see* **Note 2**).
10. Deleted edges cannot necessarily be recovered easily. Make sure to save the network regularly and approach the cutoff point in a stepwise process.
11. Edge lengths have no meaning: They are adjusted by the organic layout algorithm to create a functional network. All information about degree of sequence relationship is contained in the presence or absence of these edges, depending on the E-value/Bit-score cutoff point.

12. For the majority of stylistic representations, choose "Discrete Mapping" and assign individual characteristics defined in the information CSV file (*see* **Note 2**) different graphical features.

Acknowledgments

The authors would like to express their gratitude to Andreas Winkler (Institute of Biochemistry, Graz University of Technology) for constructive discussion and valuable feedback.

References

1. Koehler Leman J, Szczerbiak P, Renfrew PD et al (2023) Sequence-structure-function relationships in the microbial protein universe. Nat Commun 14(1):2351. https://doi.org/10.1038/s41467-023-37896-w
2. Atkinson HJ, Morris JH, Ferrin TE et al (2009) Using sequence similarity networks for visualization of relationships across diverse protein superfamilies. PLoS One 4:e4345. https://doi.org/10.1371/journal.pone.0004345
3. Altschul SF, Gish W, Miller W et al (1990) Basic local alignment search tool. J Mol Biol 215(3):403–410. https://doi.org/10.1016/S0022-2836(05)80360-2
4. Shannon P, Markiel A, Ozier O et al (2003) Cytoscape: a software environment for integrated models of biomolecular interaction networks. Genome Res 13(11):2498–2504. https://doi.org/10.1101/gr.1239303
5. Wiese R, Eiglsperger M, Kaufmann M (2001) yFiles: visualization and automatic layout of graphs. In: Proceedings of the 9th international symposium on graph drawing (GD 2001). Springer-Verlag
6. Böhm C, Gourinchas G, Zweytick S (2022) Characterisation of sequence-structure-function space in sensor-effector integrators of phytochrome-regulated diguanylate cyclases. Photochem Photobiol Sci 21(10):1761–1779. https://doi.org/10.1007/s43630-022-00255-7
7. Gourinchas G, Etzl S, Winkler A (2019) Bacteriophytochromes – from informative model systems of phytochrome function to powerful tools in cell biology. Curr Opin Struct Biol 57:72–83. https://doi.org/10.1016/j.sbi.2019.02.005
8. Takala H, Edlund P, Ihalainen J, Westenhoff S (2020) Tips and turns of bacteriophytochrome photoactivation. Photochem Photobiol Sci 19(11):1488–1510. https://doi.org/10.1039/d0pp00117a

Chapter 3

Strategies for Coexpression of Bilin-Producing Enzymes in *Escherichia coli*

Oliver Maximilian Eder, Marta Jančić, and Andreas Winkler

Abstract

Protein cofactors vastly expand the chemical space of proteins and thus allow for otherwise inaccessible functionalities. For instance, heme for oxygen transport, flavins that are key to redox reactions, or copper ions facilitating electron transport are only some examples among a plethora of other protein cofactors. In the protein family of phytochromes, bilin compounds act as cofactors and govern their core functionality—the sensing of red/far-red light. This light-sensing capability regulates not only central metabolic pathways in plants but also influences other light-regulated organisms; thus, understanding the molecular mechanisms of phytochromes is of great interest. For biochemical characterization, the phytochrome proteins are most commonly overexpressed in heterologous expression hosts. However, many hosts, like *Escherichia coli*, lack the machinery to produce sufficient levels of bilins. For the generation of biliverdin, this can be alleviated by the coexpression of a heme oxygenase and the phytochrome, most commonly in a two-plasmid (helper plasmid/expression plasmid) setup. In this chapter, we introduce two alternative strategies, bi-cistronic coexpression and genomic integration of the heme oxygenase. The generation of these systems is shown in detail, and their advantages/disadvantages are discussed compared to the well-established two-plasmid systems. We anticipate that these approaches are also feasible for other bilins, for example, phytochromobilin and/or phycocyanobilin, which will expand the toolbox for phytochrome expression and provide alternatives for challenging study designs.

Key words Photoreceptor, Pigment, Heme oxygenase, Biliverdin, CRISPR, Cas9, pEcCas, pEcgRNA, Genomic integration, Bi-cistronic, Cotranslational

1 Introduction

Photoreceptor proteins depend on specific pigments to respond to different ambient light conditions ranging from the ultraviolet (UV) region to the near-infrared (NIR) part of the electromagnetic spectrum. The employed cofactors (in the visible region) or the amino acid tryptophan itself (UV absorption) enable proteins to integrate information of surrounding light conditions and eventually regulate downstream signal integration partners for organismal adaptation [1]. In the family of phytochrome photoreceptors,

Andreas Winkler (ed.), *Phytochromes: Methods and Protocols*, Methods in Molecular Biology, vol. 2970,
https://doi.org/10.1007/978-1-0716-4791-2_3,

different linear tetrapyrroles (bilins) derived from heme are used as the light-sensing cofactors, and the degree of bilin reduction is employed to additionally tune the absorption properties in different organisms [2]. With an absorption maximum in the near-IR, biliverdin (BV)-binding phytochromes found in bacteria and fungi have the longest wavelength absorption maxima and only require the action of a heme oxygenase (HO) to generate their cofactor from heme [3]. Conversely, in plants and cyanobacteria, an additional biliverdin reduction machinery is necessary to produce their preferred phytochrome cofactors phytochromobilin (PΦB) and phycocyanobilin (PCB), respectively [2].

In order to research the diverse phytochrome family, heterologous expression is typically required to obtain the amounts needed for a detailed biochemical and biophysical characterization. To this end, apophytochromes can be expressed in established bacterial or eukaryotic expression systems and subsequently reconstituted with their corresponding cofactors. However, for some phytochromes, instability of the apo form has been reported [4], and also concerns about differences between cotranslational insertion of the bilin versus subsequent reconstitution have been raised [5]. Therefore, coexpression of bilin-producing enzymes together with the phytochrome genes is typically the preferred way of generating holoprotein preparations for biochemical purification and characterization [6–9].

Historically, versatile two-plasmid expression systems [10] are the most utilized approach with a helper plasmid encoding a heme oxygenase (HO), plus if needed other bilin-producing enzymes and a second plasmid containing a phytochrome sequence. An example of this setup is the pT7 helper plasmid containing the *Synechocystis* sp. PCC 6803 *ho1*, first established by Tarutina et al. [11], and modified by Gourinchas et al. [4], in combination with a pETM11 overexpression vector encoding a bacteriophytochrome (BphP).

In this chapter, we want to portray two alternative approaches and how they were generated in our lab. First, usage of a bi-cistronic approach was already reported in previous work [12], however not in an overexpression format as outlined in this chapter. This approach is inspired by natural BphP operon structures, where the phytochrome is often co-encoded with their respective *ho* gene. This genomic architecture is either achieved with overlapping open reading frames (ORF; *see* Fig. 1d), in line with model overlapping gene sequences like *menD/menH* in *Escherichia coli* [13], or with short nucleotide spacers between the *ho* and BphP genes (*see* Fig. 1e). Thus, the awareness for this method should be furthered, and possible strategies for the design of such an expression system are discussed in this chapter. Second, we introduce an *E. coli* BL21 (DE3)-based expression host where the *Synechocystis* sp. PCC 6803 *ho1* gene under the T7 promoter from the pT7 vector was stably

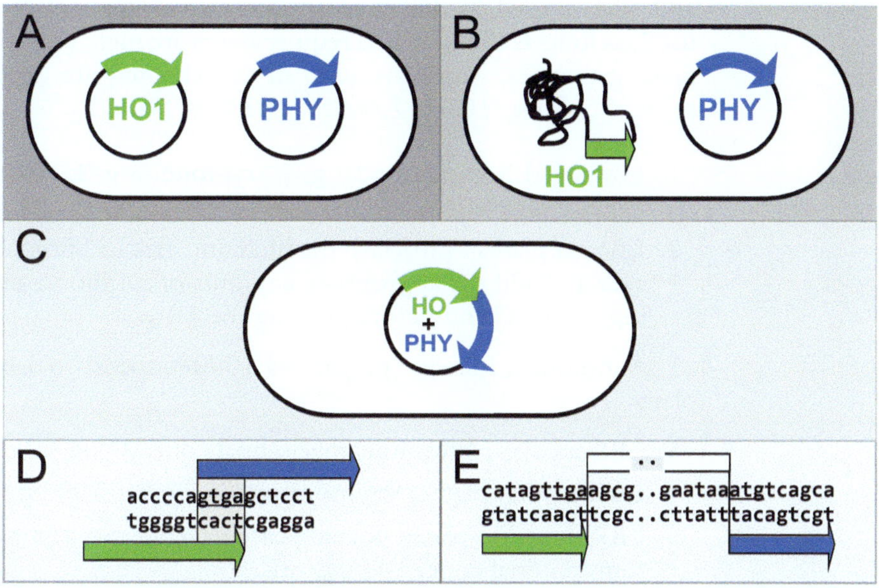

Fig. 1 Different approaches for cofactor generation during BphP expression. In all panels, green symbolizes the open reading frame (ORF) of the heme oxygenase and blue the ORF of the phytochrome. (**a**) In the classical established systems, the *ho1* gene, for example from *Synechocystis* sp. PCC 6803, is provided on a helper plasmid in parallel to an expression plasmid containing the phytochrome of interest. (**b**) Alternatively, the *ho1* gene can be stably integrated into the genome of an expression strain; thus, only the plasmid containing the phytochrome needs to be transformed into the strain. (**c**) Also, the *ho1* and phytochrome can be expressed from a single plasmid with a bi-cistronic architecture, either with overlapping open reading frames or separated by a short nucleotide spacer. (**d**) Natural genomic context of *Ms*HO (green arrow) and *Ms*PadC (blue arrow) in *Marinobacter* sp. T13-3. The gene sequence of the *ho* gene and the adjacent *Ms*PadC show overlapping stop ("tga") and start ("gtg") codons, respectively. (**e**) Natural genomic context of *Is*HO (green arrow) and *Is*PadC (blue arrow) in *Idiomarina* sp. A28L. The two ORFs are separated by an in-frame nucleotide spacer (white bar) with a length of 87 bp

integrated into the chromosomal *recA* locus by a CRISPR/Cas9 method modified from Li et al. [14].

Finally, we also want to highlight potential caveats employing the different systems and compare them to the established helper plasmid/overexpression plasmid systems. Thereby, we provide novel tools for specific use cases, a helpful introduction for those new to the field, but also provide new insights to those who sometimes struggle with the reproducibility of bilin production in bacterial systems. Though not tested specifically, these approaches should also be amenable for the production of PCB- and PΦB-binding proteins.

2 Materials

Prepare all solutions using ultrapure water and chemicals of American Chemical Society (ACS) grade or higher. For the equipment and all kits used in this chapter, alternative suppliers can also be used, as long as the specified requirements are met. Vector files and gene sequences, which are used in this chapter, are deposited at https://doi.org/10.3217/tv97k-t9f33 (*see* **Note 1**).

2.1 General Materials

1. Low salt LB medium: 10 g/L tryptone, 5 g/L yeast extract, and 5 g/L NaCl.
2. LB agar plates: Autoclave the medium after addition of 18 g/L agar and add the appropriate amounts of antibiotics after cooling to 60 °C and before pouring the plates.

Antibiotic additives (prepare as 1000× stocks) to LB media/plates:

3. LBK media/plates: Kanamycin final conc. 34 μg/mL.
4. LBSt media/plates: Streptomycin final conc. 100 μg/mL.
5. LBKSt media/plates: Kanamycin final conc. 34 μg/mL, Streptomycin final conc. 100 μg/mL.
6. LBCK media/plates: Chloramphenicol final conc. 30 μg/mL, Kanamycin final conc. 34 μg/mL.

2.2 Bi-cistronic Expression of Heme Oxygenase and Phytochrome

1. PCR polymerase: Use a proofreading polymerase of your choice, for example, Q5® High-Fidelity DNA Polymerase.
2. Coding sequence of the gene of interest: Order from your gene synthesis vendor of choice, for example, order as GeneArt Strings.
3. Overexpression target vector: here pETM11 *Ms*PadC PSM or pETM11 *Is*PadC PSM.
4. Polymerase chain reaction (PCR) cycler: for example, PEQLAB peqSTAR 2× gradient.
5. Plasmid isolation kit: for example, VWR® peqGold Plasmid MiniPrep Kit I.
6. Restriction enzyme to remove methylated plasmid DNA: for example, *Dpn*I.

2.3 CRISPR/Cas9-Mediated Genomic Insertion of ho1 from Synechocystis sp. PCC 6803 into BL21(DE3)

2.3.1 Generation of Plasmid Constructs for the CRISPR/Cas9 Approach

1. LBSt media: prepare 50 mL.
2. LBSt plates.
3. pEcCas (Addgene #73227).
4. pEcgRNA (Addgene #166581).
5. N20 sequence design web service: for example, ATUM (*see* **Note 2**); this sequence determines the target to which the Cas9 nuclease is guided.
6. PCR polymerase: Use a proofreading polymerase of your choice, for example Q5® High-Fidelity DNA Polymerase.

Table 1
Primers to assemble the pEcgRNA N20$_{recA}$ vector and the editing fragment

Primer name	Sequence (5′–3′)
N20_ecgRNA_fwd	ACTGGAAATCTGTGACGCCCGTTTTAGAGCTAGAAATAGCAAG TTAAAATAAGGCTAGTCCGTTATC
N20_ecgRNA_rev	GGGCGTCACAGATTTCCAGTCTGAAACCGCTGCGGCG
uphom_recA_fwd	TAGAGTCGACCTGCAGAAGCGCTATCATCTACAGAGAAATCCGGCG
uphom_recA_rev	CGTGGTGGGTAGCGAAACCC
downhom_recA_fwd	CGTGTTCAGCATCGATAAACGCACAG
downhom_recA_rev	CTTATGGAGCTGCACATGAAGCGTTGGCGGCAGCAC
HO1_pT7_fwd	TTTCGCTACCCACCACGATCCCGCGAAATTAATACGACTCACTATAG
HO1_pT7_rev	GTTTATCGATGCTGAACACGCGGCGGATTTGTCCTACTCA

7. Cloning strain with F′ episome: for example, XL10-Gold (*see* **Note 3**).
8. Bacterial chromosomal DNA isolation kit: for example, GeneJET™ Genomic DNA Purification Kit.
9. DNA cleanup kit: for example, Zymoclean™ Gel DNA Recovery Kit.
10. Gibson assembly kit: for example, NEBuilder® HiFi DNA assembly.
11. Plasmid isolation kit: for example, VWR® peqGold Plasmid MiniPrep Kit I.
12. Primers: *see* Table 1.

2.3.2 Genomic Insertion of ho1 *into the* recA *Locus of BL21(DE3)*

1. LBK media: prepare 200 mL.
2. LBK plates.
3. LBKSt plates.
4. *E. coli* expression strain: for example, BL21(DE3).
5. Electroporation device: for example, MicroPulser Electroporator.
6. Cuvettes for electroporation: for example, Invitrogen™ Electroporation Cuvettes, 0.2 cm (cooled on ice).
7. ddH_2O: filtered, autoclaved, ice cold.
8. 50-mL sterile tubes.
9. Centrifuge for 50-mL tubes: for example, Heraeus Multifuge 3S-R.
10. Centrifuge for 1.5-mL reaction tubes: for example, Eppendorf 5415R centrifuge.

11. 10% glycerol in ddH_2O (w/v): autoclaved, ice cold.
12. Liquid nitrogen.
13. 1 M L-arabinose stock (w/v).
14. Vector pEcgRNA $N20_{recA}$.

2.3.3 Validation of Successful Genomic Integration and Curing of pEcCas/pEcgRNA from E. coli *BL21(DE3)*

1. LB plates.
2. LBSu plates: sucrose final conc. 20 mg/mL.
3. LBSt plates.
4. LBK media: prepare 200 mL.
5. ddH_2O: sterile, room temperature (RT).
6. Inoculation loop.
7. UV crosslinker: for example, Stratagene® UV Stratalinker® 1800.
8. A plate lid cut in half, covered with aluminum foil/duct tape.
9. 1 M rhamnose stock (w/v).
10. Printed paper raster: This is used to subdivide plates for the patching experiment. Also *see* [14].
11. PCR polymerase: Use a cheap polymerase suitable for colony PCR which still has proofreading capabilities, for example, NEB OneTaq® Quick-Load 2× Master Mix.
12. DNA cleanup kit: for example, Zymoclean™ Gel DNA Recovery Kit.
13. Primers: *see* Table 2.
14. 50% glycerol (w/v): autoclaved.

2.3.4 General Phytochrome Expression Protocol

1. Expression strain (*see* Table 3).
2. Expression vectors (*see* Table 3).
3. LBK medium/plates or LBCK medium/plates depending on expression system.

Table 2
Primers to confirm correct genomic integration of *Synechocystis* sp. PCC 6803 *ho1* sequence

Primer name	Sequence
seq_genomic_fwd	CTTCGTTAGTTTCTGCTACGCCTTC
seq_genomic_rev	TTCGGAGGTGGCGAG
seq_genomic_pncC_fwd	ATGAGTGTCAACTTAGCTTCCCAG
seq_genomic_pncC_rev	TGCGTATGCATTGCAGACCTTG

Table 3
Overview of phytochrome expression systems used in our lab

Expression systems	Component 1	Component 2	Component 3
Version 1	BL21(DE3)	pT7-ho1 Cm^R helper plasmid	pET vector with BphP
Version 2	BL21(DE3)	pET vector with overlapping ORF of *ho1* and *bphp*	–
Version 3	BL21(DE3)	pET vector with ORF of *ho* and *bphp* genes separated by nucleotide spacer	
Version 4	BL21(DE3) pEc *ho1* (stable chromosomal *ho1* integration)	–	pET vector with BphP

4. 30 % w/v glucose stock: autoclave separately from media and store at RT.
5. 1 M $MgCl_2$ stock: add to LB media before autoclaving.
6. 10 mg/mL δ-aminolevulinic acid stock (δ-ALA): store at −20 °C.
7. 0.5 M Isopropyl-β-D-thiogalactopyranoside stock (IPTG): store at −20 °C.
8. 1 L flask: for example, 1 L Nalgene® baffled shake flask.
9. 2.5 L flask: for example, 2.5 L Ultra Yield® Flasks.
10. Orbital shaker: 25 mm orbital shaker achieving 130 rpm, for example, Infors HT Multitron Standard.
11. Centrifuge: for example, Beckman Coulter Avanti® J-26XP.
12. Rotor: for example, Beckman Coulter JLA-8.1000.

3 Methods

In recent years, multiple different expression systems for BphPs with simultaneous generation of the biliverdin cofactor were employed. As described, the helper plasmid-based system is the most widespread one in the literature, like the above-mentioned pT7-ho1 system. In this chapter, first, the generation of the alternative holophytochrome expression systems (*see* versions 2-4, Table 3) is outlined, followed by a general expression protocol and a discussion outlining possible hurdles when using the pT7-ho1 helper plasmid system (*see* **Note 4**) and a comparison of the benefits of the four approaches shown in Table 3 (*see* **Note 5**).

3.1 Bi-cistronic Expression of Heme Oxygenase and Phytochrome

3.1.1 Identification of a Heme Oxygenase for Bi-cistronic Expression

For many BphPs, the heme oxygenase (HO) and the phytochrome can be found in the same operon. Interestingly, in some phytochromes, the preceding heme oxygenase shares a sequence overlap of its stop codon with the start codon of the subsequent phytochrome gene, though with a relative frameshift (*see* Fig. 1d). In other phytochromes, the heme oxygenase precedes the phytochrome, but there is an in-frame nucleotide spacer separating the ORFs of both sequences. Both of these genetic architectures, overlapping ORFs or sequential ORFs, can be used for overexpressing a heme oxygenase in tandem with a phytochrome from one single plasmid. For both architectures, sequences were identified that work with our model proteins. A short discussion on their efficiency is provided below (*see* **Note 5**). In the following, a quick method to identify and confirm these heme oxygenase genes in their natural genomic context using blastp and AlphaFold2/Foldseek is shown with examples.

1. Paste the protein sequence of your phytochrome of interest into the blastp web server (https://blast.ncbi.nlm.nih.gov/Blast.cgi?PAGE=Proteins) to identify the organism of origin. In the test example *Ms*PadC, *Marinobacter* sp. T13-3 is identified. Follow the Sequence ID link (KXS53631.1) and then the DBSOURCE accession number (LSRT01000111.1) to access the whole genome shotgun sequencing information. For the test example *Is*PadC, the organism *Idiomarina* sp. A28L is identified. Follow the Sequence ID link (WP_007419415.1), then click on the "Nucleotide" link under "Related information" to access the whole genome shotgun sequence (NZ_AFPO01000007.1).
2. In the list of automatically annotated proteins, look for the accession number of your phytochrome of interest and check the locus of the coding DNA sequence (CDS). Check if there is any hypothetical protein with an overlap of the coding sequence to the phytochrome (as for the *Ms*PadC gene) or with a protein in very close proximity (as for the *Is*PadC gene). In the *Ms*PadC context, a CDS of a hypothetical protein (KXS53630.1) is identified, harboring a 4-nucleotide overlap with a relative −1 frameshift. For *Is*PadC, a CDS already annotated as "biliverdin-producing heme oxygenase" (WP_007419414.1) is identified with a nucleotide sequence spacer between them.
3. Optional: Put the translated sequence into AlphaFold2 [15] and check the quality of predictions. If a sufficiently confident structural model can be generated, run Foldseek [16] and check if structural similarity to published heme oxygenases can be observed.

3.1.2 Cloning of the Bicistronic Expression Vector

This subheading details the cloning of the identified *ho* genes into existing expression vectors of phytochromes. As examples, we use the sequences introduced in Subheading 3.1.1, *Ms*PadC [17] and *Is*PadC [4]. For the *ho* sequence retrieved from *Marinobacter*, the nucleotide sequence was partly codon optimized (*see* **Note 6**). In our case, the sequence was generated with overlaps to an existing expression plasmid containing a BphP (pETM11 *Ms*PadC). There, the *ho* gene was cloned in front/overlapping with the start codon of the His_6 tag and TEV cleavage site (*see* **Note 7**). For the *Idiomarina* derived sequences, the full *ho* sequence was codon optimized and the nucleotide spacer was left unchanged. The *ho* gene plus spacer was cloned into an expression vector (pETM11 *Is*PadC) in front of the His_6 tag and the TEV cleavage site (*see* **Note 7**). A general protocol for phytochrome expression using these vectors is shown in Subheading 3.3.

1. Perform overlap extension PCR using 3 ng of the expression vector (here, a pETM11 vector which already contains a bacteriophytochrome) as a template and the synthesized *ho* gene (plus spacer in the case of the *Idiomarina* sequence) plus overlaps to the template in 250 molar excess as megaprimer. Conduct PCR with an annealing temperature of 55 °C, an extension time sufficient to amplify the whole template plasmid and 20 cycles (*see* **Note 8**). Remove the template by *Dpn*I digest and transform into the cloning strain followed by plasmid isolation.
2. Identify correct plasmids using Sanger sequencing.

3.2 CRISPR/Cas9-Mediated Genomic Insertion of ho1 from Synechocystis sp. PCC 6803 into BL21(DE3)

3.2.1 Generation of Plasmid Constructs for the CRISPR/Cas9 Approach

To generate an expression strain with chromosomal integration of a biliverdin-producing heme oxygenase, insert the *ho1* gene from *Synechocystis* sp. PCC 6803 under the control of a T7 promoter into a chromosomal locus of choice, here the *recA* locus, (*see* **Note 9**) of *E. coli* BL21(DE3). For this purpose, the pEcCas (Addgene #73227)/pEcgRNA (Addgene #166581) system was used (*see* **Notes 3** and **9**) and modified. There are multiple important sequences which need to be generated. First, design the N20 nucleotide spacer which is complementary to the chromosomal target sequence. Clone this sequence in front of the scaffold sequence in the pEcgRNA vector. The N20 nucleotide spacer and the scaffold together form the single-guide RNA (sgRNA) which binds to the Cas9 nuclease and guides it to the chromosomal locus of choice to induce double-strand breakage. Also, a so-called editing fragment has to be designed which contains homologous sequences, for example, to the *recA* locus. In between the homologous sequences, the editing fragment contains the CDS for the protein of interest (POI) which is to be inserted into the genome, here the *ho1* gene (*see* Fig. 2). Using this technique, a stable insertion of *ho1* into *recA* (*recA*::*ho1*) is achieved which knocks

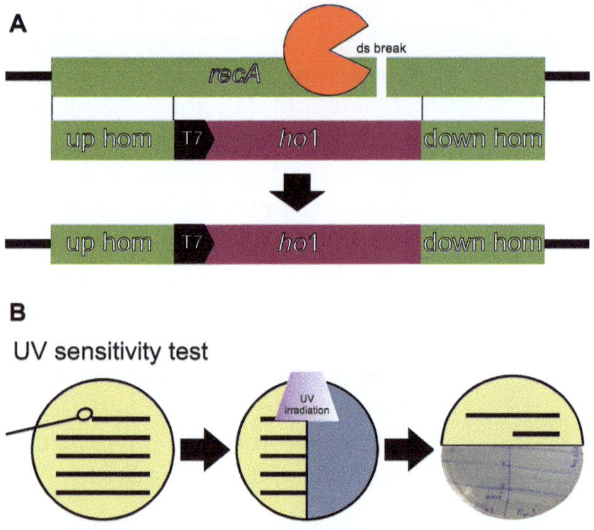

Fig. 2 Principle of CRISPR/Cas9 genomic integration and UV sensitivity tests. (**a**) Overview of events after transforming the pEcgRNA $N20_{recA}$ plasmid into recombineering cells which contain pEcCas encoding the Cas9 enzyme. The Cas9 protein (orange) is guided to its target locus by the $sgRNA_{recA}$ where it is causing double-strand cleavage, here in the *recA* locus. Aided by the λ-Red proteins, the editing fragment containing the *Synechocystis* sp. PCC 6803 *ho1* sequence is inserted into the *recA* locus, guided by the upper homology ("up hom") and downstream homology sequences ("down hom"). Thus, the *ho1* sequence replaces the *recA* sequence which is targeted by the Cas9 nuclease. The result is an *E. coli* BL21(DE3) expression strain where *ho1* is under the control of an IPTG-inducible T7 promoter. (**b**) Workflow of the UV sensitivity test. This test can be conducted after the transformation of pEcCas recombineering cells with pEcgRNA $N20_{recA}$ which results in the introduction of the *ho1* gene as described in panel A. The deletion/truncation of *recA* leads to a bacterial strain which is more sensitive to UV irradiation than *E. coli* BL21(DE3). This opens the possibility for an easy screening system for genomic integration of *ho1* into the *recA* locus. As described in Subheading 3.2.3, **step 1**, use an inoculation loop to streak clones with the potential *ho1* insertion on a LB plate. Cover half of the plate with a UV light-blocking material and irradiate the plate with UV light; the streaked cells should face the UV source. Clones which show survival under the UV protection but die upon UV exposure are *recA* negative and are highly likely to show correct integration of *ho1* into the chromosome

out the function of the recombinase gene. This insertion makes the strain sensitive to UV radiation which aids in the downstream screening for clones with successful genomic integration of *ho1*. Finally, this protocol results in a strain which generates BV upon IPTG induction which is amenable for phytochrome expression.

1. Design the N20 sequence using the online tool ATUM (*see* **Note 1**), which suggests the sequence "ACTGGAAATCTGT-GACGCCC" for the chosen *recA* locus.

2. Use overlapping primers (*see* **Note 10**) to insert the N20 nucleotide spacer sequence into the plasmid pEcgRNA before the sgRNA scaffold sequence (primers: N20_ecgRNA_fwd/rev) to generate the vector pEcgRNA $N20_{recA}$. The primers amplify the whole plasmid plus the N20 insertion. Transform the PCR mix into chemically competent XL10-Gold cells which circularize the plasmid using the introduced overhangs. Isolate the plasmid and confirm the insertion of the $N20_{recA}$ sequence by Sanger sequencing.
3. Run "killing efficiency" experiments (*see* **Note 11**) with the sgRNAs created from the N20 nucleotide spacer and the scaffold sequence in the pEcgRNA $N20_{recA}$ vector.
4. The editing fragment consists of three smaller fragments, the "upstream homology sequence," the *ho1* gene, and the "downstream homology sequence," which were together subcloned in this order into the vector pEcgRNA 34 nucleotides after the sgRNA scaffold sequence. To generate the homology sequences (*see* **Note 12**), purify genomic DNA from *E. coli* BL21(DE3) using the bacterial chromosomal DNA isolation kit. Conduct a standard PCR to amplify the 320 bp upstream homology sequence (primers: uphom_recA_fwd/rev) from the genomic DNA in parallel to the 270 bp downstream homology sequence (primers: downhome_recA_fwd/rev). PCR-amplify the *ho1* gene including the T7 promoter and the rrnB T1 terminator from the plasmid pT7-ho1 Cm^R (primers: HO1_pT7_fwd/rev). Also amplify the pEcgRNA $N20_{recA}$ vector backbone with the corresponding primers. Subject all liquid PCR samples to a DNA cleanup step.
5. Circularize the pEcgRNA $N20_{recA}$ vector backbone, the upstream homology sequence, the *ho1* fragment, and the downstream homology sequence with a Gibson assembly kit.
6. Optional: Amplify the editing fragment from the final vector pEcgRNA $N20_{recA}$ + editing fragment by PCR (primers: uphom_recA_fwd/downhom_recA_rev), followed by a DNA cleanup step using a kit to generate a highly concentrated linear editing fragment stock (~500 ng/μL) (*see* **Note 13**).

3.2.2 Genomic Insertion of ho1 *into the* recA *Locus of BL21(DE3)*

With the following protocol, the *ho1* sequence is inserted into the *recA* locus, thereby disrupting its open reading frame (*see* Fig. 2 and **Note 14**). This is achieved by transforming the plasmid pEcCas (*see* **Note 15**) encoding the Cas9 enzyme into BL21(DE3) (*see* **Note 16**). From these cells, new electrocompetent cells are prepared which are shortly treated with arabinose to induce expression of the λ-Red recombinase system co-encoded by pEcCas. Thus, the cells are ready to repair any double-strand break of the chromosome aided by the λ-Red recombinase proteins. Upon providing the pEcgRNA $N20_{recA}$ vector and the editing fragment, the

$sgRNA_{recA}$/Cas9 complex introduces a double-strand break in the *recA* locus which is repaired with the editing fragment containing the *hol* sequence. Since the sequence which is targeted by the $sgRNA_{recA}$ is deleted in this process, clones with successful genomic integration survive, whereas other clones should die due to continuous chromosomal damage exerted by the $sgRNA_{recA}$/Cas9 complex. A detailed graphical depiction of the workflow is shown in the original paper [14].

1. Prepare electrocompetent *E. coli* BL21(DE3) cells (*see* **Note 17**) by growing them in 35 mL LB media at 37 °C and 200 rpm to an OD_{600} of 0.5–0.6. Cool the culture in ice water for 10 min, transfer to a 50-mL sterile tube and centrifuge (10 min, 4000 *g*, 4 °C). After this point, all solutions in contact with the cells should be sterile and ice cold. Resuspend the pellet in 1 mL ddH_2O until homogeneous and fill up to 35 mL ddH_2O followed by centrifugation (10 min, 4000 *g*, 4 ° C). Resuspend the pellet in 1 mL ddH_2O and transfer to a sterile 1.5-mL reaction tube. Centrifuge (10,000 *g*, 30 s, 4 °C) the solution, discard the supernatant, and repeat resuspension and centrifugation in the 1.5-mL reaction tube. Finally, resuspend in 200 μL ddH_2O for immediate usage or with 10% glycerol (w/v) in ddH_2O for flash freezing in liquid nitrogen and storage.
2. Transfer 40 μL of electrocompetent *E. coli* BL21(DE3) to precooled 0.2 cm electroporation cuvettes, add 100 ng of pEcCas plasmid and conduct the electroporation at 1.8 kV followed by outgrowth for 30 min at 37 °C in 150 μL SOC media. Plate on LBK plates.
3. Grow a clone in 50 mL LBK media at 37 °C and 200 rpm until an OD_{600} of 0.35 is achieved, followed by addition of L-arabinose to a final concentration of 50 mM. Incubate at 37 °C and 200 rpm for 30 min to induce expression of the λ-Red proteins. From this culture, again electrocompetent cells should be prepared as outlined in **step 1** to generate the "recombineering cells" (*see* **Note 18**).
4. Transform the electrocompetent recombineering cells with 100 ng of the vector pEcgRNA $N20_{recA}$ and 500 ng linear editing fragment stock (*see* Subheading 3.2.1, **step 5**), according to the protocol in **step 2**, followed by plating on LBKSt plates (*see* **Note 19**).

3.2.3 Validation of Successful Genomic Integration and Curing of pEcCas/pEcgRNA from E. coli BL21(DE3)

To check the successful integration of *hol* into the *recA* locus, subject clones to UV irradiation since *ΔrecA* strains show markedly decreased survival rates under UV light. Positive clones still harbor both vectors pEcCas and pEcgRNA $N20_{recA}$ which can be removed in the last part of the protocol. pEcCas encodes a second sgRNA under the control of a rhamnose promoter which upon induction leads to the targeted degradation of pEcgRNA by Cas9. The removal of the pEcgRNA plasmid can be confirmed by simultaneous patching on LBK and LBKSt plates, where successful degradation of the vector is indicated by survival on the first and death on the second plate. Optionally, the curing of pEcgRNA can also be confirmed by colony PCR. The vector pEcCas also contains the gene sequence of *sacB* coding for the enzyme levansucrase. This enzyme produces the toxic compound levan, which turns the BL21 (DE3) pEcCas strain highly sensitive to sucrose. Thus, plating on sucrose-containing plates efficiently removes pEcCas by applying this negative selection pressure. Lastly, the genomic integration is confirmed by colony PCR using two primer pairs, where in each, one primer binds in the genome and the other within *ho1*.

1. Resuspend single colonies from the step above in 50 μL of sterile ddH_20 to create a cell stock used here in **steps 1** and **2** for each colony individually. Dip a sterile inoculation loop into the cell stock and streak a continuous line covering the full diameter of an LB plate; 6–8 samples can be fit on the same plate in parallel lines. Create a cell suspension of BL21(DE3) as negative control and a *ΔrecA* cloning strain like XL10-Gold as positive control. Cover half of the plate for later UV irradiation. To do that, cut the lid of a standard LB plate in half and cover first with a layer of aluminum foil followed by a layer of duct tape. Prewarm the UV source in the UV crosslinker for 2 min, then set to energy mode and 5000 mJ. Turn the LB plate with the lines of bacterial samples facing the UV source and cover the plate so that, for each sample streaked as a line, half of it is covered. Irradiate with UV light (workflow Fig. 2b; *see* **Note 20**) and place the LB plate at 37 °C overnight.
2. In parallel to the UV sensitivity tests, use 5 μL each of the same cell stocks as above to inoculate 4 mL of LBK media + 10 mM rhamnose and grow at 200 rpm for 6 h at 37 °C. This induces the sgRNA which targets pEcgRNA, leading to degradation of the plasmid. Spin the cultures at 4000 *g* (5 min, RT) and resuspend in 4 mL LB media and grow at 200 rpm for 2 h at 37 °C.
3. With a marker, divide an LBSu plate into four equal parts. Use a sterile inoculation loop to streak four different LB cultures per LBSu plate to induce removal of the pEcCas plasmid. Thin out the liquid cultures appropriately, as the next step requires single colonies.
4. On the next day, check the UV-treated plates from **step 1**. Samples which did grow in the shielded environment but not in the UV-exposed areas are *recA* negative and likely positive for the genomic integration. Check the LBSu plates from **step 2**; single colonies from this plate, which are UV sensitive and thus *recA negative*, are further processed. Subdivide three

plates (LB, LBK, and LBSt) into around 30 small rectangular squares with a printed template which is laid under each plate (*see* also [14]). Touch single colonies on the LBSu plate with a sterile inoculation loop and patch 0.5 cm lines on each of the three plates for each single colony (*see* **Note 21**). Incubate the three plates at 37 °C overnight. Clones which grow on the LB but not LBK or LBSt plate were successfully cured of the pEcCas and pEcgRNA vectors, meaning their complete removal.

5. Streak a clone which is sensitive to kanamycin and streptomycin on an LB plate using an inoculation loop to retrieve single colonies. Dip single colonies with a pipette tip in 10 μL of sterile ddH20 at room temperature to create a cell stock. Confirm the correct sequence of the *Synechocystis* sp. PCC 6803 *ho1* in the genome of BL21(DE3) by PCR using 1 μL of cell stock per clone. Utilize a premixed formulation containing a DNA polymerase with proofreading capabilities and two primer pairs (seq_genomic_fwd/rev; seq_genomic_pncC_fwd/rev) to confirm the successful genomic integration by standard PCR (*see* **Note 22**). Purify the PCR products of these reactions with a DNA cleanup kit. To the liquid PCR mix, add three times the volume of the column binding buffer followed by the standard DNA cleanup procedure. Send the purified PCR products for Sanger sequencing using sufficient primers to confirm the correct integration of *ho1* into the *recA* locus of BL21(DE3).
6. From the cell stock of **step 5**, inoculate 10 mL LB media and grow till an OD_{600} of 1. Mix 500 μL of cell suspension with 500 μL of 50% glycerol and store these stocks at −80 °C.
7. Optional: quick protocol to test successful *Synechocystis* sp. PCC 6803 *ho1* genomic integration in BL21(DE3). Inoculate BL21(DE3) WT and BL21(DE3) *ho1* in 10 mL LB media without any additives and grow till reaching the stationary phase (>20 h). Spin 1 mL of culture each (10,000 *g*, RT, 30 s) and check the color of the supernatant; the BL21(DE3) *ho1* sample should show green coloring. Measure the supernatants on the NanoDrop, a distinct peak around 400 nm should be visible if biliverdin is successfully produced by the genomic integration strain (*see* **Note 23**).

3.3 A General Phytochrome Expression Protocol

1. Choose your desired expression system (e.g., from Table 3) and transform the expression strain with a vector containing the phytochrome and/or a heme oxygenase. In our case, the overexpression vector is a pETM11 vector with a kanamycin-resistance cassette. Plate on LBK plates (LBCK in version 1) and incubate overnight at 37 °C.

2. Scratch off multiple colonies (20–50, *see* **Note 24**) of the plates and inoculate them in 100 mL (*see* **Note 25**) of LB media containing appropriate antibiotics, 0.3% glucose (w/v), and 8.5 mM $MgCl_2$ in a 1 L baffled flask (*see* **Note 26**). Shake at 130 rpm at 37 °C until an OD_{600} of 0.8–1 is achieved.
3. Transfer 20–30 mL of pre-culture into 1 L of LB media containing appropriate antibiotics, 0.3% glucose (w/v), and 8.5 mM $MgCl_2$ in a 2.5 L baffled flask. Shake at 130 rpm at 37 °C until an OD_{600} of 0.5 is achieved.
4. Add δ-ALA to a final concentration of 10 μg/mL and put the flasks into a shaker precooled to 18 °C. Shake at 18 °C, 130 rpm for 30 min to cool down the culture (*see* **Note 27**).
5. Add IPTG to a final concentration of 0.25 mM and shake overnight at 18 °C, 130 rpm. Harvest the bacteria by centrifugation at 4500 rpm. A strong green coloration of the pellet is usually a good indication of successful phytochrome production. Alternatively, conduct sodium dodecyl sulfate–polyacrylamide gel electrophoresis (SDS-PAGE) on samples before and after induction and check for overexpression bands on the gel. Store the pellet at −20 °C in the freezer for further processing (for comments on advantages of different expression approaches, *see* **Notes 4** and **5**).

4 Notes

1. Under the link https://doi.org/10.3217/tv97k-t9f33, nucleotide sequences of expression vectors and the modified pEcgRNA vector are deposited. The file Data_comment.pdf contains an overview of the files deposited, whereas the folder "Data" contains the deposited nucleotide sequences.
2. When using the webpage ATUM, set the parameters to *E. coli* K-12 MG1655, PAM NGG, and wild-type Cas9. Insert the sequence around the desired cutting site (approx. 200 bp) and choose the N20 nucleotide spacer with the highest scores. Alternatively, choose other webpages to generate N20 sequences or consult the table in the supplements of this paper [18], which contains a list of N20 sequences which work in *E. coli*. Carefully read this paper, as some of these sequences only work efficiently with simultaneous knockdown of the *recA* gene locus. To design the N20 nucleotide spacer, the upstream and downstream homology sequences, as well as the seq_genomic primers, use the *recA* nucleotide sequence accessible on www.ncbi.nlm.nih.gov for the reference chromosome of *E. coli* strain K-12 substr. MG1655 (Sequence ID NC_000913.3, region 2,822,708–2,823,769; this sequence

is on the negative strand). When designing the N20 sequence and the editing fragment, keep in mind that after recombineering the target site of the sgRNA should be deleted with the insert of choice to avoid continuous chromosome cutting and cell death.

3. When working with the pEcgRNA plasmid as deposited at Addgene, remember that it contains the *ccdB* gene, which allows only the usage of cloning strains to which the gene product is not toxic, like XL10-Gold. However, when transforming a mutated vector where the N20 nucleotide spacer sequence successfully replaced the *ccdB* sequence, this circumstance can be handy. Use a strain which cannot tolerate *ccdB*, like NEBalpha, so only clones where the plasmid was successfully mutated to contain the N20 nucleotide spacer can grow.

4. At least for chemically competent *E. coli* BL21(DE3) cells, cotransformation of the pT7-hol helper plasmid with the pET vector containing the POI is very inefficient. Instead, BL21(DE3) should be transformed with pT7-hol, clones selected, and then made chemically competent again before transformation with a vector containing the POI. Note that, at least in our hands, this BL21(DE3) pT7-hol cells have a limited shelf life at −80 °C and should be newly prepared at least every 4–6 months. Sometimes even clone-to-clone variation of the BL21(DE3) pT7-hol cells can occur; therefore, it pays off to pick at least four clones separately for preparation of competent cells and testing their expression capabilities. When producing a new stock for the helper plasmid pT7-hol Cm^R, be aware to not use a cloning strain with a natural antibiotic resistance against chloramphenicol like XL10-Gold.

5. Expression yields in the different systems can depend on the phytochrome of interest, but a trend is observed depending on the system used for biliverdin generation. The following list is sorted from higher to lower expression yields: (a) expression strain with genomic integration of *Synechocystis* sp. PCC 6803 *hol* (version 4); (b) pT7-hol (*Synechocystis* sp. PCC 6803) helper plasmid-based expression; (c) bi-cistronic expression plasmid using the HO/phytochrome architecture separated by a nucleotide spacer (version 3); and (d) bi-cistronic expression plasmid using the HO/phytochrome architecture with overlapping stop and start codon (version 2). It should be noted that the differences between the genomic integration strain and the pT7 helper plasmid approach are minute and might be within errors observed in batch-to-batch reproducibility. Expression yield-wise, both of these approaches outperform the bi-cistronic expression systems. It should be noted that between the two different bi-cistronic architectures—separation by a nucleotide spacer versus overlapping ORFs—

deviations in expression yields are minute. Still, cotranslational expression of overlapping coding regions is the closest to the natural situation at least observed for some BphPs. Though the yield potentially drops, this system has advantages because of its straightforward usage and its usability in circumstances where yield is less important like high-throughput approaches. Apart from slightly higher yields, genomic integration provides benefits like increased shelf life and slightly faster growth characteristics compared to the pT7 helper plasmid-based expression.

6. The majority of the *Ms*HO coding sequence was codon optimized for expression in *E. coli*; however, the last 21 nucleotides remained in its natural sequence context to preserve potentially important sequence properties for bi-cistronic expression. The decision to protect specifically 21 nucleotides of the *ho* gene was based on the idea to at least include a potential cryptic Shine–Dalgarno sequence in the upstream CDS, but the influence of not codon optimizing this region was not tested in detail. For a general review on overlapping genes, *see* [13].
7. The overlapping sequences in the *Ms*HO/*Ms*PadC expression plasmid constitute the upstream *ho* CDS stop codon and the downstream start codon corresponding to the His_6-tag. In comparison, the *Is*HO/*Is*PadC expression plasmid harbors *Is*HO, the nucleotide spacer, and then the start codon of the His_6-tag. Although the inclusion of the His-tags for the PadC proteins deviates from the genetic context found in the organisms, this setup still leads to sufficiently high expression levels. As mentioned, *Ms*HO has an overlapping stop codon with its phytochrome, whereas in *Is*HO, there is a nucleotide spacer before the adjacent phytochrome. However, at least for overexpressions, *Ms*HO can be cloned with an overlapping start codon in front of *Is*PadC, and *Is*HO can be cloned with a nucleotide spacer in front of *Ms*PadC. In both cases, the phytochrome is still produced loaded with BV. In our tests, the highest phytochrome yield was achieved for *Is*HO cloned with a nucleotide spacer in front of *Ms*PadC. However, cloning of *Is*HO with an overlap of the stop codon or *Ms*HO with a nucleotide spacer was not tested.
8. The overlap extension protocol used for the insertion of the *Ms*HO/*Is*HO into the pETM11 vector was first described in [19]. It should be noted that the molar excess range (around 250×) of the megaprimer (aka insert), the annealing temperature of 55 °C and the use of 20 cycles are pivotal for the success of the method.
9. First and foremost, it is strictly necessary to read and understand the two publications where the pEcCas/pEcgRNA [14] and an earlier version [20] are presented. The genomic

integration location for heterologous protein expression matters; thus, the *recA* locus chosen as stable integration at this chromosomal location showed decent expression levels for *E. coli* in the literature [21]. Furthermore, *recA* knockout/knockdown was also shown to aid the efficiency of CRISPR/Cas9-mediated recombineering in *E. coli* [18]. For a general overview about CRISPR/Cas9 basics and methods, *see* https://www.addgene.org/guides/crispr/.

10. The overlapping primer design for the insertion of the N20 nucleotide spacer into the pEcgRNA vector was performed according to the protocols described by Liu and Naismith et al. [22].

11. If you want to design an additional/alternative N20 nucleotide spacer sequence to target a different gene locus, experimentally try multiple (at least three) designs to decide on the most suitable one. Run preexperiments on their DNA double-strand cutting efficiency. Create electrocompetent cells of your choice harboring the induced pEcCas vector as described in Subheading 3.2.2, **steps 1** and **3**. Transform these recombineering-ready cells with pEcgRNA empty (no sgRNA) and in parallel pEcgRNA plus the novel sgRNA sequence. Count the colonies on both plates, the pEcgRNA empty plate should show a multitude of colonies, whereas the second plate should show very few if any colonies. The ratio between these plates (# colonies pEcgRNA N20/# colonies pEcgRNA empty) should usually be <5% to indicate a high efficiency of double-strand breakage by Cas9 in the desired locus.

12. Instead of amplifying the homology sequences from the chromosome and assembling the editing fragment by Gibson assembly, the editing fragment can also be ordered from a nucleotide synthesis company. If you do that and want to target a gene locus which might show variability between strains, one could think about amplifying the gene region of interest with PCR and sequencing it to confirm that the regions targeted show the expected sequence. Also remember not to codon optimize the sequence of the homology regions.

13. We initially inserted the editing fragment into the pEcgRNA + $N20_{recA}$ vector, mainly, because we thought it would be easier to handle that way. When generating the BL21(DE3) pEcCas arabinose-induced recombineering cells, transformation of the whole pEcgRNA $N20_{recA}$ + editing fragment works for genomic insertion. However, we realized that cotransformation of the pEcgRNA + $N20_{recA}$ vector with a highly concentrated linear editing fragment stock (generated in Subheading 3.2.1, **step 5**) works more efficiently in our hands. Sidenote: Due to the study context in which the pEcCas/pEcgRNA system was

intended to be used, the resistance cassette of the pEcgRNA vector was exchanged to Amp^R. If your study design does not require this, it is strongly suggested to not follow this complication. Therefore, the backbone amplification primers are not provided in the manuscript as they include the generation of overhangs to an ampicillin-resistance cassette.

14. Note that, in the shown example, not the full *recA* sequence is deleted but the start codon and roughly 300 bp of the CDS are kept, which resembles a truncation of the protein. To avoid translation of *ho1* under the *recA* promoter, the *ho1* ORF was inserted on the opposite strand compared to the reading direction of *recA*.
15. For isolating the pEcCas plasmid, it is recommended to use Midiprep-scale isolation kits, since the vector copy number is very low.
16. To generate *E. coli* BL21(DE3) pEcCas cells, also high-quality chemically competent cells can be used; however, their transformation efficiency is extremely low and retrieval of positive transformants is not guaranteed.
17. When using electrocompetent cells, it is pivotal to use plasmid preparations which were eluted with highly pure ddH_2O to prevent arcing.
18. When preparing the electrocompetent BL21(DE3) pEcCas arabinose-induced cells, it is highly recommended to immediately pursue with the workflow. Shelf life of flash-frozen cell stocks in 10% glycerol at −80 °C was observed to be less than 2 weeks.
19. After the last recombineering step (*see* Subheading 3.2.2, **step 4**), successful genomic integration can already be confirmed using the colony PCR/sequencing approach. However, this step is rather time-consuming and, since not a single clone was observed which was UV sensitive and did not have the correct insertion, might not be needed.
20. The exact dose for UV irradiation needs to be empirically determined, as it strongly depends on the *E. coli* strain used. Conduct experiments to identify the ideal dosage which kills a *ΔrecA* cloning strain but not your *recA*-positive expression strain.
21. In our hands, the least efficient step in the pEcCas/pEcgRNA workflow is the rhamnose-induced removal of the pEcgRNA plasmid. Thus, the patching experiment (*see* Subheading 3.2.3, **step 2**) should be conducted with at least 30 clones to retrieve a strain which is cured from pEcCas and pEcgRNA. Since the biliverdin-producing properties of the strain were more important to us than removal of pEcgRNA, the final generated strain

used for successful expression of phytochromes still contains this plasmid and is therefore ampicillin resistant. It was thus termed BL21(DE3) *recA::ho1* pEcgRNA. Removal of pEcCas works very efficiently with high curing success.

22. To confirm correct insertion of the gene of interest, it is recommended to use at least two primer pairs for amplification and sending the PCR product purified with a DNA cleanup kit for Sanger sequencing to check the sequence of the insert and the upstream/downstream regions. In this case, primer pair 1 (seq_genomic_fwd/rev) binds upstream of the *recA* sequence which is not part of the editing fragment and within the *Synechocystis* sp. PCC 6803 *ho1* sequence, respectively. Primer pair 2 (seq_genomic_pncC_fwd/rev) binds within the *ho1* gene and a neighboring gene to *recA*, here *pncC*.
23. For the optional quick protocol to confirm active HO1 in BL21(DE3) (*see* Subheading 3.2.3, **step** 7), it is not necessary to add additives like δ-ALA or IPTG as the leaky expression from the T7 promoter and the natural heme metabolism in the strain is sufficient to allow biliverdin accumulation. In fact, it was observed that addition of δ-ALA or IPTG in the absence of a phytochrome, which would sequester biliverdin, leads to less obvious green coloring of the culture supernatant.
24. Inoculation with multiple colonies instead of a single clone can improve the reproducibility of the expression process. Furthermore, this allows to conduct the expression protocol from inoculating the preculture to adding IPTG in 1 day which in our experience also improves reproducibility.
25. 100ml preculture is sufficient to inoculate a maximum of 4 L main culture. Scale up the preculture volume accordingly if more than 4 L of main culture is generated. For example, if 12 flasks with 1 L media each should be inoculated, generate at least 400 mL of preculture ideally in one flask to aid uniformity over all flasks.
26. Proper aeration is key; use baffled flasks and the culture volume should be maximally 20% of the flask volume in all liquid culturing steps. In addition, cultures should be shaken thoroughly; we typically use 130 rpm on a 25 mm orbit shaker.
27. At least for our proteins, cooling of the culture to 18 °C prior to IPTG addition (*see* Subheading 3.3, **step 4**) is absolutely essential for proper expression of soluble holophytochrome to the point where neglecting this step leads to complete failure of soluble protein production.

Acknowledgements

We want to express our sincere gratitude to Birgit Wiltschi who provided critical insights into the CRISPR/Cas9 method with extensive knowledge about the original pCas/pTargetF plasmid system.

O.M.E. is supported by the Austrian Science Fund (FWF) grant https://doi.org/10.55776/DOC130 and was trained within the framework of the PhD program Biomolecular Structures and Interactions (BioMolStruct). This research was funded in whole or in part by the Austrian Science Fund (FWF) grant https://doi.org/10.55776/P32022 to A.W. For open access purposes, the authors have applied a CC BY public copyright license to any author-accepted manuscript version arising from this submission.

References

1. van der Horst MA, Hellingwerf KJ (2004) Photoreceptor proteins, "star actors of modern times": a review of the functional dynamics in the structure of representative members of six different photoreceptor families. Acc Chem Res 37(1):13–20. https://doi.org/10.1021/ar020219d
2. Hughes J, Winkler A (2024) New insight into phytochromes: connecting structure to function. Annu Rev Plant Biol 75(1):153–183. https://doi.org/10.1146/annurev-arplant-070623-110636
3. Beale SI (1994) Biosynthesis of open-chain tetrapyrroles in plants, algae, and cyanobacteria. Ciba Found Symp 180:156–168; discussion 168–71. https://doi.org/10.1002/9780470514535.ch9
4. Gourinchas G, Etzl S, Göbl C et al (2017) Long-range allosteric signaling in red light-regulated diguanylyl cyclases. Sci Adv 3: e1602498. https://doi.org/10.1126/sciadv.1602498
5. Song C, Mroginski MA, Lang C et al (2018) 3D structures of plant phytochrome A as Pr and Pfr from solid-state NMR: implications for molecular function. Front Plant Sci 9:498. https://doi.org/10.3389/fpls.2018.00498
6. Landgraf FT, Forreiter C, Hurtado Picó A et al (2001) Recombinant holophytochrome in Escherichia coli. FEBS Lett 508(3):459–462. https://doi.org/10.1016/s0014-5793(01)02988-x
7. Gambetta GA, Lagarias JC (2001) Genetic engineering of phytochrome biosynthesis in bacteria. Proc Natl Acad Sci USA 98(19): 10566–10571. https://doi.org/10.1073/pnas.191375198
8. Yang X, Kuk J, Moffat K (2008) Crystal structure of *Pseudomonas aeruginosa* bacteriophytochrome: photoconversion and signal transduction. Proc Natl Acad Sci USA 105(38):14715–14720. https://doi.org/10.1073/pnas.0806718105
9. Schmidt P, Gertsch T, Remberg A et al (1998) The complexity of the Pr to Pfr phototransformation kinetics is an intrinsic property of native phytochrome. Photochem Photobiol 68(5): 754–761. https://doi.org/10.1111/j.1751-1097.1998.tb02541.x
10. Mukougawa K, Kanamoto H, Kobayashi T et al (2006) Metabolic engineering to produce phytochromes with phytochromobilin, phycocyanobilin, or phycoerythrobilin chromophore in Escherichia coli. FEBS Lett 580(5): 1333–1338. https://doi.org/10.1016/j.febslet.2006.01.051
11. Tarutina M, Ryjenkov DA, Gomelsky M (2006) An unorthodox bacteriophytochrome from *Rhodobacter sphaeroides* involved in turnover of the second messenger c-di-GMP. J Biol Chem 281(46):34751–34758. https://doi.org/10.1074/jbc.M604819200
12. Multamäki E, García de Fuentes A, Sieryi O et al (2022) Optogenetic control of bacterial expression by red light. ACS Synth Biol 11(10):3354–3367. https://doi.org/10.1021/acssynbio.2c00259
13. Wright BW, Molloy MP, Jaschke PR (2022) Overlapping genes in natural and engineered genomes. Nat Rev Genet 23(3):154–168. https://doi.org/10.1038/s41576-021-00417-w
14. Li Q, Sun B, Chen J et al (2021) A modified pCas/pTargetF system for CRISPR-Cas9-assisted genome editing in Escherichia coli. Acta Biochim Biophys Sin Shanghai 53(5): 620–627. https://doi.org/10.1093/abbs/gmab036
15. Mirdita M, Schütze K, Moriwaki Y et al (2022) ColabFold: making protein folding accessible to all. Nat Methods 19(6):679–682. https://doi.org/10.1038/s41592-022-01488-1

16. van Kempen M, Kim SS, Tumescheit C et al (2024) Fast and accurate protein structure search with Foldseek. Nat Biotechnol 42(2): 243–246. https://doi.org/10.1038/s41587-023-01773-0
17. Böhm C, Gourinchas G, Zweytick S et al (2022) Characterisation of sequence-structure-function space in sensor-effector integrators of phytochrome-regulated diguanylate cyclases. Photochem Photobiol Sci 21(10):1761–1779. https://doi.org/10.1007/s43630-022-00255-7
18. Moreb EA, Hoover B, Yaseen A et al (2017) Managing the SOS response for enhanced CRISPR-Cas-based recombineering in E. coli through transient inhibition of host RecA activity. ACS Synth Biol 6(12):2209–2218. https://doi.org/10.1021/acssynbio.7b00174
19. Bryksin AV, Matsumura I (2010) Overlap extension PCR cloning: a simple and reliable way to create recombinant plasmids. BioTechniques 48(6):463–465. https://doi.org/10.2144/000113418
20. Jiang Y, Chen B, Duan C et al (2015) Multigene editing in the Escherichia coli genome via the CRISPR-Cas9 system. Appl Environ Microbiol 81(7):2506–2514. https://doi.org/10.1128/AEM.04023-14
21. Englaender JA, Jones JA, Cress BF et al (2017) Effect of genomic integration location on heterologous protein expression and metabolic engineering in E. coli. ACS Synth Biol 6(4): 710–720. https://doi.org/10.1021/acssynbio.6b00350
22. Liu H, Naismith JH (2008) An efficient one-step site-directed deletion, insertion, single and multiple-site plasmid mutagenesis protocol. BMC Biotechnol 8:91. https://doi.org/10.1186/1472-6750-8-91

Chapter 4

Preparation of Phycocyanobilin (PCB) with and without $^{13}C/^{15}N$ Isotope Labeling for in vitro Phytochrome Autoassembly

Jon Hughes and Christina Lang

Abstract

Free phycocyanobilin (PCB) is useful in structure/function studies of phytochrome photoreceptors. Characteristically, apophytochromes can attach bilin cofactors such as PCB autocatalytically in vitro, thereby allowing chemical modification or separate isotopic labeling for vibrational spectroscopy, in particular NMR. Here we describe methods for producing, purifying, and analyzing unlabeled and universally $^{13}C/^{15}N$-labeled PCB from cyanobacteria. We also outline methods for in vitro autoassembly and analysis of holophytochromes from apoprotein and PCB.

Key words Bilin cofactor, Isotope labeling, Methanolysis, *Synechocystis* culture, Autoassembly, Photoreceptor, Tetrapyrrole

Abbreviations

ARR4	*Arabidopsis* response regulator 4
BphP	Bacteriophytochrome photoreceptor
BV	Biliverdin IXα
CHES	N-cyclohexyl-2-aminoethanesulfonic acid
Cph1	Cyanobacterial phytochrome from *Synechocystis* 6803
Cys	Cysteine
DMSO	Dimethyl sulfoxide
EDTA	Ethylenediaminetetraacetate
FR	Far-red (700–800 nm near-infrared) light
H_2O	Pure water
HEPES	4-(2-Hydroxyethyl)-1-piperazineethanesulfonic acid
HPLC/FPLC	High-/fast-performance liquid chromatography
LED	Light-emitting diode
PAR	Photosynthetically active radiation (400–700 nm)
PCB	Phycocyanobilin
Pfr	FR-absorbing phytochrome state
PIF6	Phytochrome-interacting factor 6

Andreas Winkler (ed.), *Phytochromes: Methods and Protocols*, Methods in Molecular Biology, vol. 2970,
https://doi.org/10.1007/978-1-0716-4791-2_4,

Pr	R-absorbing phytochrome state
PΦB	Phytochromobilin
R	Red light
SAR	Specific absorbance ratio ($A_{\lambda max,R}/A_{280nm}$)
SDS	Sodium dodecyl sulfate
ΔA_λ	(After R)–(after FR) difference spectrum
$\Delta\Delta A_{R\text{-}FR}$	Difference between the absorbance values at $\lambda_{max,R}$ and $\lambda_{min,FR}$ in the ΔA_λ difference spectrum
ε	Molar extinction coefficient
λ	Wavelength

1 Introduction

The apoprotein moiety of various biliproteins can attach bilin (linear tetrapyrrole) cofactors either with the help of a separate lyase or, as in the case of phytochrome photoreceptors, autonomously [1]. Whereas in bacteriophytochrome photoreceptors (BphPs), the cofactor is biliverdin IXα (BV), in both plant phytochromes and Cph1-like phytochromes from cyanobacteria [2, 3]; it can be phycocyanobilin or phytochromobilin (PCB or PΦB, respectively). The attachment process is known variously as chromophorylation, reconstitution, or, better, autoassembly. Although apophytochromes tend to aggregate at high concentrations, posttranslational autoassembly in vitro has advantages in several experimental contexts, in particular when a synthetic analog is to replace the natural cofactor, for example in the case of locked BV [4], when specifically cofactor atoms are to be labeled isotopically for spectroscopic studies, in particular nuclear magnetic resonance (NMR) [5–10], or when the host used for overproduction cannot be engineered genetically to biosynthesize the cofactor. Even in the case of *E. coli*, where BV, PCB, and PΦB biosynthetic pathways can be introduced easily [11, 12] (*see* Chapter 6 by Shenkutie and Guan in this volume), various laboratories routinely add excess free chromophore to their holophytochrome extract, presumably to allow residual apoproteins to autoassemble in vitro.

Whereas in Cph1-like cyanobacterial phytochromes (and at least also the eukaryotic alga *Mesotaenium* [13]), the native chromophore is PCB, higher plants use PΦB. λ_{max} values of both R-absorbing and FR-absorbing phytochrome states (Pr and Pfr, respectively) are thereby shifted ca. 11 nm bathochromically on account of the delocalized π-electron system extending to include the vinyl rather than ethyl D-ring side chain [14]. Surprisingly, Pfr → Pr thermal reversion in vitro is accelerated in PΦB adducts, especially in the case of plant phytochrome B (phyB), leading to the suggestion that phyB might also act as a thermosensor [15]. Whether this is relevant in vivo is unclear, however, as thermal reversion is strongly inhibited by Pfr-specific binding of cellular partners such as ARR4 [16, 17] and PIF6 [18].

Here we describe optimized procedures for the production of both unlabeled PCB from commercially available *Spirulina* tablets and universally $^{13}C/^{15}N$-labeled PCB from *Synechocystis* 6803 cultures. The original method used sucrose-density ultracentrifugation to prepurify the phycobilisome antennae complexes from lysed *Synechocystis* [6], but this nontrivial procedure is better replaced by final reverse-phase high-performance liquid chromatography (RP-HPLC) purification. The purity available from the latter is very high but seldom necessary, as the phytochrome autoassembly reaction is well able to select PCB from contaminants that pass through simple reverse-phase cartridges. If that reaction is carried out early on in the phytochrome purification procedure, excess PCB is removed quantitatively along with other contaminants.

Although *Spirulina* tablets provide an ideal source of unlabeled PCB, *Synechocystis* cells grown in liquid blue-green 11 medium (BG-11) in dim white light can also be used. However, using nonradioactive isotopes of different atoms in the medium offers valuable opportunities for research exploiting atomic mass differences or nuclear spin. The original procedure for u-^{15}N PCB production used BG-11 in which conventional nitrate was replaced by $^{15}NO_3^-$ while the carbon source remained as (unlabeled) CO_2 from humidified air bubbled through the culture [6]. In principle, u-^{13}C labeling could be achieved using gaseous N_2 with ca. 0.1% $^{13}CO_2$, except that cost factors would demand a closed system to recycle the $^{13}CO_2$ with simultaneous removal of photosynthetic oxygen (perhaps using alkaline pyrogallol [1,2,3-trihydroxybenzene]). However, carbonate ($^{13}CO_3^{2-}$, effectively bicarbonate [$H^{13}CO_3^-$] at pH 8–10, as preferred by *Synechocystis*) can be used instead, but with the associated problem of basification: The substrate of ribulose bisphosphate carboxylase in photosynthesis is CO_2, thus using HCO_4^- as a carbon source generates OH^- that rapidly kills the standard (unbuffered) BG-11 culture once toxic pH values above 11 are reached. To counter this, 60 mM N-cyclohexyl-2-aminoethanesulfonic acid (CHES) buffer is added and the pH is initially adjusted to ca. 8 with NaOH: As CHES is neither toxic to nor metabolized by *Synechocystis*, this allows the culture to grow to saturation before a toxic pH is reached [6]. The problem remains that this system cannot be exposed to (unlabeled) atmospheric CO_2 yet photosynthetic O_2 must be allowed to escape. We reasoned, however, that a semiclosed system minimizing CO_2 intrusion while allowing partial-pressure-driven O_2 escape might provide an appropriate solution. This has indeed been borne out in practice [8, 9, 19–21], as shown by the mass spectra of typical samples in Fig. 1.

In Subheadings 3.1 and 3.2, we describe phycocyanin extraction from *Spirulina* tablets (unlabeled) and from isotope-labeled

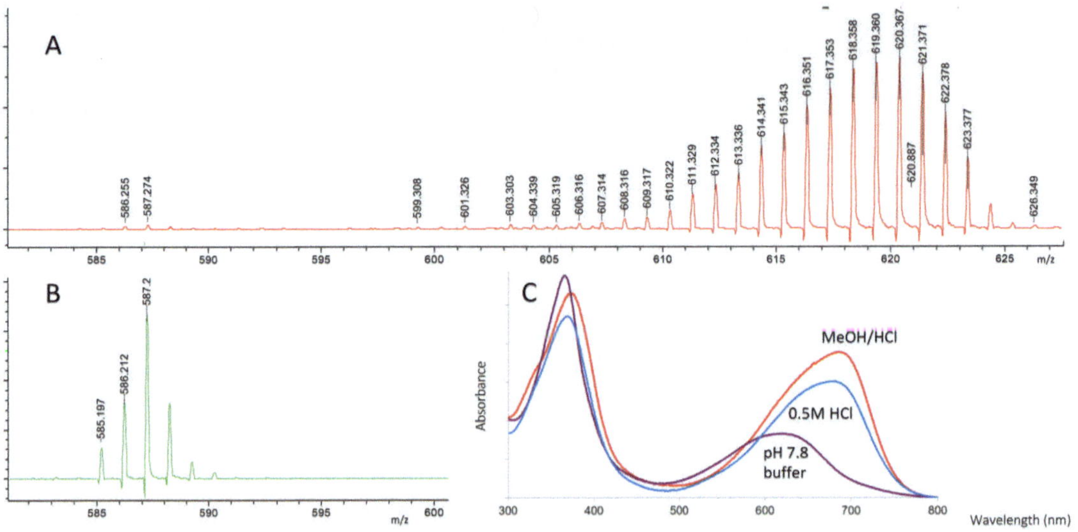

Fig. 1 Analyses of free PCB generated as described. Mass spectra: (**a**) u-$^{13}C/^{15}N$-labeled PCB from semiclosed *Synechocystis* 6803 culture (note trace amounts of unlabeled PCB). (**b**) Unlabeled PCB from *Spirulina* tablets. The *x*-axis is the m/z (mass/charge) ratio of the protonated bilin (PCB + $^{1}H^{+}$). Courtesy of Prof. Günter Lochnit, JLU Giessen. (**c**) UV-Vis absorption spectra of HPLC-purified (unlabeled) PCB from *Spirulina* tablets in various solvents at equal dilution. (Authors' unpublished data)

Synechocystis liquid culture, respectively. Subheading 3.3 describes subsequent PCB production, initial purification, and quantification. Subheading 3.4 describes optional HPLC purification. Subheading 3.5 provides an overview of phytochrome autoassembly with PCB in vitro. A laboratory with dim cyan safelight (ideally from 490 nm light-emitting diodes [LEDs]) is required.

2 Materials

2.1 Extraction of Phycocyanin from Spirulina *Tablets*

1. Large cooled centrifuge with capacity to spin >1 L at >4000 *g* at 4 °C.
2. Rotor and bottles for **item 1** above.
3. Magnetic stirrer/hot plate and ca. 4 cm stir bar.
4. Calibrated pH meter.
5. Large pestle and mortar.
6. 1 L Erlenmeyer flask.
7. *Spirulina* tablets (Green Valley, Süsel, Germany), ca. 100 g. Similar tablets are widely available from health food vendors.
8. Cyanobuffer (1 L): 100 mM K/HPO_4 buffer, pH 7.0, 5 mM Na_2EDTA. While stirring, mix 100 mM K_2HPO_4 and 100 mM KH_2PO_4 to pH 7.0; then add 1.85 mg ethylenediaminetetraacetate (EDTA) per L.

9. H_2O (pure water, Millipore or Milli-Q).
10. 1.8 mL Eppendorf tubes.
11. Micropipettes with disposable tips.

2.2 Production and Extraction of U-^{13}C/^{15}N-Labeled Phycocyanin from Synechocystis 6803 Liquid Culture

1. Large cooled centrifuge with capacity to spin >1 L at >4000 *g* at 4 °C.
2. Rotor and bottles for **item 1** above.
3. High-speed cooled centrifuge with capacity to spin >100 mL at >25,000 *g* at 4 °C.
4. Rotor and tubes for **item 3** above.
5. Simple spectrophotometer with disposable cuvettes (*see* **Note 1**).
6. pH meter, calibrated.
7. 300 mL baffled glass Erlenmeyer flasks with porous stopper, sterile.
8. 2 L glass screw-top bottle, sterile.
9. Cyanobuffer (1 L): 100 mM K/HPO_4, pH 7.0 (mix 100 mM K_2HPO_4 and 100 mM KH_2PO_4 to pH 7.0 while stirring), 5 mM EDTA (add 1.85 mg ethylenediaminetetraacetate [EDTA, disodium salt] per L).
10. Media for *Synechocystis* 6803 liquid culture: *see* Table 1 and **Notes 2** and **3**. Dissolve the components of each solution sequentially in 900 mL H_2O with stirring, then titrate to pH 8 with NaOH, and add H_2O to 1 L final volume. Sterile filter.
11. H_2O (pure water, Millipore or Milli-Q).
12. Micropipettes with disposable tips.
13. 1.8 mL Eppendorf tubes.

2.3 Preparation of Free PCB by Methanolysis of Phycocyanin

1. Large cooled centrifuge with capacity to spin >1 L at >4000 *g* at 4 °C.
2. Ultraviolet–visible (UV-Vis) spectrophotometer (250–800 nm range).
3. Methanol-resistant, 1 cm path length microcuvettes for **item 2** above.
4. Magnetic stirrer with stir bar.
5. Water bath at 54 °C with magnetic stirrer and stir bar.
6. Rotary evaporator with water bath at 50 °C, 1 L flask, condenser, and vacuum pump.
7. SpeedVac concentrator in darkness, with vacuum pump.
8. 250-mL screw-top bottle.

Table 1
Composition of BG-11 medium and variants

Main solution (1 L)	g/L	Mol. mass	mM
$NaNO_3$ or $Na^{15}NO_3$	1.500 0.38	85 86	17.6 4.36
K_2HPO_4	0.040	174	0.23
$MgSO_4{\cdot}7H_2O$	0.075	246.5	0.30
$CaCl_2{\cdot}2H_2O$	0.036	147	0.24
Citric acid	0.006	192	0.031
Ferric ammonium citrate	0.006	ca. 300	ca. 0.020
EDTA (disodium salt)	0.001	336	0.003
Na_2CO_3 or $NaH^{13}CO_3$ (semiclosed system)	0.020 4.0	106 85	0.19 47.1
CHES (semiclosed system)	12.5	207	60
Trace element solution	1 mL		
to pH 8.0 with NaOH			
to 1 L with H_2O			
Trace element solution (1 L)			
H_3BO_3	2.86	62	46.1
$MnCl_2{\cdot}4H_2O$	1.81	198	9.14
$ZnSO_4{\cdot}7H_2O$	0.22	288	0.77
$NaMoO_4{\cdot}2H_2O$	0.39	219	1.78
$CuSO_4{\cdot}5H_2O$	0.08	250	0.32
$Co(NO_3)_2{\cdot}6H_2O$	0.05	291	0.17
to 1 L with H_2O			

The components in red can be varied as shown

9. 1.5-L beaker.
10. C18 reverse-phase cartridge (Sep-Pak, Waters).
11. 10-mL disposable syringe for **item 10** above.
12. Methanol, 250 mL, technical grade.
13. Methanol, 250 mL, HPLC grade.
14. Ammonium sulfate (AmS), ca. 50 g fine powder (grind if necessary).
15. Dimethyl sulfoxide (DMSO), <1 mL.
16. Concentrated HCl, <1 mL.
17. H_2O (pure water, Millipore or Milli-Q).
18. Micropipettes with disposable tips.
19. 1.8 mL Eppendorf tubes.

2.4 Purification of PCB by HPLC

1. HPLC or fast protein liquid chromatography (FPLC) (e.g., ÄKTA purifier) system with 1 mL sample loop and ca. 80 mL preparative C18 reverse-phase column, particle diameter 7 μm (UltraSep ES PHARM RP18E, SepServ, Berlin). An 18 mL precolumn with the same matrix is recommended.
2. 1 mL syringe for **item 1** above.

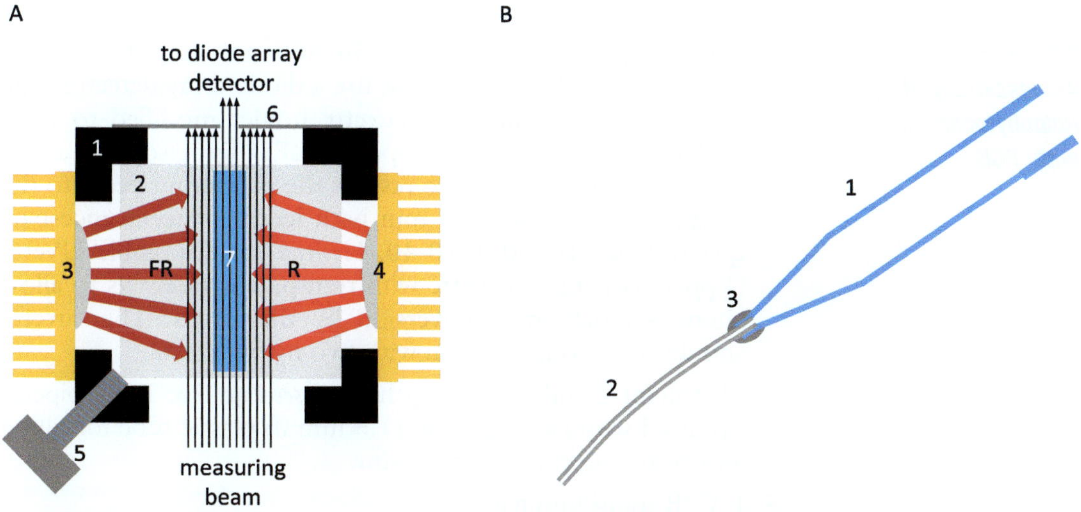

Fig. 2 Modified cuvette holder and filling aide. (**a**) Horizontal cross section of modified cuvette holder in UV-Vis spectrophotometer (8453, Agilent). The measuring beam is indicated with black arrows. (1) Workshop-modified cuvette holder (08451-60104, Agilent); (2) fluorescence-type quartz microcuvette with transparent sidewalls; (3, 4) high-power FR and R LED arrays with appropriate constant-current power supplies and heatsinks (λ_{max} 735 nm, LED735-66-60 and λ_{max} 630 nm, LED630-66-60, respectively; Roithner, Vienna); (5) nylon fixing screws to immobilize the cuvette; (6) thin plate with ca. 1.5 × 3 mm window to restrict the measuring beam to the sample in the cuvette; and (7) holophytochrome sample in cuvette. (**b**) Longitudinal cross section of Elephant Trunk filling aide allowing samples to be exchanged, accurately diluted, and mixed without moving the cuvette. A 1 mL disposable pipette tip (1) is trimmed to fit a thin (ca. 1.5 mm ∅) plastic tube (2) providing access to the cuvette baseplate. The tube is back-inserted through the pipette tip and sealed with a thin strip of parafilm (3)

3. UV-Vis spectrophotometer (250–800 nm range).
4. Methanol-resistant, 1 cm path length microcuvettes for **item 3** above.
5. High-speed (50,000 *g*) cooled centrifuge with rotor for Eppendorf tubes.
6. Calibrated pH meter.
7. Sonication bath.
8. 1 mL pipette tip fitted with ca. 3 cm thin (1.5 mm ∅) plastic tube ("Elephant Trunk," *see* Fig. 2b).
9. HPLC running buffer (1 L, sterile filtered): 70% 7.5 mM Na/HPO_4, pH 6.0 (mix 7.5 mM solutions of Na_2PO_4 and phosphoric acid to pH 6.0), 30% acetonitrile.
10. C18 reverse-phase cartridge (Sep-Pak, Waters).
11. 10 mL disposable syringe for **item 10** above.
12. H_2O (pure water, Millipore or Milli-Q).
13. Micropipettes with disposable tips.
14. 1.8 mL Eppendorf tubes.

2.5 In Vitro Autoassembly of Holophytochromes with PCB

1. UV-Vis spectrophotometer (250–800 nm range) with PC, modified as in Fig. 2a: Ideally, use a diode-array detector type (e.g., 8453, Agilent) with cuvette holder modified to carry 630 nm and 735 nm high-power LED light sources and a ca. 1.5 × 3 mm window to restrict the measuring beam to the sample region. Use a 10 mm path length, fluorescence-type quartz (e.g., 104002F-10-40, Hellma) or disposable UVette (Eppendorf) microcuvette with transparent sidewalls to allow lateral sample photoconversion by the LEDs. The cuvette should be fixed in position with two nylon screws.
2. "Elephant Trunk" pipetting helper, *see* Fig. 2b: 1 mL pipette tip fitted with ca. 3 cm thin (1.5 mm ∅) plastic tube for filling and emptying the cuvette in situ.
3. UV/B transilluminator.
4. White-light transilluminator.
5. Digital camera with macro lens fitted with ca. 620 nm interference filter.
6. HESβ buffer (1 L): 20 mM 4-(2-hydroxyethyl)-1-piperazineethanesulfonic acid (HEPES), pH 7.8, 1 mM Na_2EDTA, 50 mM NaCl, 1 mM β-mercaptoethanol.
7. Sample loading buffer, fivefold concentrate (5× SLB): 10% (w/v) sodium dodecyl sulfate (SDS), 250 mM Tris–Cl pH 6.8, 50% glycerol, 25% (v/v) β-mercaptoethanol, 0.01% (w/v) bromophenol blue (marker).
8. 50–200 kDa protein molecular weight markers.
9. SDS–polyacrylamide gel electrophoresis (SDS–PAGE) apparatus with buffers and 10% running gel.
10. 1 mM zinc acetate (200 mL).
11. Conventional Coomassie Blue stain and destain solutions.
12. H_2O (pure water, Millipore or Milli-Q).
13. 1.8 mL Eppendorf tubes.
14. Micropipettes with disposable tips.

3 Methods

3.1 Extraction of Phycocyanin from Spirulina Tablets

Large amounts of phycocyanin for PCB production can be obtained easily from dried *Spirulina* as described here (but *see* **Note 2**).

1. Grind ca. 100 g *Spirulina* tablets to a fine powder.
2. Place the powder in a ca. 1 L Erlenmeyer flask wrapped in aluminum foil, add 600–700 mL Cyanobuffer, and mix thoroughly. Stir ca. 6 h on a magnetic stirrer at ca. 30 °C.
3. Clarify the extract by centrifugation at ca. 4000 *g* for 15 min at 4 °C.
4. Carefully remove the supernatant without disturbing the pellet (that can be discarded). Proceed to Subheading 3.3.

3.2 Production and Extraction of u-$^{13}C/^{15}N$-Labeled Phycocyanin from Synechocystis 6803 Liquid Culture

Isotope-labeled PCB can easily be produced by growing *Synechocystis* 6803 in appropriate media (*see* Table 1 and **Note 3**). *Synechocystis* is sensitive to strong light: work in subdued light; never allow the cells to be exposed to sunlight. Bear in mind that **steps 1** and **2** below require ca. 3 months (*see* **Note 4**).

1. Initiate growth of an axenic *Synechocystis* 6803 start culture under sterile conditions: Suspend a colony grown on BG-11 with 1% agar in 100 μL liquid BG-11, then transfer to 50 mL medium according to Table 1 in a sterile 300 mL baffled flask stoppered with cotton wool on a shaker (130 rpm) under constant illumination with dim white light (fluence rate ca. 30 μmol/m^2 s photosynthetically active radiation [PAR]) at 26 °C. Allow to grow for ca. 5 weeks.
2. When the start culture has reached stationary phase (OD_{750} following tenfold dilution ca. 0.2–0.4 (but *see* **Note 1**), after ca. 5 weeks), use ca. 20 mL to inoculate 1 L sterile-filtered medium (*see* Table 1) in a sterile 2 L screw-top bottle with stir bar. Leave the screw cap loose (semiclosed system to allow O_2 escape). Place the culture on a magnetic stirrer with temperature and light conditions as **step 1** above. Allow to grow for ca. 5–7 weeks.
3. Harvest the cells when they reach stationary phase (after 6–8 weeks, OD_{750} as **step 2** above) by centrifugation at 4000 *g* and 4 °C for 15 min. Keep the cells on ice and work in the cold room.
4. Wash the cells: Discard the supernatant, resuspend in the cell pellet in 200 mL cold Cyanobuffer, and spin down again; then discard the supernatant. Washed cells can be shock-frozen in liquid nitrogen and kept at −80 °C until needed.
5. Resuspend the cells in 100 mL Cyanobuffer.
6. Lyse the cell slurry using a French press (three passages at 6–7 MPa).
7. Clarify the extract by centrifugation at 25,000 *g* for 30 min at 4 °C.
8. Carefully remove the supernatant without disturbing the pellet (that can be discarded). Proceed to Subheading 3.3.

3.3 Preparation of Free PCB by Methanolysis of Phycocyanin

The phycocyanin light-harvesting biliproteins that dominate the extract from Subheadings 3.1 and 3.2 are rich sources of PCB. The cofactor is attached via a covalent thioether link that can be broken by treatment with hot methanol following the simple procedure below. It is not necessary to use a Soxhlet apparatus.

1. Measure the volume of the clarified dark green supernatant (from Subheadings 3.1 or 3.2 above) with a graduated cylinder and transfer to a 1.5 L beaker.
2. While stirring, slowly add 370 mg powdered AmS per mL supernatant to achieve 70% saturation. Finally, allow at least 5 min for the salt to dissolve completely.
3. Allow the protein to precipitate overnight at 4 °C in darkness.
4. Spin down the precipitate at 4000 *g* for 20 min at 4 °C. Discard the colorless supernatant.
5. Chlorophyll removal: Resuspend the pellet in ca. 50 mL of cold, technical-grade methanol, then spin down again at 4000 *g* for 20 min at 4 °C. Remove the green methanolic supernatant (collect for redistillation). Repeat this procedure several times until the supernatant is nearly colorless.
6. Set up a water bath in darkness with magnetic stirrer to 54 °C.
7. Resuspend the washed pellet in 200 mL HPLC-grade methanol in a 250 mL screw-top bottle. Insert a stir bar and close the lid tightly. From this point on, work in dim ca. 490 nm safelight.
8. Methanolysis: Place the bottle in 54 °C water bath and stir overnight in darkness.
9. Clarify the crude extract: Centrifuge at 4000 *g* for 20 min at 4 °C. Remove the blue-green supernatant carefully without disturbing the pellet (can be subjected to a second round of methanolysis).
10. Place the clarified extract in a 1 L rotary evaporator flask (if desired, store at 4 °C in darkness until the second methanolysis is completed, then pool the extracts in the flask).
11. Set the water bath of a rotary evaporator to 50°.
12. To condense the extract to ca. 10 mL, attach the flask to the rotary evaporator and apply vacuum. Collect the methanol with a condenser for reuse and to protect the vacuum pump.
13. Carefully remove the concentrated extract, if necessary, using a small volume of methanol to salvage residual PCB.
14. Using a 10 mL syringe, wet the surface of a C18 reverse-phase cartridge with 20 mL H_2O, then activate the matrix with 10 mL HPLC-grade methanol followed by 20 mL H_2O again.

15. Dilute the concentrated PCB extract 1:1 with H_2O and apply to cartridge with syringe. Use an additional cartridge if the PCB is not bound quantitatively.
16. Wash the cartridge with 10 mL H_2O via the syringe.
17. Using the syringe, slowly elute the PCB with HPLC-grade methanol (ca. 1 mL) into an Eppendorf tube. If the first few drops appear oily, discard them. A yellow-green residue remains trapped in the cartridge. Cartridges can be reused several times: They fail to bind PCB when exhausted.
18. Dry down in a SpeedVac in darkness and redissolve in 200 μL DMSO. Alternatively, the dry PCB can be kept at −80 °C, preferably under argon.
19. Measure the PCB concentration spectrophotometrically (*see* **Notes 5** and **6**). Fix the cuvette in place, then add 500 μL of freshly prepared 95% MeOH/5% conc. HCl and record the baseline 300–800 nm. Then add exactly 2.5 μL of the extract, mixing with the help of the Elephant Trunk. Measure the UV-Vis absorption spectrum 300–800 nm. For pure PCB under these conditions, λ_{max} = 374 and 690 nm ($\varepsilon_{374nm} = 47.9$, $\varepsilon_{690nm} = 37.9\ mM^{-1}\ cm^{-1}$ [22], *see* Fig. 1c and **Note 7**).
20. Label the PCB extract in DMSO with its concentration (correct for the 200× dilution) and store at −80 °C, preferably under argon. Following the procedures outlined here, the PCB is usually pure enough for autoassembly of holophytochromes (*see* Subheading 3.5). Further purification is achieved by HPLC, as detailed in Subheading 3.4.

3.4 Purification of PCB by HPLC

The procedure in Subheading 3.3 above yields significant amounts of 3′-methoxy PCB [23]. Although this cannot form covalent phytochrome adducts, it can enter the apoprotein pocket and thus might lead to artifacts in some circumstances. As described below, high-purity PCB can be obtained via reverse-phase preparative HPLC with aqueous acetonitrile, despite rather low solubility in this medium. The PCB is then concentrated rapidly and stored in DMSO. Work in dim cyan (490 nm) safelight.

1. Dissolve ca. 500 nmol PCB in 900 μL HPLC running buffer by incubation for 15 min in a sonication bath.
2. Clarify the sample at 50,000 *g* for 5 min at 4 °C.
3. Remove the supernatant carefully, then inject into the HPLC column via a 1 mL sample loop.
4. Run the column at 6 mL/min, pressure ca. 4.2 MPa.
5. Collect the blue PCB eluate in fractions and pool. Add 0.25× the pool volume H_2O.

6. Working swiftly in the cold room (*see* **Note 8**), exchange the solvent with cold HPLC-grade methanol using a fresh, pre-wetted C18 cartridge (*see* Subheading 3.3, **steps 14–17**).
7. Dry down the pure PCB extract swiftly using a SpeedVac at room temperature in darkness (*see* **Note 8**).
8. Dissolve the sample in 200 μL DMSO.
9. Measure the PCB concentration spectrophotometrically (*see* **Notes 5** and **6**). Fix the cuvette in place, then add 500 μL of freshly prepared 95% MeOH/5% conc. HCl and record the baseline 300–800 nm. Then add exactly 2.5 μL of the extract, mixing with the help of the Elephant Trunk. Measure the UV-Vis absorption spectrum 300–800 nm. For pure PCB under these conditions, $\lambda_{max} = 374$ and 690 nm ($\varepsilon_{374nm} = 47.9$, $\varepsilon_{690nm} = 37.9$ mM^{-1} cm^{-1} [22], *see* Fig. 1c and **Note 7**).
10. Label the PCB extract in DMSO with its concentration (correct for the 200× dilution) and store at −80 °C, preferably under argon.

3.5 *In Vitro* Autoassembly of Holophytochromes with PCB

Plant phytochromes and Cph1-like cyanobacterial phytochromes can autoassemble with PCB (or PΦB). If the expression host lacks the appropriate bilin biosynthetic pathway, apophytochromes can be produced and subsequently autoassembled in vitro. The following gives a general indication of the latter procedure with PCB. Work in dim cyan (490 nm) safelight.

1. The apophytochrome sample should be in 1–20 mL HESβ buffer. As apophytochromes tend to aggregate, it is often wise to initiate the autoassembly reaction at cell lysis, that is, in the crude sample. This poses the problem, however, that the concentration of apoprotein is unknown and therefore that the amount of PCB to add is also unknown. This can be avoided by lysing a known small fraction of the washed cells and adding PCB stepwise while monitoring the spectrophotometric difference signal ($\Delta\Delta A_{R\text{-}FR}$, see below). When the difference signal no longer increases, sufficient PCB has been added. Calculate the equivalent volume of PCB required for the rest of the sample.
2. Add a twofold excess of PCB to the clarified apoprotein sample, mix carefully, and allow the autoassembly reaction to proceed at 30 °C for 30 min.
3. Meanwhile, mount the cuvette in the holder of the spectrophotometer and fix it in position with the nylon screws (*see* Fig. 2a).
4. Use an Elephant Trunk (*see* Fig. 2b and **Notes 5** and **6**) to add 500 μL of HESβ to the cuvette. Measure and store the baseline spectrum 250–800 nm.

5. Remove the buffer with the Elephant Trunk, refill with fresh buffer and measure the spectrum: It should be flat with a noise level <0.005 absorbance units. If not, repeat from **step 4**.
6. Using the Elephant Trunk, remove the buffer and place a 500 μL aliquot of the sample (or an appropriate dilution thereof) in the cuvette.
7. Check the UV-Vis absorption spectrum to ensure that $A_{\lambda max}$ is <2. If the value exceeds 2 (<1% transmission) at any wavelength, dilute appropriately.
8. Irradiate the sample with actinic red light (R) for 10 s, measure the spectrum, and then repeat. If the fluence rate is high enough, there should be no change. Store the final "after R" spectrum (Pr + Pfr photoequilibrium, *see* **Note 9**). Measure the spectrum again a few minutes later to check for thermal Pfr → Pr reversion.
9. Repeat the procedure with far-red (FR) light and store the final "after FR" spectrum (Pr).
10. Derive and store the (after R)–(after FR) difference spectrum (ΔA_{λ}). The difference between the maximum and minimum ($\Delta\Delta A_{R\text{-}FR}$) is a quantitative parameter of the phytochrome concentration and is independent of purity.
11. Add a small (ca. 1 μL) additional aliquot of PCB to the sample in the cuvette and mix with the Elephant Trunk.
12. Repeat **steps 7–10**. If the $\Delta\Delta A_{R\text{-}FR}$ difference signal has increased, the sample contains unassembled apoprotein—in which case, redetermine the amount of additional PCB necessary for twofold excess and repeat from **step 2**.
13. Purify the holoprotein (typically by affinity and/or ion-exchange and size-exclusion chromatography) to remove contaminants, including the excess PCB. Measure $\Delta\Delta A_{R\text{-}FR}$ before and after each step, noting the total eluate volume and any dilution: Calculate the % yield through the purification process from these data. Calculate also the $A_{\lambda max,R}$:A_{280nm} specific absorbance ratio (SAR) after each step—this is a quantitative parameter of the holophytochrome purity (*see* also **Note 10** and Fig. 3 for typical spectra of pure phytochrome). When spectra are measured during purification, remove ca. 1–20 μL from the cuvette and mix with 5× SLB, *see* **Note 11**.
14. When purification is complete, warm the samples in SLB to 70 °C for 20 min to denature, then spin down any debris at ca. 13,000 *g* for 5 min at room temperature. Load appropriate volumes of each in a total of 10 μL SLB onto a standard SDS–10% polyacrylamide gel (with a lane of marker proteins). Following electrophoresis, soak the gel for 5 min in 1 mM zinc

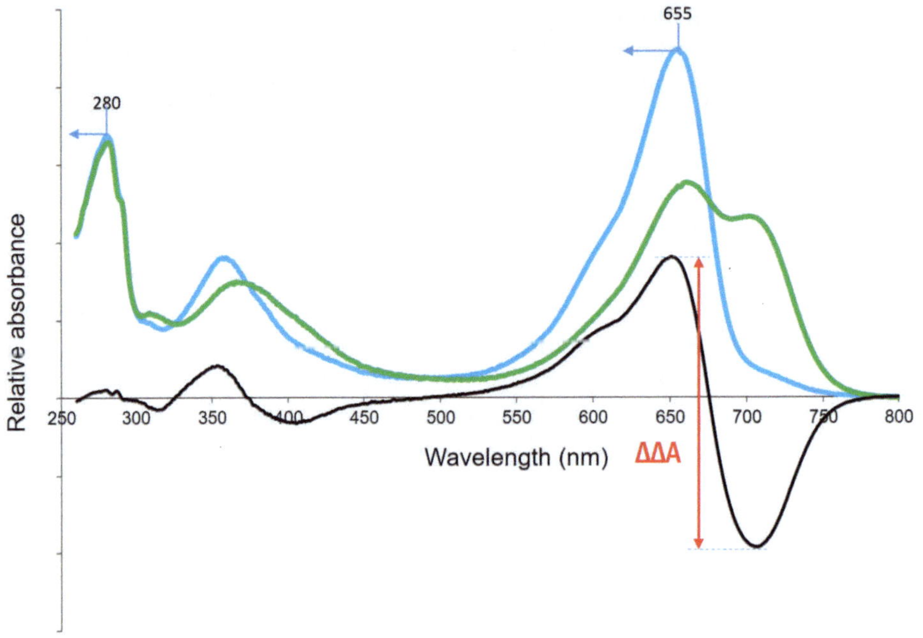

Fig. 3 UV-Vis absorption spectra of Cph1Δ2 holophytochrome (photosensory module, residues 1–512 with C-terminal His_6 affinity tag) autoassembled with PCB in vitro and purified by Ni-affinity and size-exclusion chromatography. The "after FR" (Pr), "after R" (Pr/Pfr photoequilibrium), and difference (ΔA_λ) spectra are shown in cyan, green, and black, respectively. The vertical red arrow indicates the $\Delta\Delta A_{R\text{-}FR}$ difference signal that is proportional to the phytochrome concentration, independently of purity. The specific absorbance ratio (SAR) of the $Pr_{\lambda max}$ (here at 655 nm) and protein (at 280 nm) peaks is a useful parameter of purity. (Courtesy of Dr. Joel Mailliet, JLU Giessen)

acetate, then visualize the orange-fluorescent bilin bands using a UV/B transilluminator (photograph the gel using a digital camera fitted with a ca. 620 nm interference filter—the filter is essential). Then stain the gel conventionally with Coomassie Blue and photograph again on a white-light transilluminator.

4 Notes

1. Cell densities are commonly defined on the basis of optical density (OD) units measured spectrophotometrically. This is profoundly unscientific, however, as the measured value relates to light scattering rather than absorption, and thus is dependent on the geometry of the spectrophotometer as well as the cell density.
2. Unlabeled PCB can also be produced from *Synechocystis* cultures using conventional BG-11 medium either assembled with standard reagents (*see* Table 1) or using commercially available concentrate. For this, follow Subheading 3.2.

3. u-^{15}N-labeled PCB can be produced with BG-11 in which $NaNO_3$ is replaced by $Na^{15}NO_3$ (*see* Table 1; the lower concentration relative to standard BG-11 is to reduce costs). In both cases, the carbon source is usually CO_2 from sterile-filtered air that is humidified by passing through a sterile water column before being bubbled through the medium. With such aeration, a stirrer is not necessary. Alternatively, the carbon source can also be unlabeled (HCO_3^-) or labeled ($H^{13}CO_3^-$) bicarbonate, in which case both the standard carbonate component and aeration are omitted and CHES is added to buffer the released hydroxyl ions (*see* Table 1; note that technical-grade CHES sometimes contains traces of oligo (vinylsulfonic acid), a potent toxin—analytical-grade CHES should therefore be used). The system is semiclosed (to allow photosynthetic oxygen to escape), thus mixing is provided by a stirrer rather than aeration.
4. The cultures are grown to saturation in continuous dim white light. Growth rate is thereby light-limited but with the advantage that extensive phycobilisomes are developed, thereby increasing the proportion of phycocyanin (and thus PCB) in the crude protein extract.
5. For accurate measurement, switch on the spectrophotometer to allow 30 min for warm-up.
6. Even slight movements of the cuvette have significant effects in spectrophotometry, thus accurate measurement requires that samples be exchanged by pipette while the cuvette remains fixed in place (for example with nylon screws, *see* Fig. 2a). Sample exchange and mixing can be carried out using a plastic Pasteur pipette or, better, a 1 mL micropipette with a tip fitted with a thin plastic tube ("Elephant Trunk," *see* Fig. 2b), also allowing accurate dilution. Never use a glass (Pasteur) pipette to fill a cuvette—the slightest scratch will cause light scattering.
7. From the ε_{690nm} value, a solution of protonated PCB in methanol with $A_{690nm} = 1.00$ in a 1 cm path length cuvette has a concentration of 26.3 μM. If λ_{max} is below 675 nm, the sample is grossly contaminated with chlorophyll.
8. Avoid renewed formation of 3′-methoxy PCB during the concentration procedure. Work quickly with the cartridge in the cold room using ice-cold methanol, then dry down the eluate swiftly at room temperature.
9. Because the Pfr absorbance spectrum overlaps throughout with that of Pr, actinic irradiation generates a Pfr/Pr photoequilibrium rather than pure Pfr. If the photoconversion quantum efficiencies of Pr and Pfr are approximately equal, as is usually the case, the highest proportion of Pfr is generated at the

wavelength where the quotient $\varepsilon_{Pr}/(\varepsilon_{Pr} + \varepsilon_{Pfr})$ is maximal [24]. That is usually significantly hypsochromic of $\lambda_{max,Pr}$.

10. The concentration of a sample of a known protein in pure form can generally be estimated from A_{280nm} under denaturing conditions (e.g., in 8 M urea) and ε_{280nm} calculated from the peptide sequence (*see* https://web.expasy.org/protparam/). However, in the case of holophytochromes, the contribution of the bilin at 280 nm must be taken into account: For PCB, this is ca. 10 mM^{-1} cm^{-1} under denaturing conditions [25]. Thus, in the case of a *pure* holophytochrome, $\varepsilon_{\lambda max,R}$ of the Pr state can be derived from the SAR and ε_{280nm}. Compare that with the value derived from the absorbance of the holophytochrome formed when a known amount of free PCB is added to excess apophytochrome [25].
11. It is worth visualizing the holophytochrome purification process by retaining samples at each step for SDS–PAGE. First, incubate for a few minutes with 1 mM zinc acetate, then visualize the bilin by fluorescence at ca. 620 nm on a UV/B transilluminator [26] (by chance, the same setup is used conventionally for ethidium bromide–DNA imaging). Then stain the gel conventionally with Coomassie Blue to show all proteins.

Acknowledgments

We gratefully acknowledge the support of the *Deutsche Forschungsgemeinschaft* (DFG) and the contributions of Drs. Holger Strauss, Janina Hahn (both FMP Berlin), Genady Kozhukh, Günter Lochnit (both JLU Giessen), and Wolfgang Gärtner (University of Leipzig) to the development of these methods.

References

1. Scheer H, Zhao KH (2008) Biliprotein maturation: the chromophore attachment. Mol Microbiol 68(2):263–276. https://doi.org/10.1111/j.1365-2958.2008.06160.x
2. Hughes J, Lamparter T, Mittmann F, Hartmann E, Gartner W, Wilde A, Borner T (1997) A prokaryotic phytochrome. Nature 386(6626):663. https://doi.org/10.1038/386663a0
3. Yeh KC, Wu SH, Murphy JT, Lagarias JC (1997) A cyanobacterial phytochrome two-component light sensory system. Science 277:1505–1508
4. Inomata K, Noack S, Hammam MA, Khawn H, Kinoshita H, Murata Y, Michael N, Scheerer P, Krauss N, Lamparter T (2006) Assembly of synthetic locked chromophores with *Agrobacterium* phytochromes Agp1 and Agp2. J Biol Chem 281(38):28162–28173. https://doi.org/10.1074/jbc.M603983200
5. Hahn J, Strauss HM, Schmieder P (2008) Heteronuclear NMR investigation on the structure and dynamics of the chromophore binding pocket of the cyanobacterial phytochrome Cph1. J Am Chem Soc 130(33):11170–11178
6. Strauss HM, Hughes J, Schmieder P (2005) Heteronuclear solution-state NMR studies of the chromophore in cyanobacterial phytochrome Cph1. Biochemistry 44(23):8244–8250

7. Stöppler D, Song C, van Rossum BJ, Geiger MA, Lang C, Mroginski MA, Jagtap AP, Sigurdsson ST, Matysik J, Hughes J, Oschkinat H (2016) Dynamic nuclear polarization provides new insights into chromophore structure in phytochrome photoreceptors. Angew Chem 55(52):16017–16020. https://doi.org/10.1002/anie.201608119
8. Song C, Mroginski MA, Lang C, Kopycki J, Gärtner W, Matysik J, Hughes J (2018) 3D structures of plant phytochrome A as Pr and Pfr from solid-state NMR: implications for molecular function. Front Plant Sci 9:498. https://doi.org/10.3389/fpls.2018.00498
9. Song C, Psakis G, Kopycki J, Lang C, Matysik J, Hughes J (2014) The D-ring, not the A-ring, rotates in Synechococcus OS-B′ phytochrome. J Biol Chem 289(5): 2552–2562. https://doi.org/10.1074/jbc.M113.520031
10. Song C, Rohmer T, Tiersch M, Zaanen J, Hughes J, Matysik J (2013) Solid-state NMR spectroscopy to probe photoactivation in canonical phytochromes. Photochem Photobiol 89(2):259–273. https://doi.org/10.1111/php.12029
11. Gambetta GA, Lagarias JC (2001) Genetic engineering of phytochrome biosynthesis in bacteria. Proc Natl Acad Sci USA 98(19): 10566–10571. https://doi.org/10.1073/pnas.191375198
12. Landgraf FT, Forreiter C, Hurtado PA, Lamparter T, Hughes J (2001) Recombinant holophytochrome in *Escherichia coli*. FEBS Lett 508(3):459–462
13. Wu SH, McDowell MT, Lagarias JC (1997) Phycocyanobilin is the natural precursor of the phytochrome chromophore in the green alga *Mesotaenium caldariorum*. J Biol Chem 272(41):25700–25705
14. Elich TD, Lagarias JC (1989) Formation of a photoreversible phycocyanobilin-apophytochrome adduct in vitro. J Biol Chem 264(22):12902–12908
15. Burgie ES, Gannam ZTK, McLoughlin KE, Sherman CD, Holehouse AS, Stankey RJ, Vierstra RD (2021) Differing biophysical properties underpin the unique signaling potentials within the plant phytochrome photoreceptor families. Proc Natl Acad Sci USA 118(22): e2105649118. https://doi.org/10.1073/pnas.2105649118
16. Mira-Rodado V, Sweere U, Grefen C, Kunkel T, Fejes E, Nagy F, Schäfer E, Harter K (2007) Functional cross-talk between two-component and phytochrome B signal transduction in *Arabidopsis*. J Exp Bot 58: 2595–2607. https://doi.org/10.1093/jxb/erm087
17. Sweere U, Eichenberg K, Lohrmann J, Mira-Rodado V, Bäurle I, Kudla J, Nagy F, Schäfer E, Harter K (2001) Interaction of the response regulator ARR4 with phytochrome B in modulating red light signaling. Science 294(5544):1108–1111
18. Smith RW, Helwig B, Westphal AH, Pel E, Borst JW, Fleck C (2017) Interactions between phyB and PIF proteins alter thermal reversion reactions in vitro. Photochem Photobiol 93(6):1525–1531. https://doi.org/10.1111/php.12793
19. Song C, Psakis G, Lang C, Mailliet J, Gärtner W, Hughes J, Matysik J (2011) Two ground state isoforms and a chromophore *D*-ring photoflip triggering extensive intramolecular changes in a canonical phytochrome. Proc Natl Acad Sci USA 108(10):3842–3847. https://doi.org/10.1073/pnas.1013377108
20. Rohmer T, Lang C, Gärtner W, Hughes J, Matysik J (2010) Role of the protein cavity in phytochrome chromoprotein assembly and double-bond isomerization: a comparison with model compounds. Photochem Photobiol 86(4):856–861. https://doi.org/10.1111/j.1751-1097.2010.00740.x
21. Röben M, Hahn J, Klein E, Lamparter T, Psakis G, Hughes J, Schmieder P (2010) NMR spectroscopic investigation of mobility and hydrogen bonding of the chromophore in the binding pocket of phytochrome proteins. ChemPhysChem 11(6):1248–1257. https://doi.org/10.1002/cphc.200900897
22. Cole WJ, Chapman DJ, Siegelman HW (1967) The structure of phycocyanobilin. J Am Chem Soc 89:3642–3645. https://doi.org/10.1021/ja00990a055
23. Lindner I, Braslavsky SE, Schaffner K, Gärtner W (2000) Model studies of phytochrome photochromism: protein-mediated photoisomerization of a linear tetrapyrrole in the absence of covalent bonding. Angew Chem Int Ed Engl 39(18):3269–3271
24. Mancinelli A (1994) The physiology of phytochrome action. In: Kendrick RE, Kronenberg GHM (eds) Photomorphogenesis in plants, 2nd edn. Kluwer Academic Publishers, Dordrecht, pp 211–269
25. Lamparter T, Esteban B, Hughes J (2001) Phytochrome Cph1 from the cyanobacterium *Synechocystis* PCC6803. Purification, assembly, and quaternary structure. Eur J Biochem 268(17):4720–4730. https://doi.org/10.1046/j.1432-1327.2001.02395.x
26. Berkelman TR, Lagarias JC (1986) Visualization of bilin-linked peptides and proteins in polyacrylamide gels. Anal Biochem 156(1): 194–201

Chapter 5

Production of Plant Phytochromes by Combining Insect Cells and Bacterial Expression Systems

Cheng Chi, Zhengdong Wang, Ruihan Wang, and Jizong Wang

Abstract

Phytochromes, acting as red and far-red light receptors, orchestrate the developmental plasticity of plants to acclimate to the ambient environment throughout their entire life cycle. Recent structural biological studies have provided significant insights into the molecular bases of phytochrome-mediated light signaling in plants. Distinct approaches have been employed in these studies for phytochrome expression and purification. In our research, we utilized both bacteria and insect cells to separately synthesize the phytochromobilin (PΦB) molecule and express apophytochrome protein and subsequently purify the holophytochrome protein from the mixtures of the two expression systems. Here, we detail the procedures of this strategy, including plasmid construction, PΦB biosynthesis, apophytochrome expression, holophytochrome purification, and characterization of the phytochrome proteins. This approach can serve as an essential supplement to existing phytochrome purification methodologies, and we hope that investigators will benefit from it in achieving successful purification of the phytochromes they are interested in.

Key words Phytochrome, PΦB biosynthesis, Apophytochrome expression, Holophytochrome purification, Bacteria, Insect cells

1 Introduction

Among the photoreceptors in plants, phytochromes (phys) primarily perceive red and far-red light. These phytochromobilin (PΦB)-containing proteins exist in two states: a dark-adapted, red-light-absorbing state designated as Pr, localized in the cytoplasm, and a photo-activated, far-red-light-absorbing state designated as Pfr, which translocates into the nucleus to induce light signaling, guiding the plant's adaptation to its changing environment [1, 2]. Although extensive studies have been conducted on the signaling mechanisms of phytochromes, the molecular bases of their actions remain poorly understood. Recent structural

Cheng Chi and Zhengdong Wang contributed equally.

Andreas Winkler (ed.), *Phytochromes: Methods and Protocols*, Methods in Molecular Biology, vol. 2970,
https://doi.org/10.1007/978-1-0716-4791-2_5,

biological studies on phytochrome A (phyA) and phytochrome B (phyB) have provided valuable insights into this aspect [3–7], with high-quality protein samples serving as the primary prerequisite for these advancements.

In plants, the *ELONGATED HYPOCOTYL 1* (*HY1*) and *HY2* genes encode two enzymes responsible for the biosynthesis of the bilin molecule PΦB [8–10]. HY1 is a heme oxygenase, and HY2 is a bilin reductase. These enzymes cooperate sequentially to convert heme to PΦB in the plastids. PΦB molecules enter the cytosol by simple diffusion or through unknown transport mechanisms. The cytoplasmic apophytochrome protein covalently incorporates PΦB via its intrinsic lyase activity to assemble the Pr form of phytochrome (phy-Pr). A thioester linkage is catalyzed between a conserved cysteine residue and the A-ring ethylidene substituent of PΦB. Based on the well-studied biosynthesis of phytochrome, genetic engineering in *Escherichia coli* (*E. coli*) and yeast strains has been exploited to express phytochrome proteins [11–16]. These expression approaches have made significant contributions to the biochemical and structural analyses of phytochromes. In our recent studies on the structures of phyA and phyB [6, 7], we developed another strategy for phytochrome biosynthesis, utilizing *E. coli* to synthesize PΦB and insect cells to express apophytochromes. The phy-Pr proteins were then purified from their mixtures. Using this method, we successfully purified *Arabidopsis* phyA and phyB, as well as a maize phyA. Here, we present this strategy in detail and hope it provides an alternative approach for the successful purification of phytochrome in the research community.

2 Materials

For clarity and to avoid duplication with published data, we use a phytochrome species from tomato, *Solanum lycopersicum* phyA (SlphyA), as an example to illustrate our purification approach in the following sections.

2.1 Plasmid Construction

1. cDNA library of *Solanum lycopersicum* (Courtesy of Prof. Jingquan Yu, Zhejiang University).
2. Primers for the PCR amplification of *SLPHYA* (sense primer: 5′--tttcaggggcccctgggatccATGTCGTCTTCAAGACCTAGCCA-3′; antisense primer: 5′--cttggtaccgcatgcctcgagTCAGCTTGAGTTGGTAGCCACT-3′).
3. Synthesized DNA products of the *HO1* (*see* **Note 1**) and *HY2* (AT3G09150) genes.
4. Synthesized DNA product of the *SUMO* (*see* **Note 2**) gene.

5. pRSFDuet-1 plasmid (Novagen).
6. pFastBac-1 plasmid (Invitrogen).
7. *E. coli* DH5α competent cells.

2.2 PΦB Biosynthesis

1. *E. coli* BL21 (DE3) competent cells.
2. Luria-Bertani (LB) medium: 10 g/L tryptone, 5 g/L yeast extract, 10 g/L NaCl (*see* **Note 3**). Autoclave at 121 °C for 20 min.
3. Kanamycin: 60 mg/mL in water. Sterilize by filtration through a 0.22-μm filter.
4. IPTG: 0.6 M isopropyl-β-D-thiogalactoside in water. Sterilize by filtration through a 0.22-μm filter.
5. 5-ALA: 0.1 M 5-aminolevulinic acid in dimethyl sulfoxide (DMSO). Sterilize by filtration through a 0.22-μm filter.

2.3 Expression of apo-SlphyA

1. *E. coli* DH10Bac competent cells.
2. LB agar plate: Supplement the LB medium with 15 g/L agar. Autoclave at 121 °C for 20 min. Cool the medium to 50 °C and add 50 μg/mL kanamycin, 10 μg/mL tetracycline, 7 μg/mL gentamicin, 40 μg/mL IPTG, and 100 μg/mL X-gal (*see* **Note 4**). Pour the mixture into the plates and store them at 4 °C once solidified.
3. Plasmid mini kit.
4. Sf21 insect cells (Invitrogen).
5. Cell transfection reagent: Cellfectin® II Reagent (Invitrogen).
6. Cell culture medium: Sf-900 II SFM (Invitrogen). Supplement the Sf-900 II SFM medium with penicillin/streptomycin (Invitrogen) to a final concentration of 50 units/mL penicillin and 50 μg/mL streptomycin. Store at 4 °C.

2.4 Purification of SlphyA-Pr

1. Gravity fast flow Nickel (Ni^{2+}) Sepharose column (Cytiva).
2. Ion-exchange chromatography column: RESOURCE Q6 (Cytiva).
3. Size-exclusion chromatography column: Superose 6 Increase 10/300 GL (Cytiva).
4. PreScission protease (*see* **Note 5**).
5. Lysis buffer: 25 mM Tris-HCl, 150 mM NaCl, 5% (v/v) glycerol, 20 mM imidazole, pH 8.0. Prepare and store at 4 °C.
6. Ion-exchange chromatography Buffer A: 25 mM Tris-HCl, pH 8.0. Prepare and store at 4 °C.
7. Ion-exchange chromatography Buffer B: 25 mM Tris-HCl, 1 M NaCl, pH 8.0. Prepare and store at 4 °C.

8. Size-exclusion chromatography Buffer C: 25 mM Tris-HCl, 150 mM NaCl, 2% glycerol, 3 mM dithiothreitol (DTT), pH 8.0. Prepare and store at 4 °C.

2.5 Characterization of SlphyA Proteins

1. SDS-PAGE gel: Comprises 5% stacking gel and 10% resolving gel.
2. Coomassie brilliant blue staining solution: 2 g/L Coomassie brilliant blue, 40% (v/v) ethanol, 10% (v/v) acetic acid, 50% (v/v) water.
3. Zinc solution: 150 mM Tris-HCl, 20 mM zinc acetate, pH 7.0. Prepare and store at room temperature.
4. GelDoc Go Imaging System (BIO-RAD).
5. Cary 60 spectrophotometer (Agilent).

3 Methods

3.1 Plasmid Construction

1. Amplify the DNA products of *SlPHYA* (UniProt entry: Q41331; residues 1–1123) from the cDNA library of *Solanum lycopersicum* using the specific primers detailed in Subheading 2.1.
2. Use molecular cloning technology (*see* **Note 6**) to construct the pFastBac-SUMO-SlphyA plasmid (*see* Fig. 1a and **Note 7**) for apo-SlphyA expression in insect cells, and the pRSFDuet-1-HO1-HY2 plasmid (*see* Fig. 1b and **Note 8**) for PΦB biosynthesis in *E. coli*.

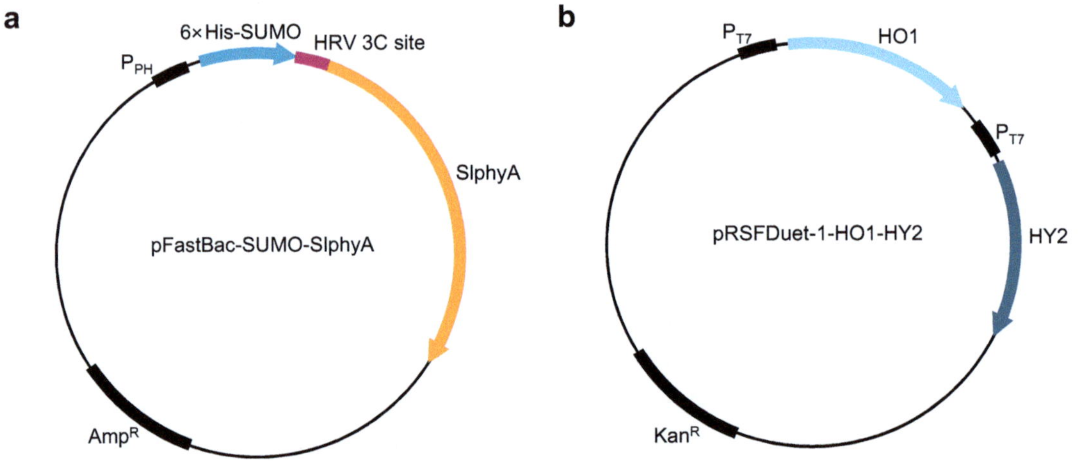

Fig. 1 Plasmid constructs for apo-SlphyA expression and PΦB biosynthesis. (**a**) Plasmid map of the pFastBac-SUMO-SlphyA construct used for apo-SlphyA expression in insect cells. P_{PH}: polyhedrin promoter. HRV 3C site, also known as the Pre Scission protease cleavage site. Amp^R: ampicillin resistance. (**b**) Plasmid map of the pRSFDuet-1-HO1-HY2 construct used for PΦB biosynthesis in *E. coli*. P_{T7}: T7 promoter. Kan^R: kanamycin resistance

3. Transform the plasmids into *E. coli* DH5α competent cells. Screen positive clones containing the pFastBac-SUMO-SlphyA plasmid by ampicillin resistance and those containing the pRSFDuet-1-HO1-HY2 plasmid by kanamycin resistance. Sequencing verification is required for both plasmids.
4. Extract the two plasmids and store them at −20 °C.

3.2 PΦB Biosynthesis

1. Transform the pRSFDuet-1-HO1-HY2 plasmid into *E. coli* BL21 (DE3) competent cells.
2. Culture the BL21 (DE3) strain in LB medium supplemented with 60 μg/mL kanamycin at 37 °C with continuous shaking at 200 rpm.
3. When the absorbance at 600 nm (OD_{600}) of the culture reaches approximately 1.0, cool the culture to 16 °C.
4. Add 0.8 mM IPTG and 100 μM 5-ALA to the culture to induce PΦB biosynthesis (*see* **Note 9**). Continue culturing for 10 h at 16 °C with continuous shaking at 170 rpm, and maintain the culture in darkness (*see* **Note 10**).
5. Harvest the *E. coli* cells by centrifugation at 4300× *g* for 10 min at 4 °C in darkness or under dim green light.
6. Proceed to Subheading 3.4 or store the cell pellets protected from light at −80 °C.

3.3 Expression of apo-SlphyA

1. Transform the pFastBac-SUMO-SlphyA plasmid into *E. coli* DH10Bac competent cells to generate bacmid DNA containing the recombinant *SlPHYA* (*see* **Note 11**).
2. Inoculate a white colony from the LB agar plate into 2 mL of LB medium supplemented with 50 μg/mL kanamycin, 10 μg/mL tetracycline, and 7 μg/mL gentamicin. Culture at 37 °C with continuous shaking at 200 rpm for 12 h.
3. Extract the recombinant bacmid DNA from the culture using the Plasmid Mini Kit and dissolve in 40 μL of sterilized distilled water. Store the bacmid DNA at 4 °C for short-term use or at −20 °C for long-term preservation (*see* **Note 12**).
4. Transfection of Sf21 insect cells: Dilute 6 μL of bacmid DNA and 6 μL of Cellfectin® II separately in 100 μL of Sf-900 II SFM without antibiotics. Mix the two solutions gently and incubate for 30 min at 28 °C. Add the transfection mixture to Sf21 insect cells (1×10^6 cells per 35-mm well of a 6-well plate) and incubate for 5 h at 28 °C. Remove the transfection mixture and replace it with 2 mL of Sf-900 II SFM containing antibiotics. Incubate the cells at 28 °C for 72 h (*see* **Note 13**).
5. Harvest and amplify the recombinant baculovirus: After 72 h post-transfection, harvest the virus by centrifugation at 500× *g* for 5 min. Preserve the virus-containing supernatant and mark

it as the P1 viral stock. To amplify the viral stock, infect a monolayer culture of Sf21 cells (2×10^7 cells and 25 mL of Sf-900 II SFM with antibiotics per 15-cm plate) with 0.5 mL of P1 viral stock at 28 °C. After 72 h post-infection, harvest the virus and mark it as the P2 viral stock. Store all viral stocks protected from light at 4 °C (*see* **Note 14**).

6. Expression of apo-SlphyA in Sf21 insect cells: Culture 1 L of Sf21 cells in Sf-900 II SFM with antibiotics to a density of 2×10^6 cells/mL at 28 °C with continuous shaking at 120 rpm. Infect the culture with 25 mL of P2 viral stock. After 48 h post-infection at 28 °C with continuous shaking at 120 rpm, harvest the cells by centrifugation at 2200× *g* for 10 min (*see* **Note 15**).
7. Proceed to Subheading 3.4 or store the cell pellets at −80 °C.

3.4 Purification of SlphyA-Pr

All the steps in this section are conducted in darkness or dim green light.

1. Resuspend cell pellets from 1 L of *E. coli* culture in 30 mL of lysis buffer, and cell pellets from 1 L of Sf21 culture in 60 mL of lysis buffer. Mix the two suspensions thoroughly (*see* **Note 16**).
2. Lyse the cells by sonication on ice at 150 W for 3 min with a 2-s on/4-s off cycle.
3. Clarify the lysate by centrifugation at 23,000× *g* for 1 h at 4 °C. Collect the supernatant.
4. Load the supernatant onto a gravity column containing 1 mL of Ni^{2+} Sepharose for affinity chromatography at 4 °C. Repeat this step once for maximal purification.
5. Wash the column with 5 mL of lysis buffer twice at 4 °C.
6. Resuspend the Ni^{2+} Sepharose in 2 mL of lysis buffer and add 10 μL of PreScission protease to the suspension. Cleave the His-SUMO tag from the recombinant SlphyA proteins at 4 °C for 8 h. Resuspend the Ni^{2+} Sepharose once or twice during the enzyme cleavage (*see* **Note 17**).
7. Collect the supernatant and the lysis buffer used to wash the Ni^{2+} Sepharose (2 mL per wash, performed twice at 4 °C).
8. Dilute the 6 mL protein sample with Buffer A to a total volume of 30 mL. Load the dilution onto a RESOURCE Q6 column and perform ion-exchange chromatography at 4 °C. Elute protein samples using a gradient of NaCl from the mixture of Buffer A and B. Characterize the peak fractions of protein samples as described in Subheading 3.5 (*see* **Note 18**).
9. Load 2 mL of purified SlphyA-Pr proteins from the ion-exchange chromatography onto a Superose 6 Increase 10/300 GL column and perform size-exclusion chromatography with Buffer C at 4 °C and a flow rate of 0.5 mL/min.

Characterize the peak fractions of protein samples as described in Subheading 3.5.

3.5 Characterization of SlphyA Proteins

1. Load the SlphyA protein samples onto the SDS-PAGE gel and perform electrophoresis (*see* **Note 19**).
2. Characterize the purity of the SlphyA-Pr protein using Coomassie brilliant blue staining (Fig. 2a) and the PΦB incorporation of the SlphyA-Pr protein by zinc-induced fluorescence under 302 nm UV light using the GelDoc Go Imaging System (Fig. 2a).
3. Illuminate 50 μL of SlphyA proteins with far-red light (740 nm, 50.4 $\mu mol \cdot m^{-2} \cdot s^{-1}$) on ice for 30 min. Then, maintain the sample in darkness on ice for 1 h.
4. Measure the absorption spectrum of SlphyA-Pr using a Cary 60 spectrophotometer with a sub-micro quartz cell (50 μL) at 22 °C and a scan rate of 4800 nm/min. Collect the absorbance data from 200 to 900 nm (Fig. 2b).
5. Illuminate the same SlphyA sample with red light (630 nm, 340.1 $\mu mol \cdot m^{-2} \cdot s^{-1}$) on ice for a saturated irradiation (e.g., 30 min; *see* **Note 20**). Measure the absorption spectrum of SlphyA-Pfr as described in **step 4** (Fig. 2b).
6. Subtract the Pfr spectrum from the Pr spectrum to generate a difference spectrum (ΔAbsorbance; Fig. 2b).

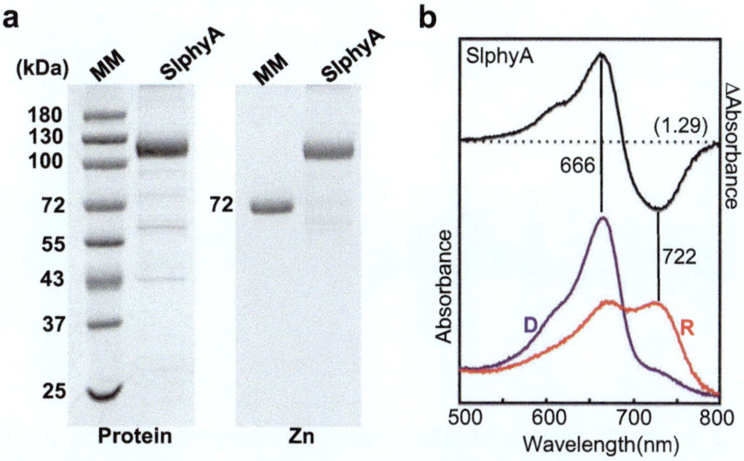

Fig. 2 Characterization of SlphyA proteins. (**a**) SDS-PAGE analysis of full-length holo-SlphyA proteins. Gels were stained with Coomassie brilliant blue (left) and assayed by zinc-induced fluorescence (right). MM, molecular mass standards. (**b**) UV-vis absorption spectra of SlphyA proteins in darkness (D) and after saturating irradiation with 630 nm red light (R). The value of spectral change ratio (SCR) is shown in parentheses

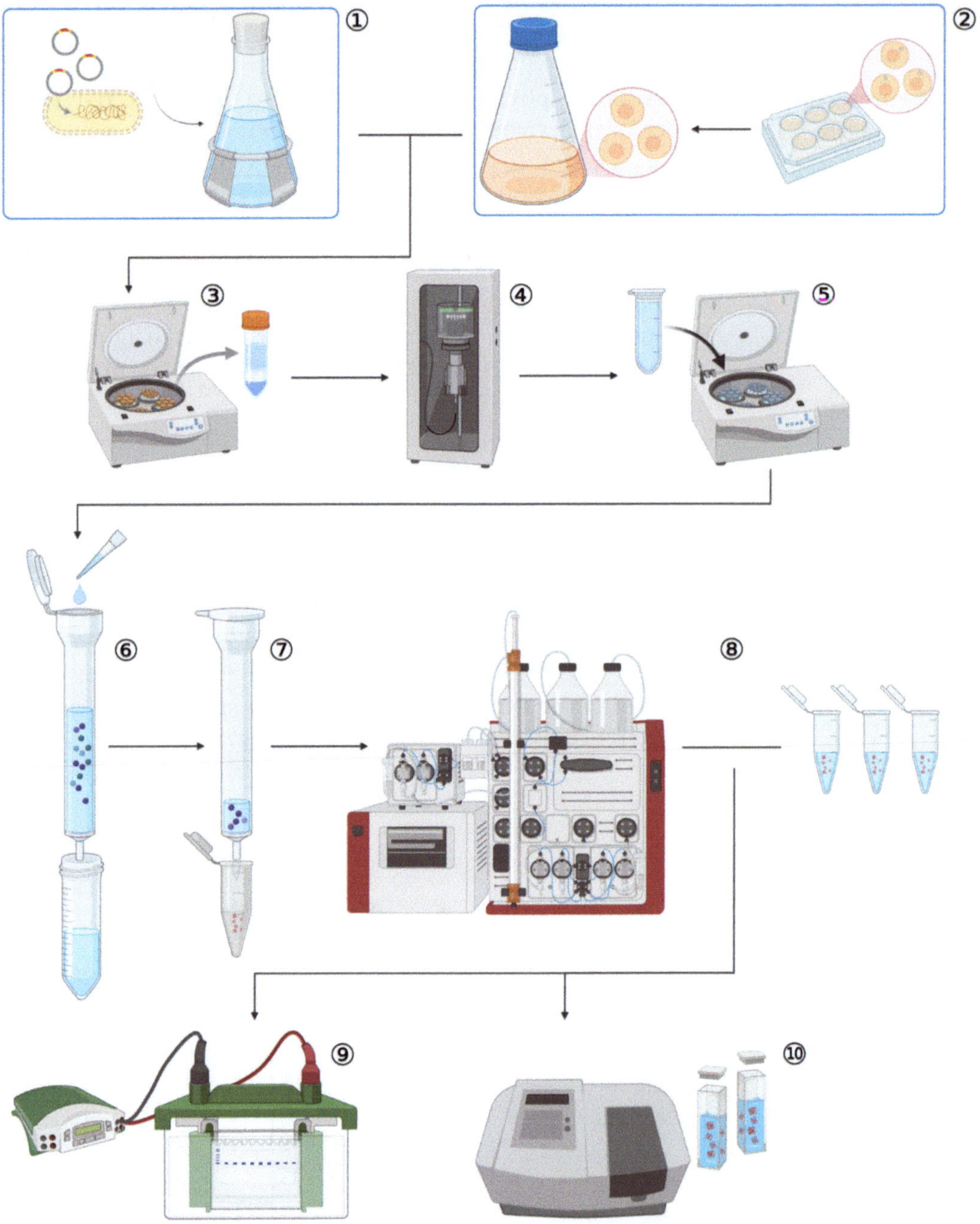

Fig. 3 Schematic diagram of plant phytochrome biosynthesis using bacteria and insect cells. ①: PΦB biosynthesis in *E. coli*. ②: Expression of apophytochrome in insect cells. ③: Harvest *E. coli* cells and insect cells by centrifugation at 4 °C and resuspend the cell pellets in lysis buffer. ④: Lyse cells by sonication on ice in darkness or dim green light. ⑤: Clarify the lysate by centrifugation at 4 °C and collect the supernatant. ⑥: Purify the holophytochrome proteins from the supernatant using the Ni^{2+} Sepharose column. ⑦: Remove the N-terminal 6 × His-SUMO tag with PreScission protease. ⑧: Further purify the holophytochrome proteins through ion-exchange and subsequent size-exclusion chromatography. Characterize the holophytochrome proteins by SDS-PAGE assays (⑨) and spectral analyses (⑩). (This figure was created with BioRender.com)

A schematic view of all procedures in this strategy for plant phytochrome biosynthesis is shown in Fig. 3.

4 Notes

1. The *HO1* (UniProt entry: P72849) gene, identified in the cyanobacterium *Synechocystis* sp. PCC 6803 [17], is the homologue of *HY1* in *Arabidopsis*. It has been widely utilized in genetic engineering for bilin molecule biosynthesis in bacteria [3, 4, 6, 7, 12–14]. Moreover, the synthesized DNA products of the *HO1* and *HY2* genes are codon-optimized for the *E. coli* expression system.
2. The SUMO (UniProt entry: Q12306) protein from *Saccharomyces cerevisiae* is widely used as a solubility-promotion tag for target proteins.
3. All reagents mentioned in this and subsequent sections can be purchased from Sigma-Aldrich or other preferred sources.
4. X-gal is an abbreviation for 5-bromo-4-chloro-3-indolyl-β-D-galactoside. It is widely used in the blue-white screening of genetic engineering. When preparing a 20 mg/mL stock solution, dissolve X-gal in dimethylformamide (DMF). Filtration sterilization is not required for this stock solution. However, stock solutions of tetracycline (10 mg/mL in ethanol) and gentamicin (7 mg/mL in water) should be sterilized by filtration through a 0.22-μm filter.
5. Pre Scission protease is a recombinant protein of human rhinovirus (HRV) type 14 3C protease, incorporating an N-terminal Glutathione-S-transferase (GST) tag. Consequently, it is also referred to as HRV 3C protease. This protease exhibits high recognition specificity for an octapeptide sequence, LEVLFQGP, or the core pentapeptide, LFQGP, and efficiently cleaves the amide bond between G and P at 4 °C. Typically, the protease is expressed and purified from *E. coli*, allowing for both in-house purification and commercial procurement.
6. The homologous recombination cloning technology was used for plasmid construction with a commercial kit. However, any other molecular cloning method could be used as preferred.
7. In this expression construct, a recombinant *SlPHYA* gene encoding six tandem histidines (6 × His), SUMO (residues 26–121), a PreScission protease cleavage site (LEVLFQGP), and SlphyA is inserted into the multiple cloning sites of pFastBac-1 vector under the control of the polyhedrin promoter.
8. In this expression construct, two PΦB biosynthesis genes encoding full-length HO1 (residues 1–240) and HY2 (residues

46–329) with the truncation of the plastid localization sequence are inserted into the multiple cloning sites of the pRSFDuet-1 vector, both under the control of T7 promoters.

9. Previous studies have demonstrated that the genetically engineered *E. coli* cells can successfully synthesize PΦB molecules [3, 4, 6, 7, 12–14]. This is presumably due to the presence of heme, a substrate for PΦB biosynthesis, in the cytoplasm of prokaryotic cells. In eukaryotic cells, heme molecules are typically compartmentalized within plastids or mitochondria. Additionally, supplementing 5-ALA, a vital substrate for heme biosynthesis [8, 10], can significantly enhance PΦB synthesis in *E. coli* by increasing the cytoplasmic heme level.
10. The free PΦB molecule is extremely light-sensitive; therefore, it is essential to keep cultures containing PΦB molecules in darkness.
11. The bacmid, which propagates in the DH10Bac strain, is a large plasmid containing the baculovirus genome. Consequently, it is also referred to as the baculovirus shuttle vector. The recombinant gene (*SlPHYA*, in this case) on the pFastBac plasmid can be transferred to bacmid via a transposition mechanism in the DH10Bac strain, thereby generating white colonies that contain the recombinant bacmid. Transfection of insect cells with the recombinant bacmid DNA can generate the recombinant baculovirus, which can be used to infect insect cells for protein expression. For a detailed description of the principles and procedures of the transposition process occurring in the DH10Bac strain, refer to the official user guide "Bac-to-Bac® Baculovirus Expression System (Catalog Numbers 10359-016, 10360-014, 10584-027, 10712-024)" available on the Invitrogen website.
12. A detailed procedure for bacmid DNA extraction can be found in the instruction manual of the purchased kit or the user guide mentioned in **Note 11**.
13. The Sf9 insect cell line can serve as an alternative choice for cell transfection and protein expression of apo-SlphyA. Both Sf9 and Sf21 cell lines are derived from the pupal ovarian tissue of the fall armyworm, *Spodoptera frugiperda*. A detailed procedure for insect cell transfection can be found in the user guide mentioned in **Note 11**.
14. The strategy for amplifying the recombinant baculovirus can be adjusted based on actual protein expression. A detailed description of the amplification of baculovirus stocks can be found in the user guide mentioned in **Note 11**.
15. To achieve optimal expression of apo-SlphyA, small-scale expression tests (e.g., 50 mL insect cell culture) with varied

volumes of inoculated virus and infection time courses are recommended.

16. To ensure a saturated assembly of SlphyA-Pr, we recommend mixing equal volumes of cell cultures dedicated to PΦB biosynthesis and apo-SlphyA expression. However, the required amounts of *E. coli* cells containing PΦB can be adjusted based on the spectral characterization of the purified SlphyA proteins. Specifically, the ratio A666/A280 (A666: absorbance at 666 nm; A280: absorbance at 280 nm) calculated from the spectrum of SlphyA-Pr in Subheading 3.5, can roughly estimate the proportion of holo-SlphyA protein in the total protein sample. A ratio close to 1.0 indicates high purity of the holo-SlphyA protein, whereas a ratio much less than 1.0 suggests that more *E. coli* cells are needed for the subsequent purification.
17. An alternative strategy involves eluting the recombinant SlphyA proteins with a high concentration of imidazole (e.g., 250 mM) and cleaving the His-SUMO tag in the solution for a shorter duration (e.g., 2 h). Although this expedites the purification process, protein precipitation may occur in the buffer containing high imidazole concentrations.
18. It is crucial to carefully characterize the purified SlphyA-Pr protein before further use. Detailed procedures for ion-exchange and subsequent size-exclusion chromatography can be found in the instruction manual of the purchased columns.
19. All characterizations in this section involve routine assays of phytochrome research, and more detailed information on these assays can be found in the references [3–7, 11–16].
20. The saturated red light irradiation of SlphyA proteins is necessary to achieve a maximal number of SlphyA-Pfr proteins. Although the specified time course for saturated irradiation serves as an empirical reference, a shorter duration, such as 15 min, is likely sufficient, as indicated in the previous study [16].

Acknowledgments

This work was supported by the National Key R&D Program of China (2024YFA1306701 to J.W.), the Shandong Provincial Natural Science Foundation (ZR2024QC259 to C.C.), and the Taishan Scholars Program of Shandong Province (C.C.).

References

1. Hughes J, Winkler A (2024) New insight into phytochromes: connecting structure to function. Annu Rev Plant Biol 75:153–183. https://doi.org/10.1146/annurev-arplant-070623-110636
2. Cheng MC, Kathare PK, Paik I et al (2021) Phytochrome signaling networks. Annu Rev Plant Biol 72:217–244. https://doi.org/10.1146/annurev-arplant-080620-024221
3. Li H, Burgie ES, Gannam ZTK et al (2022) Plant phytochrome B is an asymmetric dimer with unique signalling potential. Nature 604(7904):127–133. https://doi.org/10.1038/s41586-022-04529-z
4. Burgie ES, Li H, Gannam ZTK et al (2023) The structure of *Arabidopsis* phytochrome A reveals topological and functional diversification among the plant photoreceptor isoforms. Nat Plants 9(7):1116–1129. https://doi.org/10.1038/s41477-023-01435-8
5. Wang J, Zhou C, Guan Z et al (2023) Plant phytochrome A in the Pr state assembles as an asymmetric dimer. Cell Res 33(10):802–805. https://doi.org/10.1038/s41422-023-00847-7
6. Zhang Y, Lin X, Ma C et al (2023) Structural insights into plant phytochrome A as a highly sensitized photoreceptor. Cell Res 33(10):806–809. https://doi.org/10.1038/s41422-023-00858-4
7. Wang Z, Wang W, Zhao D et al (2024) Light-induced remodeling of phytochrome B enables signal transduction by phytochrome-interacting factor. Cell 187(22):6235–6250. https://doi.org/10.1016/j.cell.2024.09.005
8. Muramoto T, Kohchi T, Yokota A et al (1999) The Arabidopsis photomorphogenic mutant *hy1* is deficient in phytochrome chromophore biosynthesis as a result of a mutation in a plastid heme oxygenase. Plant Cell 11(3):335–347. https://doi.org/10.1105/tpc.11.3.335
9. Davis SJ, Kurepa J, Vierstra RD (1999) The *Arabidopsis thaliana HY1* locus, required for phytochrome-chromophore biosynthesis, encodes a protein related to heme oxygenases. Proc Natl Acad Sci USA 96(11):6541–6546. https://doi.org/10.1073/pnas.96.11.6541
10. Kohchi T, Mukougawa K, Frankenberg N et al (2001) The Arabidopsis *HY2* gene encodes phytochromobilin synthase, a ferredoxin-dependent biliverdin reductase. Plant Cell 13(2):425–436. https://doi.org/10.1105/tpc.13.2.425
11. Wahleithner JA, Li L, Lagarias JC (1991) Expression and assembly of spectrally active recombinant holophytochrome. Proc Natl Acad Sci USA 88(23):10387–10391. https://doi.org/10.1073/pnas.88.23.10387
12. Gambetta GA, Lagarias JC (2001) Genetic engineering of phytochrome biosynthesis in bacteria. Proc Natl Acad Sci USA 98(19):10566–10571. https://doi.org/10.1073/pnas.191375198
13. Landgraf FT, Forreiter C, Picó AH et al (2001) Recombinant holophytochrome in *Escherichia coli*. FEBS Lett 508(3):459–462. https://doi.org/10.1016/S0014-5793(01)02988-X
14. Zhang J, Stankey RJ, Vierstra RD (2013) Structure-guided engineering of plant phytochrome B with altered photochemistry and light signaling. Plant Physiol 161(3):1445–1457. https://doi.org/10.1104/pp.112.208892
15. Shin AY, Han YJ, Song PS et al (2014) Expression of recombinant full-length plant phytochromes assembled with phytochromobilin in *Pichia pastoris*. FEBS Lett 588(17):2964–2970. https://doi.org/10.1016/j.febslet.2014.05.050
16. Han YJ, Cho JY, Kim JI (2019) Expression, purification, and spectral characterization of phytochromes. Methods Mol Biol 2026:95–111. https://doi.org/10.1007/978-1-4939-9612-4_7
17. Cornejo J, Willows RD, Beale SI (1998) Phytobilin biosynthesis: cloning and expression of a gene encoding soluble ferredoxin-dependent heme oxygenase from *Synechocystis* sp. PCC 6803. Plant J 15(1):99–107. https://doi.org/10.1046/j.1365-313X.1998.00186.x

Chapter 6

Overcoming the Challenges of Producing Plant Phytochromes in *Escherichia coli*

Sintayehu Manaye Shenkutie and Kaoling Guan

Abstract

Biochemical and structural studies of plant phytochromes have been limited until recently by the difficulty of producing them in large amounts and as chromophore-bound, photochemically competent holoproteins in bacterial expression systems. This chapter provides protocols to achieve high-level production of the photosensory modules of plant phytochrome B in *Escherichia coli*. Protocols for codon optimization, designing multiple constructs, coexpression of the bilin chromophore synthesizing enzymes, and optimizing expression vectors, expression strains, and growth conditions are also included. These protocols can be easily adopted to overcome frequently encountered problems during heterologous production of eukaryotic phytochromes in *Escherichia coli*.

Key words Phytochrome, Recombinant protein, Bacterial expression system, *Sorghum bicolor*, *Glycine max*, *Escherichia coli*

1 Introduction

Phytochromes are photochromic biliprotein photoreceptors found in plants, bacteria, and fungi, characterized by bilin lyase activity, red/far-red photoreversibility, and signal transduction following photoactivation [1]. Through their autocatalytic bilin lyase activity, plant apophytochromes autoassemble to form holophytochromes in which a phytochromobilin (PΦB) or phycocyanobilin (PCB) chromophore is covalently linked to a conserved cysteine residue in the GAF domain [2]. Holophytochromes undergo photoconversion between a red light absorbing Pr state and a far-red light absorbing Pfr state [3]. As light-regulated signaling proteins, phytochromes are components of various signal transduction pathways controlling complex developmental and physiological responses.

Plant phytochromes consist of two distinct modules: an N-terminal photosensory module (PSM) and a C-terminal module (CTM)—*see* Fig. 1. In the context of multidomain proteins, a

Andreas Winkler (ed.), *Phytochromes: Methods and Protocols*, Methods in Molecular Biology, vol. 2970,
https://doi.org/10.1007/978-1-0716-4791-2_6,

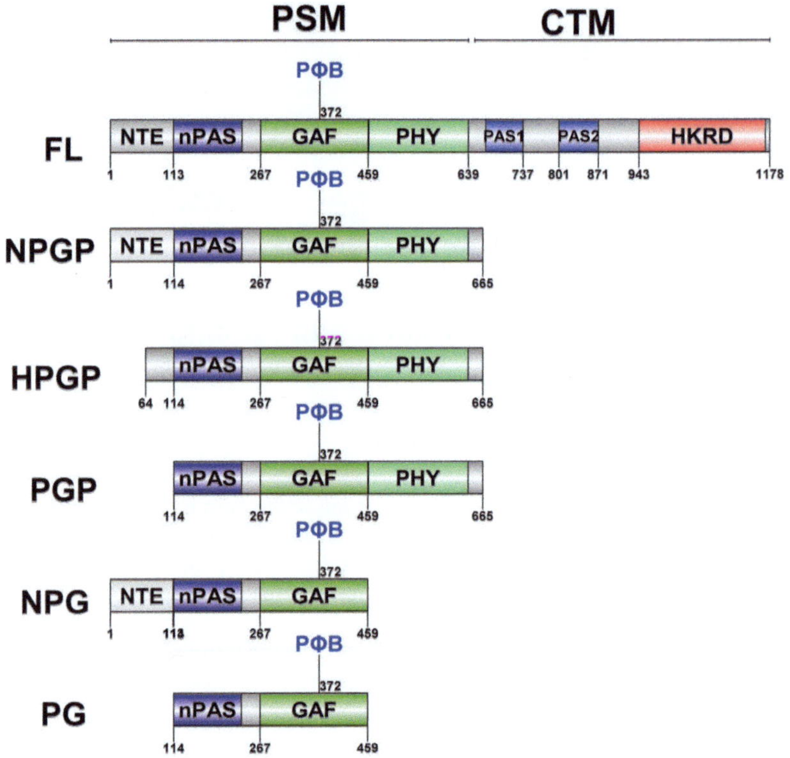

Fig. 1 Domain architecture of the full-length (FL) phytochrome B from Sorghum bicolor and selected constructs. The domain organization of phytochrome B is illustrated, highlighting its modular architecture. The PSM consists of the NTE, nPAS, GAF, and PHY domains. The CTM consists of two consecutive PAS domains and an HKRD domain. Five constructs are depicted, designated as NPGP, HPGP, PGP, NPG, and PG, reflecting specific combinations of domains retained from the FL protein. HPGP contains a putative helix-forming motif in the NTE

"module" refers to a conserved sequential arrangement of domains. The PSM includes an N-terminal extension (NTE), an N-terminal Per/Arnt/Sim (nPAS) domain, a cyclic guanosine monophosphate (c GMP)-specific phosphodiesterase/adenylyl cyclase/FhlA (GAF) domain, and a phytochrome-specific (PHY) domain. The CTM, on the other hand, contains three consecutive domains: PAS1, PAS2, and a histidine kinase-related domain (HKRD).

Historically, plant phytochromes were isolated from etiolated plant tissues. This was facilitated by two rapid and sensitive analytical techniques: spectrophotometric difference assay developed by Butler et al. [3] and zinc-induced fluorescence after SDS-PAGE developed by Berkelman et al. [4]. These methods enabled holophytochrome quantification in crude cell lysates without requiring prior protein purification. However, isolating endogenous phytochromes remains laborious and often yields insufficient quantities of high-purity material for biophysical studies. The development of recombinant phytochrome production in the *Escherichia coli* expression system has since overcome these limitations, enabling substantial progress in structure-function studies of plant phytochromes.

Escherichia coli is the most common host for the mass production of recombinant proteins for structural biology. It is well characterized, easy to handle and requires inexpensive equipment for growth, storage, and genetic manipulation [5]. However, the soluble production and proper folding of multi-domain eukaryotic proteins in *E. coli* present technical challenges. Toxicity, incorrect folding, poor solubility, and the formation of inclusion bodies are the most common problems. Several strategies have been proposed to alleviate these issues. Codon usage bias can be addressed by either employing a synthetic gene with an improved codon adaptation index or utilizing engineered strains that possess additional copies of genes encoding rare tRNAs [6]. Low copy number plasmids can be used to improve plasmid stability by minimizing background expression and toxicity, while poor solubility can be resolved by coexpression of indispensable cofactors such as chaperones in *E. coli* or solubility-enhancing fusion tags. Besides, with genetic engineering, different constructs can be generated and tested to improve bacterial production and purification [7]. The present chapter describes detailed protocols for optimization of codons, protein constructs, expression vectors, and expression strains to improve the production of the photosensory module of plant phytochromes in *E. coli*. Readers interested in expression of full-length plant phytochromes should refer additionally to protocols of the Vierstra lab [8, 9].

2 Materials

The following methods necessitate the provision of the specified equipment, consumables, reagents, buffers, growth media, and host strains. Standard microbiology and molecular biology laboratory essentials, including autoclaves, French presses, water baths, heat blocks, −80 °C freezers, 0.22- and 0.45-μm membrane filters, restriction enzymes, DNA polymerase, DNA ligase, personal computers, thermal cyclers, ice makers, weighing balances, Petri dishes, various culture tubes, centrifuge bottles, pH meters, and magnetic stirrers, are required. Access to the Internet, liquid nitrogen, Milli-Q water, DNA synthesis, and DNA sequencing services is also essential.

2.1 Equipment

1. Incubator: Heraeus B 5042.
2. Incubator shaker: Eppendorf Brunswick Innova 42/42R.
3. Triple-baffled shake flasks: Sigma-Aldrich or Erlenmeyer reusable 0.5-L and 2-L flasks.
4. Centrifuges: Refrigerated with fixed angle rotors Beckman Coulter or Sorvall.
5. Size exclusion chromatography column: Superdex 200 26/60 prep grade column.
6. HPLC glass syringe: 0.25 mL volume.
7. Fluorescent lamp: Vilber Lourmat TCP-20-LM.
8. Digital camera fitted with a ~630-nm interference filter.
9. UV/B transilluminator.
10. Fast Protein Liquid Chromatography (FPLC) system: ÄKTA or Cytiva.
11. Vertical electrophoresis cells: Protein electrophoresis equipment, Bio-Rad.
12. Benchtop spectrophotometer and cuvettes: for optical density measurement.
13. UV-Vis spectrophotometer: 8453Agilent, with cuvette holder modified to allow efficient Pr/Pfr photoconversion and micro-cuvettes with transparent side-walls (*see* Chapter 4 by Hughes and Lang).
14. Light-emitting diodes (LEDs): 660 and 730 nm, Roithner.
15. White light box: for SDS-PAGE gel documentation.

2.2 Consumables and Reagents

1. Antibiotic stock 1000×: 100 mg/mL Ampicillin, 50 mg/mL Kanamycin, 50 mg/mL Spectinomycin and 10 mg/mL Tetracycline in H_2O (pure water) and sterile-filtered.
2. Isopropyl β-D-1-thiogalactopyranoside (IPTG) stock 1000×: 1 M in H_2O and sterile-filtered.
3. Arabinose stock 100×: 20% (w/w) in H_2O and sterile-filtered.
4. IMAC column: Ni-NTA Superflow Agarose resin (Qiagen).
5. Zinc acetate solution: 1 mM.
6. Media components: yeast extract, tryptone, agar, KCl, $MgCl_2$, $MgSO_4$, and glucose.
7. Buffer component: NaCl, Tris, 37% (w/w) HCl, NaOH, HEPES, $(NH_4)_2SO_4$, KH_2PO_4, K_2HPO_4, DTT, SDS, bromophenol blue, EDTA, 14.3 M β-mercaptoethanol, DNase I, PMSF, 100% glycerol, iminodiacetic acid (IDA), and imidazole.
8. Desalting columns: PD-10 columns packed with Sephadex G-25 resin.

9. Concentrators: Centrifugal Vivaspin 15R or stirred pressure cell with 25-kDa cutoff.
10. Low molecular weight SEC standards: e.g., Cytiva, thyroglobulin (669 kDa), ferritin (440 kDa), catalase (232 kDa), aldolase (158 kDa), albumin (67 kDa), ovalbumin (43 kDa), chymotrypsinogen A (25 kDa) and, RNAse A (14 kDa).
11. SDS-PAGE components: Tris-HCl pH 6.8 and pH 8.8, 40% acrylamide/bisacrylamide solution (29:1), ammonium persulfate (APS), and TEMED.
12. Tris/Glycine/SDS 10× buffer: 0.25 M Tris base, 1.923 M glycine, and 1% (w/v) SDS.
13. SDS-PAGE high range protein marker, e.g., PM-III, AppliChem.
14. Hot Coomassie stain: 0.5% (w/v) Coomassie Brilliant Blue R-250, 25% (v/v) isopropanol and 10% (v/v) acetic acid.
15. Destaining solution: 10% acetic acid.

2.3 List of Primer (Table 1)

2.4 Bacterial Growth Media

1. SOC-medium: 0.5% (w/v) yeast extract, 2% (w/v) tryptone, 10 mM NaCl, 2.5 mM KCl, 10 mM $MgCl_2$, 10 mM $MgSO_4$, and 20 mM glucose. Sterile-filtered. *See* **Note 1**.
2. Luria Bertani (LB) medium: 1% (w/v) tryptone, 0.5% (w/v) yeast extract, and 1% (w/v) NaCl adjusted to pH 7.5 with NaOH. Autoclaved, then sterile antibiotic added when the

Table 1
Primer sequences for PCR amplifications of genes encoding multiple constructs

Constructs	Primers
NPGP	Fwd. CCCGGTACCGAATTCATTAAAGAGGAGAAATTAACCATGGCTTCTGGTTCT Rev. GTAGGGAAGCTTTTAGTGATGATGATGATGATGACCCAGCTGAACCTGACC
HPGP	Fwd. CGGGAATTCTTAAAGAGGAGAAATTAACCATGGctACCGAATCTGTTTCTAAAG Rev. as NPGP above
PGP	Fwd. CGGGAATTCTTAAAGAGGAGAAATTAACCATGGctGCTGCTTACCTGTCTCGTA Rev. as NPGP above
NPG	Fwd. CCCGGTACCGAATTCATTAAAGAGGAGAAATTAACCATGGCTTCTGGTTCT Rev. GTAGGGAAGCTTTTAGTGATGATGATGATGATGAGCCAGCTGCAGTTCCAT
PG	Fwd. CGGGAATTCTTAAAGAGGAGAAATTAACCATGGctGCTGCTTACCTGTCTCGTA Rev. as NPG above

This table details the primers utilized for PCR amplification of genes encoding the constructs NPGP, HPGP, PGP, NPG, and PG. These primers were designed to include restriction sites and stop codons to facilitate subsequent cloning and expression studies

medium has cooled to ca. 50 °C. LB is used as a standard growth medium for plasmid propagation and starter cultures.

1. LBAK agar: 1.5% agar in LB medium, autoclave, cool to ca. 50 °C, then add sterile 100 μg/mL ampicillin (Amp) and 50 μg/mL kanamycin (Kan) and pour into Petri dishes.
2. LBAS agar: 1.5% agar in LB medium, autoclave, cool to ca. 50 ° C, then add sterile 100 μg/mL ampicillin (Amp) and 50 μg/mL spectinomycin (Spec) and pour into Petri dishes.
3. Super Broth (SB): 3.2% (w/v) tryptone, 2% (w/v) yeast extract and 0.5% (w/v) NaCl adjust to pH 7.0 with NaOH, autoclave, cool to ca. 50 °C, then add sterile antibiotic. SB is used as a main culture for large-scale protein production.
4. Terrific broth (TB): 1.2% (w/v) tryptone, 2.4% (w/v) yeast extract, and 0.4% glycerol autoclaved, cool to ca. 50 °C, then add sterile-filtered 17 mM KH_2PO_4, 72 mM K_2HPO_4, and sterile antibiotic.

2.5 Buffers

1. 4X SDS Sample Loading Buffer: 0.25 M Tris-HCl pH 6.8, 0.4 M DTT, 8.0% (w/v) SDS, 40% glycerol, and 0.03% (w/v) bromophenol blue.
2. TES-β: 50 mM Tris-HCl pH 7.8, 5 mM EDTA, 300 mM NaCl and 1 mM β-mercaptoethanol; include DNase I (1 μg/mL) and PMSF (2 mM) for use as extraction buffer. Filtered and degassed prior to use.
3. HES-β: 50 mM HEPES-NaOH pH 7.8, 5 mM EDTA, 300 mM NaCl and 1 mM β-mercaptoethanol. Add DNase I (1 μg/mL) and PMSF (2 mM) for use as extraction buffer. Filtered and degassed prior to use.
4. $HISI^{10}$-β: 50 mM HEPES-NaOH pH 7.8, 1 mM IDA, 500 mM NaCl, 10 mM imidazole, and 1 mM β-mercaptoethanol. Filtered and degassed prior to use. $HISI^{50}$-β and $HISI^{250}$-β contain 50 and 250 mM imidazole, respectively.
5. Ammonium sulphate buffer (AmS): 50 mM Tris–HCl pH 7.8, 1 mM IDA, and 3.3 M $(NH_4)_2SO_4$. Filtered and degassed prior to use. Store in a cold room.

2.6 Plasmid Vectors and Host Strains

1. pPROLar.A122 (Clontech): has the p15A origin and the kanamycin resistance gene (Kan^R) with N-terminal myc tag driven by a hybrid promoter (lac/ara).
2. pCDFDuet-1 (Novagen): has the CloDF13 origin and the spectinomycin resistance gene ($Spec^R$). The vector carries a T7 promoter upstream of two multiple cloning sites (MCS) with an N-terminal His-tag (MCS1) and the other with a C-terminal S-tag (MCS2).

3. p171 plasmid: carries heme oxygenase-1 (*ho1*) and phycocyanobilin: ferredoxin oxidoreductase (*pcyA*) from *Synechocystis 6803* for phycocyanobilin (PCB) synthesis.
4. p183 plasmid: utilizes *HO1* from *Synechocystis 6803* and PΦB synthase (hy2) from *Arabidopsis thaliana* for phytochromobilin (PΦB) synthesis. This plasmid was improved as pPYC01 (*see* step 5 in Sect. 3.3).
5. Library strain: DH5α (Novagen) or XL1-Blue (Stratagene) can be used for cloning and propagation of plasmids.
6. Expression strain: $BL21_{PRO}$ (Clontech) or BL21 (DE3) (Novagen) can be used for protein production.

3 Methods

A three-phase purification protocol, designed and optimized for this study, utilizes sequential ammonium sulphate precipitation for clarification, immobilized metal affinity chromatography (IMAC) for capture, and size-exclusion chromatography (SEC) for the final polishing step. Ammonium sulphate precipitation is used for sample clarification and concentration. IMAC utilizes specific metal ions attached to a resin or beads to capture hexahistidine-tagged proteins. SEC separates proteins based on molecular size and is particularly suited for final polishing steps after sample volume reduction.

3.1 Generic Protocol for Gene Optimization and Cloning

1. Obtain the native cDNA sequence of the phytochrome of interest (e.g., *Sorghum bicolor PHYB*, accession no. AF182394) from GenBank.
2. Perform codon optimization for improved expression in *E. coli* using free online tools (e.g., OPTIMIZER [10]) or specialized software (e.g., Gene Designer 2.0).
3. Evaluate the optimized sequences in terms of codon adaptation index (CAI), GC content, and mRNA "head" folding free energy (*see* **Notes 2–4**).
4. Define optimal domain boundaries based on sequence similarity and domain border, disorder, and structure predictions (*see* **Note 5**).
5. Utilize a commercial gene synthesis service to synthesize the optimized gene encoding the selected domains of phytochrome (*see* **Note 6**).
6. Test at least two bacterial expression plasmids, either sequentially or in parallel, to optimize protein production. We typically use the pPROLarA and pCDFDuet-1 systems sequentially (*see* **Note 7**).

7. Design at least three constructs as part of a multi-construct approach to identify soluble protein constructs. We tested five different constructs (*see* **Note 8**).
8. Employ standard molecular biology techniques (Polymerase Chain Reaction (PCR) using primers listed in Table 1, restriction digestion, and ligation) for construct generation and cloning (*see* **Note 9**).
9. Transform the ligation mixture into competent library strains of *E. coli*.
10. Screen for the presence and orientation of inserts using colony PCR and restriction enzyme digestion.
11. Verify the reading frame of each construct using Sanger DNA sequencing.

3.2 Expression Test from pPROLar Plasmid Vector

1. Cotransform different pPROLar *Sb*phyB constructs (pPROLar-*Sb*NPGP, pPROLar-*Sb*HPGP, and pPROLar-*Sb*PGP) with the p183 plasmid into heat-shock-competent BL21$_{PRO}$ cells for the production of holophytochromes via in vivo assembly with phytochromobilin (PΦB).
2. Plate the transformed cells on LBAK agar plates.
3. Incubate the plates overnight at 37 °C to allow colony growth.
4. Select three independent transformant colonies for each construct for expression trials.
5. Inoculate 100 mL of LB medium containing Ampicillin (100 μg/mL) and Kanamycin (50 μg/mL) in 500-mL baffled Erlenmeyer flasks with a single colony.
6. Incubate the cultures at 37 °C with shaking at 140–180 rpm until the optical density at 600 nm (OD_{600}) reaches typically ~0.6–0.8 (*see* **Note 10**).
7. Induce expression by adding 0.2% arabinose and 1 mM isopropyl β-D-1-thiogalactopyranoside (IPTG) to the culture (*see* **Note 11**).
8. Continue incubation overnight at 37 °C with shaking at 140–180 rpm.
9. Harvest induced cells by centrifugation at 3000 *g* and 4 °C for ca. 45 min.
10. Resuspended the cells in cold TES-β lysis buffer. Use 10–20 mL of lysis buffer for 1 g of biomass.
11. Lyse the resuspended cells using a precooled French pressure cell with two passages at 100 MPa.
12. Remove the cell debris by centrifugation at 4 °C and 15,000 *g* for 30–45 min.

13. Check holophytochrome production using zinc-induced fluorescence after SDS-PAGE or by measuring difference spectra.

For zinc-induced fluorescence, incubate the SDS-PAGE gel in 100 mL of 1 mM zinc acetate solution for 20 min, wash three times in H_2O, and photograph the stained gel with a digital camera fitted with a ~630 nm interference filter using a UV/B transilluminator.

To obtain the FR-R difference spectrum, measure the UV-Vis absorption spectra after saturating red and then far-red irradiation, then subtract the spectra using a PC (*see* **Note 12**). The ΔΔA value (calculated from the difference spectrum by subtracting the difference minimum in the far-red region from the difference maximum in the red region) is a quantitative parameter of the holophytochrome concentration, independent of purity.

3.3 Optimization of Expression

1. Ideally, use a full or fractional factorial design or Box–Behnken design (BBD) for optimization (*see* **Note 13**). Here, we demonstrate the use of a Box–Behnken design with three factors: Induction OD_{600} (Factor A: 0.5, 0.75, 1.0); arabinose inducer concentration (Factor B: 0.2%, 0.4%, 0.6%); and induction temperature (Factor C: 18, 28, 37 °C).
2. Include three replicate experiments at the center point (0.75 OD_{600}, 0.4% arabinose concentration, 28 °C induction temperature) to evaluate the impact of uncontrolled variations.
3. Perform 15 total expression trials in 100 mL LB medium in a 500-mL baffled flask inoculated with 100 μL of the same starter culture under different combinations of the selected factors according to Box–Behnken design (*see* Table 2).
4. Identify the optimal culture conditions based on the measured difference spectra or zinc-induced fluorescence of the expressed protein.
5. To test alternative vectors, cotransform heat-shock-competent BL21 (DE3) *E. coli* cells with one of the pCDFDuet constructs (*Sb*NPGP, *Sb*HPGP, *Sb*PGP, *Sb*NPG, or *Sb*PG) and the p183 or pPYC01 plasmids for chromophore synthesis (*see* **Note 14** and Fig. 5).
6. Plate transformed cells on LBAS agar plates.
7. Incubate plates overnight at 37 °C to allow colony growth.
8. Use colonies from freshly transformed cells for expression trials following the protocol outlined in Subheading 3.2 (expression test).
9. Continue optimization for different expression systems (pPROLar and pCDFDuet), different expression strains ($BL21_{PRO}$ and BL21 (DE3)), different constructs (NPGP, HPGP, PGP, PG, and NPG), different IPTG concentrations (0.1–1 mM) and growth media (LB, TB, and SB) until the

Table 2
Analysis of ΔΔA from difference spectra for soluble fractions during Box–Behnken optimization for PGP holoprotein expression

Trial	Induction OD	Arabinose (%)	Temperature (°C)	ΔΔA (mOD)
BB1	0.5	0.2	28	26.2
BB2	1	0.2	28	34.9
BB3	0.5	0.6	28	25.5
BB4	1	0.6	28	71.3
BB5	0.5	0.4	18	15.0
BB6	1	0.4	18	12.3
BB7	0.5	0.4	37	9.2
BB8	1	0.4	37	26.4
BB9	0.75	0.2	18	45.3
BB10	0.75	0.6	18	35.4
BB11	0.75	0.2	37	22.9
BB12	0.75	0.6	37	22.3
BB13	0.75	0.4	28	23.3
BB14	0.75	0.4	28	23.7
BB15	0.75	0.4	28	32.7

This table presents the ΔΔA values calculated from the difference spectra of soluble fractions containing PGP holoprotein. Expression was performed using the pPROLar-*Sb*PGP and p183 plasmids under a range of conditions according to the Box–Behnken design

phytochrome yield is sufficient for the intended application. Subheading 3.4 below represents the final optimized protocol following extensive refinement.

3.4 Large-Scale Production from pCDFDuet-SbPGP

1. Identify the optimal clone for each construct based on the results of extensive optimization experiments in Subheading 3.3.
2. Prepare a glycerol stock by growing the optimal clone in LB medium with appropriate antibiotics and adding glycerol to ca. 20% (*see* **Note 15**).
3. Inoculate 2 mL of LB medium with the appropriate antibiotics in a 10-mL culture tube using colonies from the glycerol stock.
4. Incubate at 37 °C and 140 rpm.
5. Monitor the optical density of the starter culture (OD_{600}) to avoid overgrowth (>0.8).

6. Inoculate 400 μL of the starter culture into 400 mL of SB medium supplemented with the appropriate antibiotics in a 2 L baffled flask (*see* **Notes 16** and **17**).
7. Incubate the main culture at 37 °C and 140 rpm until the optimized OD_{600} is reached (*see* **Notes 18–21**).
8. Cool the main cultures in an ice-water bath for 30 min.
9. Induce protein expression by adding 1 mM of IPTG.
10. Incubate the induced cultures at the optimal temperature for 16 h.
11. Harvest the induced cells by centrifugation at 6000 *g* for 30 min at 4 °C using an appropriately large centrifuge. The cell pellet should be noticeably blue-green in color.
12. Flash-freeze the cell pellets in liquid nitrogen and store the frozen pellets at −80 °C until further use.

3.5 Immobilized Metal Affinity Chromatography (IMAC)

1. Thaw frozen cell pellets at 4 °C.
2. Resuspend the pellets in cold TES-β lysis buffer. To maximize the extraction of soluble protein, the ratio of cell mass to TES-β lysis buffer should be maintained within a range of 10–20 mL of lysis buffer per gram of wet cell mass.
3. Transfer the resuspended cells to a precooled 50 mL French pressure cell.
4. Lyse the cells by passing the suspension through the pressure cell twice at 100 MPa.
5. Clarify the lysate at 15,000 *g* and 4 °C for 30 min using an appropriate centrifuge to remove cell debris.
6. Transfer the supernatant to clean 1 L centrifuge bottles.
7. Take 0.6 mL of the supernatant for UV-Vis spectral measurement.
8. Add ice-cold ammonium sulphate buffer to the supernatant to achieve 60% saturation (*see* **Note 22**).
9. Mix gently and incubate on ice for 30 min.
10. Transfer the mixture to appropriate precooled centrifuge tubes (e.g., Oak Ridge tubes).
11. Centrifuge at 4 °C and 5000 *g* for 30 min using an appropriate centrifuge.
12. Remove and discard as much of the supernatant as possible. Dissolve the dark green pellet in a minimal volume of $HISI^{10}$-β Ni-NTA binding buffer.
13. Centrifuge the holophytochrome solution at 4 °C and 15,000 *g* for 20 min.

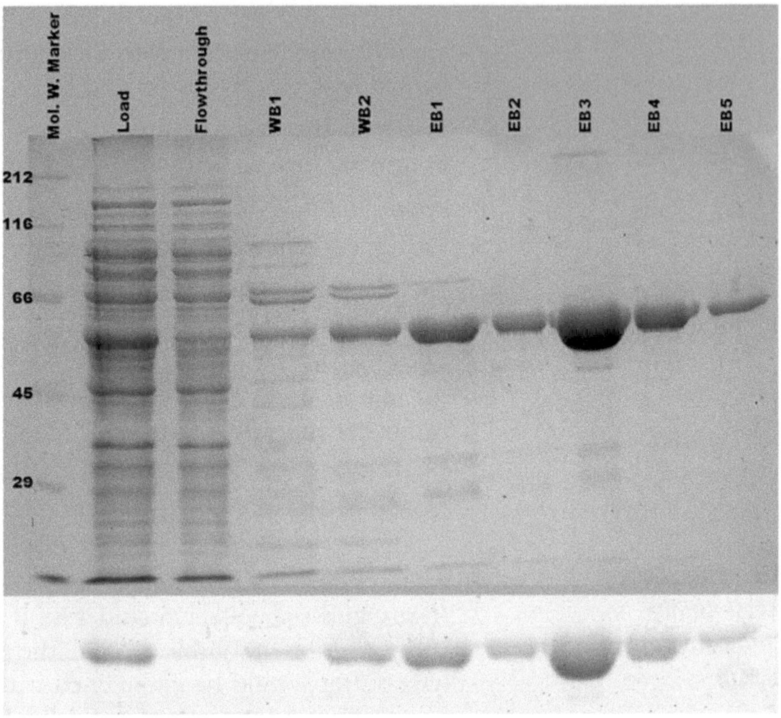

Fig. 2 SDS-PAGE analysis of large-scale Ni-NTA purification of PGP holoprotein. The figure shows SDS-PAGE analysis of PGP purification visualized using Coomassie Brilliant Blue staining (top) and zinc-induced fluorescence (bottom). The purification was performed in a binding buffer containing 50 mM HEPES (pH 7.8), 500 mM NaCl, 1 mM iminodiacetic acid (IDA), 1 mM β-mercaptoethanol (β-ME), and 10 mM imidazole. Molecular weight markers (in kDa) are shown for reference. WB and EB stands for washing buffer and elution buffer, respectively

14. Carefully remove the supernatant without disturbing the pellet, then filter through a 0.45-μm filter to remove suspended particles.
15. Pre-equilibrate a 5 mL Ni-NTA resin with five column volumes (CV) of binding buffer, then apply the filtered holophytochrome sample (*see* **Note 23**).
16. Wash the column with 2 CV of the binding buffer, followed by 3 CV $HISI^{50}$-β.
17. Elute bound phytochrome with ca. 5 CV of $HISI^{250}$-β (but *see* **Note 24**). Measure the UV-Vis spectra of the elution fractions and calculate the specific absorbance ratio (*see* Fig. 2 and **Note 25**).

3.6 Size Exclusion Chromatography (SEC)

1. Equilibrate a Superdex 200 26/60 prep grade column with HES-β buffer on an FPLC system (e.g., Äkta, Cytiva) in a dark laboratory (*see* **Note 26**).

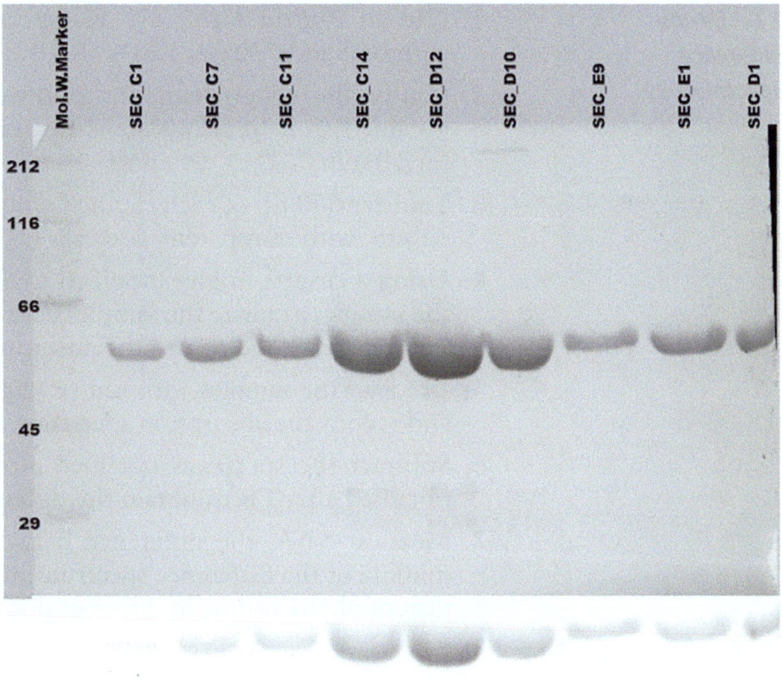

Fig. 3 SDS-PAGE analysis of PGP purified by Ni-NTA chromatography and size-exclusion chromatography. SDS-PAGE analysis of PGP holoprotein purified using Ni-NTA affinity chromatography followed by size exclusion chromatography (SEC). The purified protein was visualized using Coomassie Brilliant Blue staining (top) and zinc-induced fluorescence (bottom). Molecular weight markers (in kDa) are included for reference

2. Calibrate the column using low molecular weight standards.
3. Concentrate the sample to 5–10 mg/mL holophytochrome using a stirred pressure cell with a 25-kDa cutoff membrane (Bio-Rad), depending on the construct.
4. Photoconvert the sample to Pr with a saturating far-red (FR) light (*see* **Note 27**) and load the column via an appropriate sample loop.
5. Perform SEC at room temperature with a flow rate of 1 mL/min (in darkness to prevent photoconversion).
6. Monitor elution at 280 nm (for all proteins) and 690 nm (near the holophytochrome isosbestic point) or at 650 and 730 nm (corresponding to Pr and Pfr λ_{max}).
7. Collect eluates in 0.6-mL fractions.
8. Analyze the collected fractions using UV-Vis spectroscopy and SDS-PAGE (*see* Subheadings 3.7 and 3.8). Identify and pool fractions containing ca. 90% of the phytochrome (Fig. 3).
9. Store the pooled fractions at 4 °C until further use.

3.7 UV-Vis Difference Spectroscopy

1. Use an Agilent 8453 UV-Vis spectrophotometer equipped with 660 and 730 nm LEDs (Roithner) for actinic irradiation.
2. Clarify the supernatants or purified phytochrome samples obtained from previous steps by centrifugation at 15,000 *g* for 10 min.
3. Transfer 600 μL of each clarified sample to a fluorescence-type cuvette with transparent sidewalls (Hellma).
4. Using a cuvette holder modified to allow lateral irradiation of the sample, irradiate the samples with far-red (FR) light using a 730 nm LED and record the absorption spectrum.
5. Irradiate the samples with red (R) light using a 650 nm LED and record the absorption spectrum.
6. Subtract the spectrum obtained after R from the spectrum obtained after FR to obtain the difference spectrum.
7. Measure ΔΔA, the difference between the maxima and the minima of the difference spectrum, to quantify the concentration of phytochrome in different pooled samples (Fig. 4).

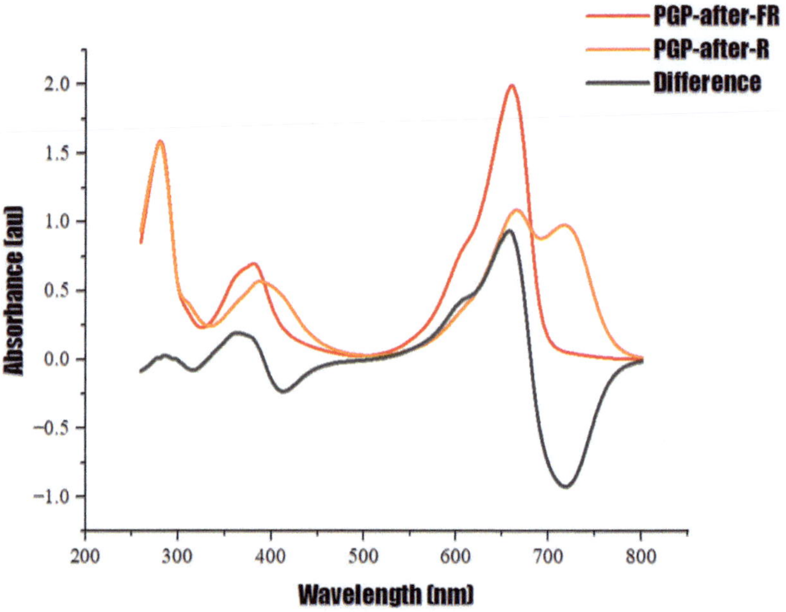

Fig. 4 Absorption spectra of PGP under red and far-red light irradiation. Absorption spectra of PGP holoprotein following saturating red light (R) irradiation (red spectrum) and far-red light (FR) irradiation (light red spectrum). The difference spectrum (black) was calculated by subtracting the spectrum obtained under red light irradiation from that under far-red light irradiation, revealing the spectral changes associated with photoconversion

3.8 Protein Analysis

1. Add 10 μL of 4× sample loading buffer to 30 μL of each protein sample, then boil at 95 °C for ca. 5 min.
2. Spin the heated samples at 12,000 *g* for 30 s.
3. Load 10 μL of each sample and the high-range protein marker onto a 10% SDS-PAGE gel prepared with the Laemmli buffer system.
4. Perform electrophoresis until the dye front reaches the bottom of the gel.
5. Stain the gel with zinc acetate solution to visualize the holo-phytochrome bands with a UV/B transilluminator, then with Coomassie Brilliant Blue R250 to visualize the protein bands with a white light box (*see* **Note 28**).
6. Estimate molecular masses of protein bands using the protein marker.

4 Notes

1. Add glucose after autoclaving the solution with the remaining ingredients, and letting it cool down. Sterilize the final solution by passing it through a 0.2 μm filter.
2. Aim for a CAI value greater than 0.80 to ensure compatibility with *E. coli*. Maintain a GC content between 40% and 60% for optimal stability and transcription, and translation efficiency. Modify the folding free energy of the mRNA head region (5′ UTR plus the initial 16 codons) to an unstable structure with a predicted free energy above −15 kcal/mol [11].
3. We employed Gene Designer 2.0 (DNA2.0, ATUM, and Newark, CA, USA) to optimize codon usage and GC content.
4. The free energy of folding of the head plus 5′-UTR was predicted using free online web servers for RNA secondary structure prediction [12].
5. ClustalW can be used to compare the sequence of the target phytochrome to phytochromes with known 3D structures. SMART-EMBL [13] can be utilized to identify conserved protein domains. Intrinsically disordered regions can be predicted from protein sequences using PONDR-FIT [14]. AlphaFold-2 or -3 can be employed to predict protein structures and identify regions with low predicted aligned error (PAE) [15]. In *Sorghum* phyB, PONDR-FIT predicted the first 113 residues to be intrinsically disordered. For protein production intended for structural studies (e.g., crystallography or cryo-EM), truncation of highly flexible and intrinsically disordered regions may be beneficial. SMART-EMBL assigned conserved domains for residues 114–639 of *Sb*phyB. The

AlphaFold structure database predicted residues 63–646 of *Sb*phyB to have low PAE values. The AlphaFold-3 server predicted residues 67–654 to have low PAE values.

6. The optimized gene for the first 655 amino acid residues of *Sorghum bicolor PHYB*, corresponding to the complete photosensory module, was synthesized with an additional ribosome binding site (TTAAAGAGGAGAAA). The synthesized gene was cloned into the EcoRV site of the pUC57 vector by the supplier (Eurofins Genomics, Germany).
7. The pPROLar.A expression plasmid is a component of the PRO Bacterial Expression System [16]. The pPROLar-based plasmids can be used in BL21 and $BL21_{PRO}$ expression *E. coli* cells but are incompatible with the T7 expression system such as BL21 (DE3) cells. All *E. coli* B strains are deficient in the lon and ompT proteases. $BL21_{PRO}$ cells constitutively produce the Tet repressor, enabling tight regulation of gene expression. The system uses endogenous RNA polymerase, achieving an induction rate of ~2% of total protein. The Tet repressor, endogenous RNA polymerase, and the low copy number minimize protein leakage, making it suitable for expressing even toxic proteins. The pPROLar system has been effectively used in our laboratory to characterize cyanobacterial phytochrome [17]. The pCDFDuet-1 plasmid vector can be expressed in BL21 (DE3) cells, which harbors a λ prophage carrying the T7 RNA polymerase gene under the control of the strong lacUV5 promoter and lacIq repressor.
8. Five different constructs or DNA inserts were generated from the photosensory module of *Sorghum bicolor* PHYB: NPGP, HPGP, PGP, NPG, and PG.
9. We used *Eco*RI and *Hind*III sites for cloning into pPROLar and *Nco*I and *Hind*III sites for cloning into pCDFDuet-1. Primer sequences (Table 1, GATC Biotech) were designed for this purpose. We used Phusion High-Fidelity DNA Polymerase (NEB) for PCR amplification.
10. The OD_{600} of a bacterial culture measures the scattering, not the absorbance, associated with the cell density and is therefore dependent on the geometry of the spectrophotometer optics. Detector collection window size and distance from the sample affect the measured OD; thus, very different values will be obtained for the same sample using different instruments. This is a significant but seldom-recognized factor when comparing culture densities in different laboratories.
11. For lac/ara hybrid promoters, Ara dominates the induction process because the AraC protein acts as both a repressor (in the absence of arabinose) and an activator (in presence of arabinose). For the lac operon, it is very important to test different concentrations of IPTG.

12. Ideally, use a fluorescence cuvette with transparent sidewalls and an Agilent 8453 UV-Vis spectrophotometer in which the cuvette holder is modified to allow lateral actinic irradiation from high-power LED arrays (ca. 640 and 730 nm; Roithner, Vienna, Austria)—*see* Chapter 4 of this edition by Hughes and Lang.
13. A full factorial design is an experiment whose design consists of two or more factors, each with discrete possible values, and whose experimental units consider all possible combinations of these values across all such factors. A complete factorial optimization for each expression vector, expression strain, and construct is not feasible. A fractional factorial design involves using a subset of conditions. The Box–Behnken design (BBD) is an independent quadratic design for a class of quantitative variables with three levels. For three variables, the treatment combinations can be graphically represented by the midpoints of the edges and the center of a cube, which denotes the process space. During optimization, each trial receives equal volumes of the starter culture grown from a single colony.
14. The phytochrome plasmids need to be cotransformed with bilin synthesizing plasmids (p171 or p183) into a suitable *E. coli* expression host. Since the bilin chromophore is indispensable for the optical and biological activity of the protein, we introduce the cofactor as early as possible. The *Synechocystis ho1* (*sll1184*) and *pcyA* (*slr0116*) genes were sequentially cloned into pPROLar.A122 to generate p171, enabling polycistronic mRNA expression. The *ho1* gene was inserted using *Hin*dIII and *Kpn*I restriction sites, while *pcyA* was cloned using *Bam*HI at the 5′ terminus and *Not*I at the 3′ terminus [18]. In the p183 plasmid, *pcyA* was replaced with a *HY2* cDNA from *Arabidopsis* [19]. hy2 contains a chloroplast transit peptide comprising the first 45 amino acid residues [20], which is suboptimal for PΦB synthase activity in *E. coli.* To evaluate the effect of its removal, *Glycine max* phyA PG was used to compare the activity of both the original p183 and its modified version, pPYC01, which contains an appropriately truncated *HY2* sequence. Deleting the transit, peptide coding sequence significantly enhanced holophytochrome production, as demonstrated by increased yields in bacterial cell pellets and the difference spectrum (Fig. 5). Although this optimized plasmid improves the production of PΦB-adducted plant phytochromes in *E. coli*, in our laboratory, plant phytochromes generally express better as PCB- than as PΦB-adducts [21, 22].
15. In many instances, the highest yields of recombinant proteins are achieved by directly inoculating starter cultures with freshly transformed cells. This avoids clonal selection and so is bad microbiology, but it saves time and is widely practiced.

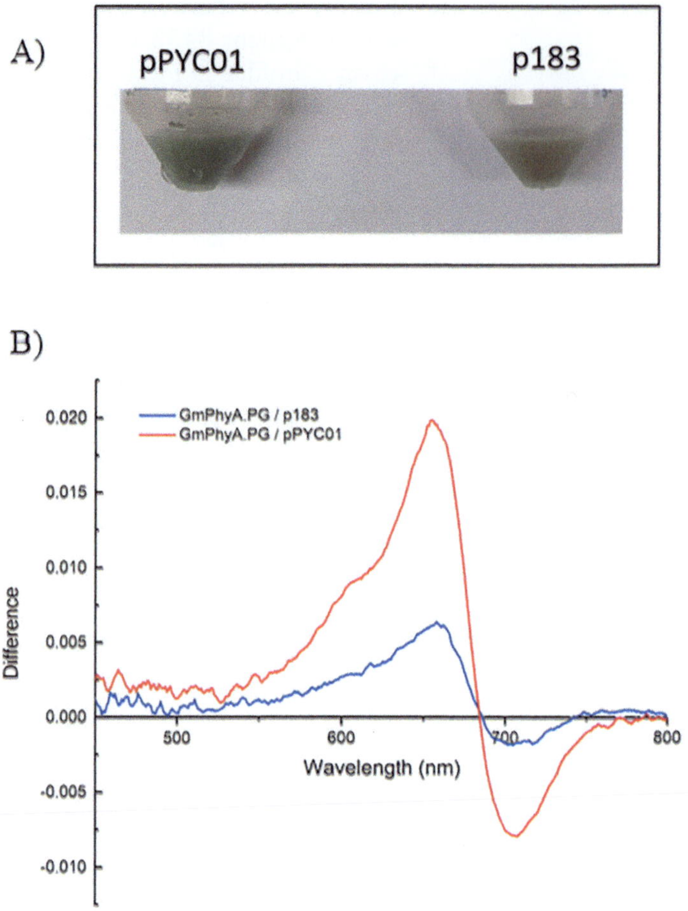

Fig. 5 Enhanced phytochromobilin production by removing the transit peptide. (**a**) Cell pellets of *E. coli* expressing *Glycine max* phyA PG with the pPYC01 plasmid (truncated hy2) compared to the p183 plasmid (wild-type hy2). (**b**) Corresponding difference spectra

16. In a batch culture, the diffusion of oxygen into the medium only occurs at the surface of the liquid that is exposed to the atmosphere; anaerobic growth will not produce holophytochrome since heme oxygenase requires molecular oxygen. It is important to avoid anoxia by maximizing oxygen access relative to the culture volume. The three commonly used methods for increasing the surface area for oxygen uptake are low volumes in large flasks, vigorous culture agitation (250–300 rpm), and flasks with baffled walls (indentations (baffles) create turbulence, improving oxygen transfer rates (OTR) within the liquid phase).
17. It is recommended to use a flask with a volume of at least five times greater than the volume of the medium. The starter culture should be diluted 1/500–1/1000 into a larger volume of the main culture.

18. The pCDFDuet constructs in conjunction with p183 in BL21 (DE3) cells grown in SB medium gave a maximum yield of 1, 8 and 12 mg of holophytochrome per litter of culture for NPGP, PGP, and PG, respectively [22].
19. The optimum temperatures of 24, 28, and 30 °C were determined for NPGP, PGP, and PG, respectively, in SB medium with cell density at induction equivalent to OD_{600} of 0.8 (but *see* **Note 13** above), and inducer concentration of 1 mM IPTG.
20. Optimal NPGP production was attained using SB medium supplemented with glucose and glycerol and the following conditions: temperature of 24 °C, cell density at induction equivalent to OD_{600} of 0.8, and inducer concentration of 1 mM IPTG. The same induction OD and inducer concentration were used for maximum PGP and PG production.
21. An important factor to achieve reproducible production of NPGP from the pCDFDuet-*Sb*NPGP construct was associated with the geometry of culture flasks. Adequate aeration was achieved with <400 mL culture volume in broad-bottom 2 L baffled flasks.
22. Certain phytochrome constructs and mutants may not tolerate ammonium sulphate precipitation.
23. In chromatography, column volume (CV) refers to the total volume occupied by the packed resin within the column bed.
24. High concentrations of imidazole slowly denature proteins; therefore, exchange the eluate into HES-β buffer (without imidazole) using a disposable Sephadex G25 NAP column (Cytiva).
25. A key quality measure for phytochrome purity is the specific absorbance ratio (SAR), defined as the ratio of absorbance at the absorbance maximum of the Pr form (ca. 660 nm) to absorbance at 280 nm.
26. Because some phytochrome constructs show different mobilities as Pr and Pfr, SEC should be carried out with Pr in a dark laboratory under safelight from a cyan (490 nm) LED.
27. Photoconversion of the sample to Pr is non-trivial: phytochromes have very high extinction coefficients (ca. 100 $mM^{-1}\,cm^{-1}$) and the sample is very concentrated—thus self-shading is a problem. From experience, photoconversion to Pr is optimally achieved by loading the column using a 250 μL HPLC glass syringe surrounded by a tubular array of about twenty 730–740 nm LEDs (e.g., ELD-740-354; Roithner, Vienna) directed toward the sample in the syringe. The short path length through the sample from all directions ensures complete photoconversion within about 10 s, even at very high concentrations.

28. Alternatives include the colloidal Coomassie method [23], Coomassie Blue dye (ProBlue Safe Stain, Giotto Biotech), and Stain-Free Gels (Bio-Rad).

Acknowledgments

We thank Peng-Yuan Chen for providing Fig. 5 and construction of the pPYC01 plasmid. We also thank Dr. Soshichiro Nagano and Christina Lang for their valuable contributions and assistance, and Professor Jon Hughes for all sorts of things. We are grateful for the financial support of the *Deutsche Forschungsgemeinschaft* (DFG).

References

1. Hughes J, Winkler A (2024) New insight into phytochromes: connecting structure to function. Annu Rev Plant Biol 75:153–183
2. Fry KT, Mumford FE (1971) Isolation and partial characterization of a chromophore-peptide fragment from pepsin digests of phytochrome. Biochem Biophys Res Commun 45: 1466–1473
3. Butler WL, Norris KH, Siegelman HW, Hendricks SB (1959) Detection, assay, and preliminary purification of the pigment controlling photoresponsive development of plants. Proc Natl Acad Sci 45:1703–1708
4. Berkelman TR, Lagarias JC (1986) Visualization of bilin-linked peptides and proteins in polyacrylamide gels. Anal Biochem 156:194–201
5. Gileadi O, Burgess-Brown NA, Colebrook SM et al (2008) High throughput production of recombinant human proteins for crystallography. In: Structural proteomics high-throughput methods. Springer, pp 221–246
6. Burgess-Brown NA, Sharma S, Sobott F et al (2008) Codon optimization can improve expression of human genes in Escherichia coli: a multi-gene study. Protein Expr Purif 59:94–102
7. Gräslund S, Sagemark J, Berglund H et al (2008) The use of systematic N-and C-terminal deletions to promote production and structural studies of recombinant proteins. Protein Expr Purif 58:210–221
8. Li H, Burgie ES, Gannam ZTK et al (2022) Plant phytochrome B is an asymmetric dimer with unique signalling potential. Nature 604: 127–133
9. Zhang J, Stankey RJ, Vierstra RD (2013) Structure-guided engineering of plant phytochrome B with altered photochemistry and light signaling. Plant Physiol 161:1445–1457
10. Puigbo P, Guzmán E, Romeu A, Garcia-Vallvé S (2007) OPTIMIZER: a web server for optimizing the codon usage of DNA sequences. Nucleic Acids Res 35:W126–W131
11. Boël G, Letso R, Neely H et al (2016) Codon influence on protein expression in E. coli correlates with mRNA levels. Nature 529:358–363
12. Reuter JS, Mathews DH (2010) RNAstructure: software for RNA secondary structure prediction and analysis. BMC Bioinformatics 11:1–9
13. Letunic I, Khedkar S, Bork P (2021) SMART: recent updates, new developments and status in 2020. Nucleic Acids Res 49:D458–D460
14. Xue B, Dunbrack RL, Williams RW et al (2010) PONDR-FIT: a meta-predictor of intrinsically disordered amino acids. Biochim Biophys Acta 1804:996–1010
15. Jumper J, Evans R, Pritzel A et al (2021) Highly accurate protein structure prediction with AlphaFold. Nature 596:583–589
16. Lutz R, Bujard H (1997) Independent and tight regulation of transcriptional units in Escherichia coli via the LacR/O, the TetR/O and AraC/I1-I2 regulatory elements. Nucleic Acids Res 25:1203–1210
17. Essen L-O, Mailliet J, Hughes J (2008) The structure of a complete phytochrome sensory module in the Pr ground state. Proc Natl Acad Sci 105:14709–14714

18. Landgraf FT, Forreiter C, Pico AH et al (2001) Recombinant holophytochrome in Escherichia coli. FEBS Lett 508:459–462
19. Gambetta GA, Lagarias JC (2001) Genetic engineering of phytochrome biosynthesis in bacteria. Proc Natl Acad Sci 98:10566–10571
20. Kohchi T, Mukougawa K, Frankenberg N et al (2001) The Arabidopsis HY2 gene encodes phytochromobilin synthase, a ferredoxin-dependent biliverdin reductase. Plant Cell 13: 425–436
21. Nagano S, Guan K, Shenkutie SM et al (2020) Structural insights into photoactivation and signalling in plant phytochromes. Nat Plants 6:581–588
22. Shenkutie SM, Nagano S, Hughes J (2024) Expression, purification and crystallization of the photosensory module of phytochrome B (phyB) from Sorghum bicolor. Acta Crystallogr F Struct Biol Commun 80:59–66. https://doi.org/10.1107/S2053230X24000827
23. Neuhoff V, Arold N, Taube D, Ehrhardt W (1988) Improved staining of proteins in polyacrylamide gels including isoelectric focusing gels with clear background at nanogram sensitivity using Coomassie Brilliant Blue G-250 and R-250. Electrophoresis 9:255–262

Part II

Biophysical and Computational Characterization of Phytochromes

Chapter 7

Characterizing Phytochrome Photochemistry Using UV–Visible Spectroscopy in Host Cell Extracts and Purified Protein

Jimena Rinaldi, Lisandro H. Otero, and Hernán R. Bonomi

Abstract

Phytochromes are red and far-red photoreceptors widespread among life, which bind a bilin chromophore molecule. They photoswitch between two main or parent photostates: the Pr state, which absorbs red light, and the Pfr state, which absorbs far-red light. Apart from the photoconversion, there is another process called dark conversion. The phytochrome photocycle can be easily followed by ultraviolet (UV)–visible or visible spectroscopy. Through the acquisition and the later data analysis of serial UV–visible absorption spectra in either unpurified or purified protein samples, many important aspects of the phytochrome photocycle can be learned.

Key words Phytochromes, Red/far-red light photoreceptors, Photocycle dynamics, UV–visible absorption spectroscopy, Biliverdin IXα chromophore, Bilin chromophore, Purified protein samples, Bacterial crude extracts, Spectral decomposition, Pr/Pfr abundance analysis, Photocycle kinetics modeling

1 Introduction

Phytochromes are red and far-red photoreceptors found across a wide range of organisms, including plants, fungi, algae, and bacteria [1]. They are involved in the modulation by light of many important cellular processes, playing a crucial role in regulating various cellular processes in response to light. Phytochromes consist of a photosensory module, which includes the PAS-GAF-PHY domain triad responsible for sensing light, and a variable output module that carries out biological functions such as enzymatic activity or transcription regulation. The photosensory module binds a bilin chromophore molecule in a 1:1 stoichiometric ratio [2].

There are three different bilin molecules that bind to phytochromes: biliverdin (BV) IXα, phytochromobilin (PCB), and phycocyanobilin (PΦB) [1]. These molecules are synthesized from

Andreas Winkler (ed.), *Phytochromes: Methods and Protocols*, Methods in Molecular Biology, vol. 2970,
https://doi.org/10.1007/978-1-0716-4791-2_7,

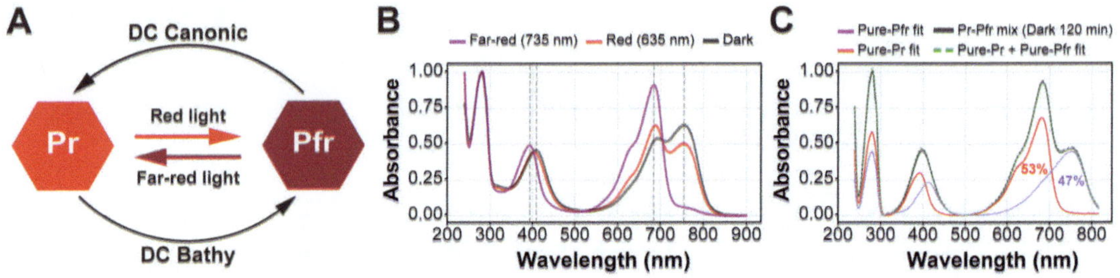

Fig. 1 The phytochrome photocycle. (**a**). Schematic representation of the phytochrome alternation between the red-light-absorbing species (Pr) and the far-red-light-absorbing species (Pfr). This alternation occurs through photoconversion triggered by red and far-red light, as well as dark conversion (DC). (**b**) Absorption spectra of the bacteriophytochrome from *Xanthomonas campestris* pv. *campestris* (*Xcc*BphP) at pH 7.5 and 25 °C. These spectra serve as a representative model after red or far-red illumination with LED light sources, as well as at equilibrium in the dark (after 24 h). (The data were extracted from Conforte et al. [24]). The four characteristic peaks are indicated with dashed lines. Soret-Pr: 394 nm, Soret-Pfr: 406 nm, Q-Pr: 684 nm, Q-Pfr: 753 nm. (**c**) A representative *Xcc*BphP spectrum at pH 7.5 and 30 °C (unpublished data) showing a Pr/Pfr mix, presented as a weighted sum of Pr and Pfr contributions. The numbers indicate the estimated proportions of Pr and Pfr in the mix

heme in two steps. First, heme oxygenase (HO) converts heme into BV [1]. Then, ferredoxin-dependent bilin reductases (FDBRs) further convert BV into either PCB or PΦB [1]. The chromophore is covalently linked via a thioether bond between a conserved cysteine residue and the vinyl group of the bilin molecule's A-ring. These photoreceptors exhibit bilin lyase activity, enabling autocatalytic assembly into functional holo-phytochromes [2].

Phytochromes alternate between two main photostates: the Pr state, which absorbs red light, and the Pfr state, which absorbs far-red light (Fig. 1a). Phytochromes naturally auto-assemble into the Pr state. Red light induces the conversion of Pr to Pfr, while far-red light triggers the reverse photoconversion from Pfr to Pr. This rapid transition occurs on the millisecond timescale [3]. In addition to photoconversion, there is a thermal relaxation process known as dark conversion (Fig. 1a), which progresses more slowly, occurring over seconds to hours [3]. For canonical phytochromes, dark conversion restores the Pr state, whereas for bathy phytochromes, it favors the Pfr state. The proportion of Pr and Pfr at equilibrium varies among phytochromes. Based on the proportion of molecules in the Pfr state at equilibrium, phytochromes are classified into four categories: bathy-type (>90%), bathy-like (50%–90%), canonical-like (10%–50%), and canonical-type (<10%) [4]. Regarding the biological function, in plant phytochromes, the Pfr photostate is generally the biologically active form, whereas the Pr form is inactive [5–8]. Phytochromes that contain a GGDEF domain (a diguanylate cyclase output module) follow a similar pattern, with Pfr being the enzymatically active state [9, 10]. In contrast, certain bacteriophytochromes—notably some with histidine kinase output domains—are active in their Pr state [11–14]. These differences underscore the importance of the Pr–Pfr balance in regulating phytochrome function in vivo, as phytochromes essentially act as light-switchable systems that respond to the ratio of red and far-red light by toggling between these two states.

The phytochrome photocycle can be readily studied using ultraviolet (UV)–visible or visible spectroscopy. These absorption spectra display distinct features: a peak at 280 nm corresponding to the protein, Soret bands in the UV/blue range, and Q bands in the red/far-red range, reflecting the bilin chromophore in the protein context [15]. Characteristic wavelengths include Soret-Pr (~390 nm, 387–398 nm), Soret-Pfr (~410 nm, 388–416 nm), Q-Pr (~680 nm, 656–703 nm), and Q-Pfr (~750 nm, 725–775 nm) (Fig. 1b). By measuring serial UV–visible absorption spectra over time following irradiation with red or far-red light, various aspects of the phytochrome photocycle can be analyzed. The literature describes various methods for extracting information on the Pr/Pfr balance and the kinetics of their interconversion [4, 12, 13, 16–20].

In this chapter, we describe the acquisition of UV–visible and visible absorption spectra, as well as the data analysis of two sample types: pure holo-phytochrome preparations and clarified extracts from bacteria overexpressing phytochrome proteins, supplemented with the bilin chromophore. The data analysis described here is based on the spectral decomposition, which enables each spectrum to be resolved into a weighted sum of pure-Pr and pure-Pfr spectra (Fig. 1c). This approach can estimate the Pr/Pfr relative abundance by integrating the spectra under different dark/light treatments or other experimental conditions. Key parameters analyzed in this chapter include (i) the wavelength and intensity of the four characteristic peaks (Soret-Pr, Soret-Pfr, Q-Pr, and Q-Pfr), (ii) the dark conversion rate, (iii) Pr/Pfr abundance at dark equilibrium, and (iv) classification into one of four categories (bathy-type, bathy-like, canonical-like, or canonical-type).

2 Materials

2.1 Samples

The sample can be either a pure holo-phytochrome sample or a clarified extract from bacterial cells overexpressing phytochrome proteins.

1. Pure holo-phytochrome sample (*see* **Note 1**) in working buffer (*see* Subheading 2.2).
2. Clarified extract from phytochrome-overexpressing bacterial cells (*see* **Note 2**).

2.2 Reagents

1. Working buffer 50 mM Tris/Cl, pH 7.5, 0.25 M NaCl.
2. Stock solution of bilin chromophore (*see* **Note 3**).

2.3 Equipment and Consumables

1. UV–visible spectrophotometer (*see* **Note 4**).
2. Quartz cuvette (*see* **Note 5**).
3. UV–visible microplate spectrophotometer (*see* **Note 6**).
4. Plastic 96-well plate, suitable for the wavelength range of 400–900 nm.
5. LED red light. For instance, 625–635 nm, 1–5 watts, and light fluence of 0.2–1 μmol m^{-2} s^{-1}.
6. LED far-red light. For instance, 750–780 nm, 1–5 watts, and light fluence of 0.2–1 μmol m^{-2} s^{-1}.
7. Data analysis software. Any suitable program can be used, such as Excel, MATLAB, or Python.

3 Methods

In this section, we describe the acquisition of UV–visible or visible absorption spectra and the subsequent data analysis of both pure holo-phytochrome preparations and clarified extracts from phytochrome-overexpressing bacterial cells. Sample preparation details are not provided here (*see* **Note 7**).

We will outline two different methods for data acquisition:

A single-sample mode using a quartz cuvette in the wavelength range of 240–900 nm for pure protein preparations.

A multi-sample mode using a 96-well plastic plate in the wavelength range of 450–900 nm for clarified bacterial extracts.

The pure protein method in the UV–visible range is more accurate and provides higher-quality data. The method using bacterial extracts in the visible range is suitable for screenings, as it allows for initial and simultaneous characterization of different phytochrome molecules—for example, various mutant versions of a single phytochrome or different phytochrome open reading frames from the same or different organisms. However, it is worth noting that the multi-sample method can also be used for pure protein samples, and the single-sample method can be applied to clarified bacterial extracts.

3.1 Single-Sample Acquisition Mode in the UV–Visible Range of a Pure Holo-Phytochrome

3.1.1 Data Acquisition

1. Configure the spectrophotometer for absorbance measurements at a fixed temperature (25 °C recommended), with a wavelength range of 200–900 nm, a resolution of 1 nm, and scanning rate of 4800 nm/min (*see* **Note 8**). Set the time sampling interval (after irradiation) to every 5 min from 0 to 180 min and then every 30 min from 180 to 1020–1200-min (*see* **Note 10**). Include a baseline correction in the settings.

2. Place 700 μL of the working buffer in the quartz cuvette and measure the baseline.
3. Remove the working buffer from the cuvette and add 700 μL of the holo-protein sample at 1 mg/ml, 10–20 μM (*see* **Note 11**). For long measurements, it is important to seal the cuvette with a lid, Parafilm, or similar material. Also, take an initial 10 μL sample of the holo-protein solution and mix it with sodium dodecyl sulfate–polyacrylamide gel electrophoresis (SDS-PAGE) loading buffer for future stability checks.
4. Irradiate the sample with red light for an appropriate duration (*see* **Note 12**). Immediately after irradiation, start the data acquisition with the settings described in **step 1** of this subheading (25 °C, 200–900 nm with a resolution of 1 nm, sampling every 5 min from 0 to 180 min and then every 30 min from 180 to 1020–1200 min, with baseline correction). Collect a final 10 μL sample of the holo-protein solution and mix it with SDS-PAGE loading buffer for further stability checks.
5. Perform an SDS-PAGE analysis to evaluate the protein's integrity at the beginning and the end of the procedure (initial and final SDS-PAGE samples).
6. Repeat the entire procedure (from **step 1–5** of this subheading) with a fresh sample; this time irradiating with far-red light.

3.1.2 First Processing of the Datasets (Fig. 2a)

Any suitable program can be used, such as Excel, MATLAB, or Python.

1. For each spectrum, calculate the baseline and subtract this baseline from the corresponding spectrum (*see* **Note 9**).
2. Normalize the spectra to the dataset's maximum value within the 270–290 nm range.

3.1.3 Calculating the Pure-Pfr Spectrum of the Dataset (Fig. 3a)

This method presents some similarities with a previous study by Assafa et al. [18]. The method is based on the fact that in all Pfr-enriched spectrum, there is always some Pr remaining. The

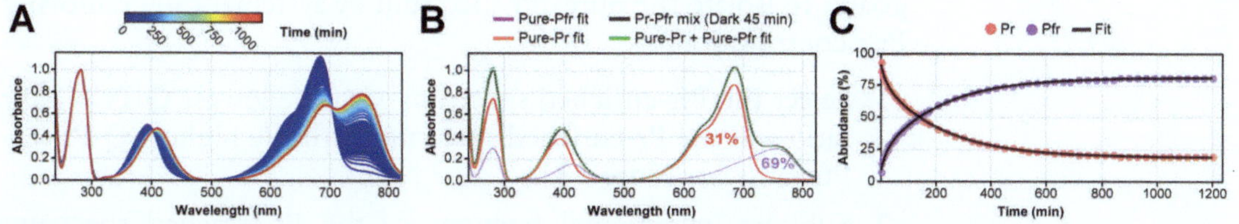

Fig. 2 Data analysis of a single pure holo-phytochrome sample of UV–visible spectra. The data shown correspond to the far-red irradiated *Xcc*BphP wild-type variant at pH 7.5 and 30 °C as an example (unpublished data). (**a**) Data after the first processing of the dataset. (**b**) Calculation of the Pr/Pfr abundance from a single spectrum. (**c**) Kinetic analysis showing Pr/Pfr abundance over time

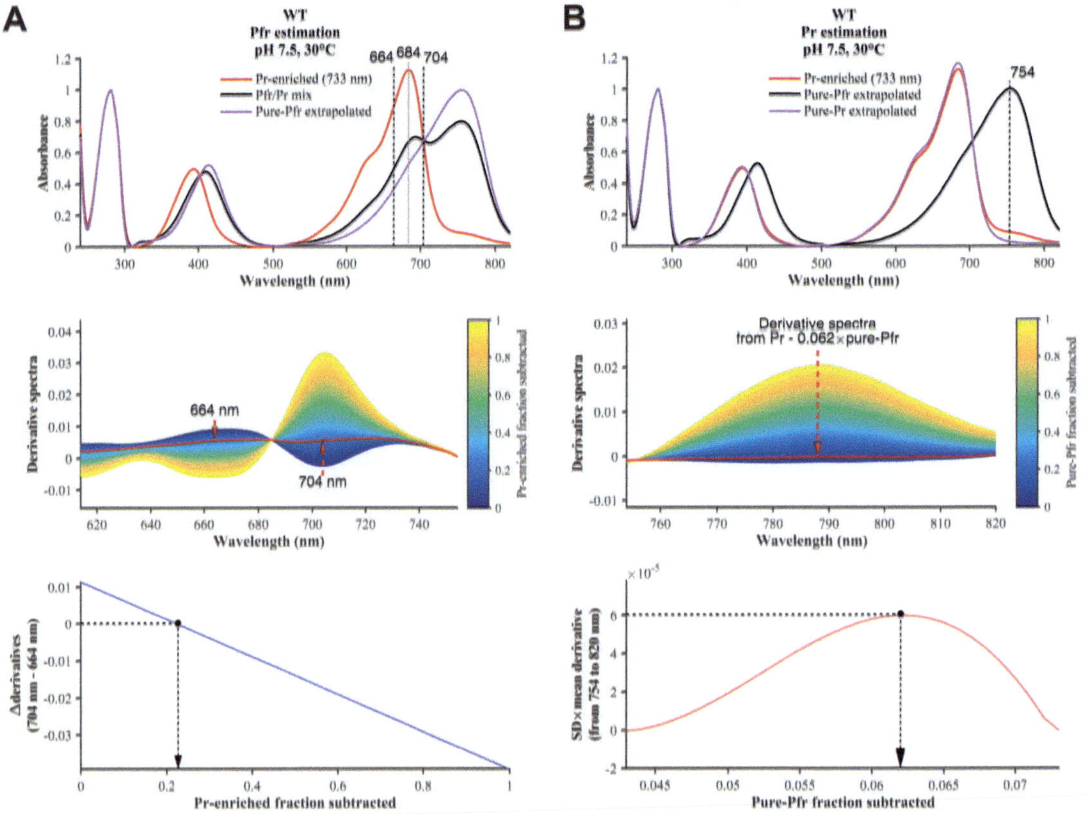

Fig. 3 Calculation of pure-Pr and Pfr spectra. (**a**) Pure-Pfr calculation. (**b**) Pure-Pr calculation. The data correspond to the *Xcc*BphP wild-type variant at pH 7.5 and 30 °C (unpublished data). Calculations were derived from dark conversion datasets initially illuminated with far-red and/or red light. The *upper panels* show Pr-enriched, Pfr/Pr mix, pure-Pr, and pure-Pfr spectra. The dashed lines indicate the wavelength ranges used for the first derivative calculations of the subtraction series, displayed in the *middle panels*. Minima and maxima around the Q-Pr peak are highlighted with red arrows in the middle panels. The *bottom panels* illustrate the criteria for determining pure-Pfr and pure-Pr spectra from the subtraction series. For pure-Pfr, the criterion is based on the minimization of the magnitude of difference between the abovementioned minima and maxima. In contrast, the pure-Pr spectrum is selected at the point where the product of the mean and standard deviation of the derivative in a low-variability region reaches a local maximum: This condition corresponds to an SD × mean value that is nearly zero and the "flattest tail" in the spectrum

goal is to isolate the pure-Pfr spectrum by systematically removing Pr's contribution.

1. Select the Pfr-enriched spectrum with the lowest Q-Pr/Q-Pfr ratio and the Pr-enriched spectrum with the highest Q-Pr/Q-Pfr ratio (*see* **Note 13**).
2. Subtract incremental fractions of the Pr-enriched spectrum (ranging from 0 to 1, with increment steps of 0.001) from the Pfr-enriched spectrum (*see* **Note 14**).

Table 1
Parameters analyzed in the spectroscopic characterization of phytochromes

Parameters	*Xcc*bphp_WT Ph 7.5/30 °C Pure	*Xcc*BphP_WT pH 7.5/25 °C Clarified extract
Soret-Pr wavelength (nm)	394	–
Soret-Pfr wavelength (nm)	414	–
Soret-Pr intensity	0.52	–
Soret-Pfr intensity	0.51	–
Q-Pr wavelength (nm)	684	682
Q-Pfr wavelength (nm)	755	755
Q-Pr intensity	1.17	–
Q-Pfr intensity	0.99	–
Pfr abundance @ equilibrium	81.3%	78%
Type	Bathy-like	Bathy-like
Half-life 1 (min)	136.0	309.4
Half-life 2 (min)	91.5	–

3. For each subtraction spectrum, compute its first derivative, which captures changes in slope across the spectrum (*see* **Note 15**).
4. Identify the minima and maxima in the derivative spectra around the wavelength corresponding to the Q-Pr peak and determine the pure-Pfr spectrum as the subtraction spectrum that minimizes the magnitude of the difference between these two values (*see* **Note 16**).
5. Fit each peak using a double Gaussian function, which reduces bias from asymmetrical peak shapes and minimizes errors caused by noisy data or insufficient wavelength sampling. This approach ensures accurate determination of both the intensities and wavelengths of the Soret and Q-Pfr bands (Table 1).

3.1.4 Calculate the Pure-Pr Spectra for the Dataset (Fig. 3b)

This method follows a similar approach to that used for calculating the pure-Pfr spectrum. To isolate the pure-Pr spectrum, the contribution of residual Pfr from a Pr-enriched spectrum must be systematically removed.

1. Select two spectra: one Pr-enriched spectrum, which is dominated by Pr but contains some Pfr, and the pure-Pfr spectrum, calculated in the previous step. Select the Pr-enriched one with the highest Q-Pr/Q-Pfr ratio.

2. Subtract incremental fractions of the pure-Pfr spectrum (0–1 with increment steps of 0.001) from the Pr-enriched spectrum (*see* **Note 17**).
3. For each subtraction spectrum, compute the first derivative in the longer wavelength region beyond the Q-Pfr peak (typically 725–900 nm) (*see* **Note 18**).
4. Calculate the mean and standard deviation (SD) for each derivative.
5. Find the optimal subtraction spectrum. Multiply the mean and SD for each subtraction spectrum and identify the subtraction where the product (mean × SD) first reaches a local maximum within the range of small incremental subtractions (*see* **Note 19**).
6. Determine the Soret and the Q-Pr bands peaks intensities and wavelengths from the pure-Pr spectrum. Use the same approach described in **step 5** from the section *Calculating the pure-Pfr spectrum of the dataset* (Table 1).

3.1.5 Calculation of the Pr/Pfr Abundance from Each Spectrum (Fig. 2b)

1. Calculate the linear combinations of the pure-Pr and pure-Pfr spectra that best fit the experimental spectrum. This step allows the estimation of Pr and Pfr abundance values at each time point (*see* **Note 20**).

3.1.6 Kinetic Analysis (Fig. 2c)

1. Reconstruct the experimental Pr/Pfr abundance as a function of time by plotting these values in an XY scatter plot.
2. Fit the data from the last step by optimizing the parameters of a monoexponential model (Eq. 1) or a double-exponential model (Eq. 2) to fit the Pr/Pfr abundance data (*see* **Note 21**).

$$f(t) = A + B \cdot e^{-t/C} \tag{1}$$

$$f(t) = D + E \cdot e^{-t/F} + G \cdot e^{-t/H} \tag{2}$$

3. Estimate the Pr/Pfr abundance in equilibrium by calculating Pr/Pfr abundance from parameter A (Eq. 1) or D (Eq. 2, Table 1).
4. Based on the equilibrium Pr/Pfr abundance, classify the phytochrome into one of the following categories: bathy-type: >90% Pfr, bathy-like: 50%–90% Pfr, canonical-like: 10%–50% Pfr, canonical-type: <10% Pfr (Table 1).
5. Estimate the dark conversion rate by calculating the half-life ($t_{1/2}$) or mean lifetime (τ) values using parameter C (Eq. 1) or F and H (Eq. 2) (Table 1). Parameters C (Eq. 1), F and H (Eq. 2) correspond to the mean lifetime. The half-life can be calculated using the following equation:

$$t_{1/2} = \tau . \ln(2)$$

3.2 Multi-sample Acquisition Mode in the Visible Range of Phytochrome-Overexpressing Crude Extracts

3.2.1 Data Acquisition

1. Prepare the acquisition settings; configure the microplate spectrophotometer for absorbance measurements at a fixed temperature (25 °C is recommended). Set the wavelength range to 450–900 nm with a resolution of 2 nm (*see* **Note 22**). For time sampling (post-irradiation), measure every 5 min from 0 to 60 min, then every 60 min from 60 to 1020–1200 min (*see* **Note 10**).
2. Pipette 200 μL of different clarified extracts into each well of a 96-well plate. If working with strains that co-express HO1, the FDBR HY2, and/or the FDBR PcyA synthases, the assembly into holoprotein will occur intracellularly without the need to add biliverdin. In this case, skip **steps 3** and **4**.
3. Add free bilin chromophore from a concentrated stock solution to each well to achieve a predefined final concentration (*see* **Note 23**).
4. Incubate the plate in the dark inside the spectrophotometer at 25 °C for an appropriate duration (*see* **Note 24**). This step ensures the completion of holo-protein assembly.
5. Illuminate the plate with red light for a specified duration (*see* **Note 12**). Immediately following irradiation, begin data acquisition as described in **step 1** of this section (25 °C, 450–900 nm with a resolution of 2 nm, time sampling every 5 min from 0 to 60 min, then every 60 min from 60 to 1020–1200 min).
6. Repeat the entire procedure (**steps 1–5** from this section) with fresh samples. This time, illuminate the plate with far-red light instead of red light.

3.2.2 First Processing of the Datasets (Fig. 4a, b)

1. Trim the data for all spectra along the *x*-axis wavelength range, considering background absorbance in the blue region (use 540–900 nm as the reference working range).
2. For each spectrum, calculate a linear baseline fit (with a non-zero slope) in the near-infrared range (use 850–900 nm as the reference working range), and subtract this baseline from the corresponding spectrum.
3. Normalize each spectrum to the maximum value in the dataset (*see* **Note 25**). The maximum value may correspond to either the Q-Pr or the Q-Pfr peak.
4. Apply a smoothing function to all spectra to reduce noise while preserving critical spectral features.

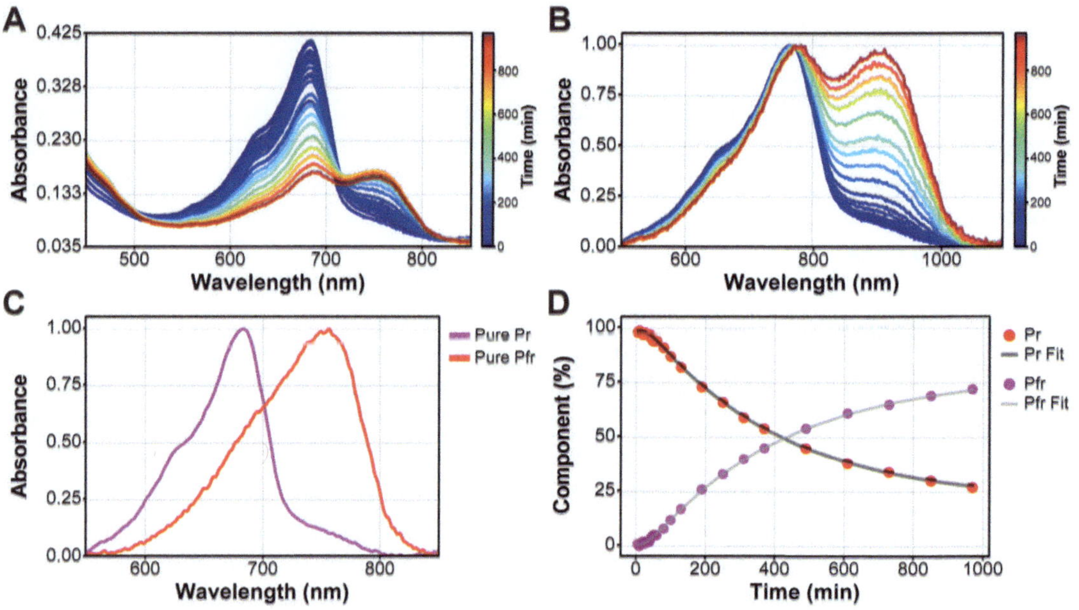

Fig. 4 Data analysis of the visible spectra from multi-phytochrome-overexpressing bacterial extract samples. The data presented correspond to the far-red irradiated *Xcc*BphP wild-type variant at pH 7.5 and 25 °C as an example [15]. For simplicity, a single dataset is shown. However, a complete screening experiment typically includes multiple datasets for both red- and far-red-illuminated samples. (**a**) Raw data. (**b**) Spectra after the first processing of the dataset. (**c**) Pure-Pr and Pfr spectra calculated. Q-Pr: 684 nm; Q-Pfr: 757 nm. (**d**) Kinetic analysis showing Pr/Pfr abundance over time

3.2.3 Calculation of the Pure-Pr and the Pure-Pfr Spectral Components (Fig. 3)

1. Follow the procedures outlined in Subheadings 3.1.3 and 3.1.4. Alternatively, the most Pr-enriched spectrum can be used as the pure-Pr spectrum. This approximation is sufficiently accurate for this type of noisy dataset and renders characterization results consistent with phytochrome proteins in pure preparation.
2. Determine the Q-Pr and Q-Pfr bands' peak intensities and wavelengths from the pure-Pr and pure-Pfr spectra (Fig. 4b). Follow the same approach described in Subheading 3.1.3, **step 5** (Table 1).

3.2.4 Calculation of the Pr/Pfr Abundance from Each Spectrum (Fig. 4c)

Calculate the linear combinations of the pure-Pr and pure-Pfr spectra that best fit the experimental spectrum, as described in the previous homonym section. This step allows estimation of Pr and Pfr abundance values at each time point.

3.2.5 Kinetic Analysis (Fig. 4d)

Perform kinetic analysis using the same procedure detailed in the previous homonym section (Table 1).

4 Notes

1. The ideal working concentration for the pure preparation is 1 mg/ml, 10–20 μM. The volume needed is 1 ml or less, depending on the cuvette geometry.
2. The required volume is 200 μl. One possibility is that the phytochrome-overexpressing bacterial extracts coexpress heme oxygenase HO1, the FDBR HY2, and/or the FDBR PcyA synthases to produce BV, PΦB, or PCB [1]. Alternatively, BV and PCB, which are commercially available, can be added externally. In some cases, PCB is used for PΦB containing phytochromes due to their similarities.
3. BV and PCB should be dissolved in dimethyl sulfoxide (DMSO) at concentrations of 20 mM and 100 mM, respectively.
4. The spectrophotometer should be capable of measuring absorption spectra in the range of 240–900 nm over time. The temperature should remain stable throughout the experiment, which may last from minutes to days, depending on the nature of the phytochrome being studied. The spectrophotometer should include a holder for a 1-cm pathlength cuvette. For example, the Agilent Technologies Cary 60 spectrophotometer with an installed temperature-controlled Peltier module allows stable measurements with high-speed scanning over extended periods (up to a few days).
5. A 1-cm pathlength quartz cuvette, ideally 2 mm wide to minimize sample usage. This cuvette size accommodates a total volume of 700 μL.
6. The device should be capable of measuring absorption spectra in the range of 240–900 nm or 400–900 nm over time in a 96-well plate, with temperature control. The temperature should remain stable throughout the experiment, which may last from minutes to days. For example, the Thermo Scientific Multiskan GO microplate spectrophotometer is suitable.
7. For pure protein preparation methods, refer to Klinke S et al. [21]. For clarified phytochrome-overexpressing bacterial extracts supplemented with the bilin chromophore, refer to Antelo GT et al. [17].
8. A suitable starting point for the scanning rate is 4800 nm/min. However, this value can be adjusted based on the spectrophotometer's characteristics and the sample's photochemical behavior. For example, the Cary60 spectrophotometer from Agilent can achieve scanning rates of up to 24,000 nm/min while maintaining good data quality. If the dark decay of the studied phytochrome is fast, with a half-life in the range of seconds or a

few minutes, it may be necessary to increase the scanning rate and/or narrow the wavelength range, focusing on the Soret and Q bands.

9. Baseline correction beyond simple blank subtraction is essential for removing drifting baselines, scattering signals, and other background offsets in UV–visible absorption spectra. The choice of correction method depends on the severity and nature of the distortion. Typically, wavelength intervals with no absorption, often in the green and near-infrared regions, are used to fit suitable functions. Common baseline estimation methods include linear offset correction and adjustments for Rayleigh (or Mie) scattering effects [22, 23].
10. The time sampling is optimized to capture the kinetics of *Xcc*BphP and its variants and is a good starting point to gain information on a broad range of phytochromes. However, faster or slower kinetics may require adjustments to the sampling intervals, either shorter or longer.
11. The protein concentration can be slightly lower, depending on the characteristics of the spectrophotometer (e.g., 0.5 mg/ml and 5–10 μM).
12. To determine the illumination time, place a suitable volume of the sample solution in a 1-cm pathlength cuvette and measure serial spectra in the range 200–900 nm for pure holo-protein or 450–900 nm for crude extracts. Perform measurements every 20 s at a controlled temperature of 25 °C, while illuminating the sample with red or far-red light from above, leaving the spectrophotometer lid open. The spectra may appear noisy, but it should provide sufficient information to establish the required illumination time for each light source. The working illumination time should correspond to the third of three consecutive overlapping spectra. Another important factor to consider is the distance between the light source and the sample. In subsequent experiments, this distance should be equal to or smaller than the one used in this initial determination in order to ensure the complete illumination of the sample. For plate illumination, the light source should be centered relative to the plate, ensuring that the distance between the light source and the outermost wells—the farthest from the center—is equal to or smaller than the one established in this experiment. This setup ensures that even if the illumination is not perfectly uniform, it will still be sufficient to fully illuminate every sample.
13. Pr and Pfr spectra are enriched in their respective states but retain a fraction of the other.
14. This process generates a series of subtraction spectra, each reflecting a slightly different assumption about the Pr

contribution to the measured spectrum. The goal is to obtain a Pfr spectrum that is as free from Pr traces as possible. As a result, the process will produce under-depleted, over-depleted, and correctly depleted spectra. In the following steps, the correctly depleted spectrum will be identified.

15. These derivative spectra exhibit a zero value at the wavelength corresponding to the Q-Pr peak. Under-depleted spectra will show positive slope values below and negative slope values above the Q-Pr peak wavelength. Conversely, over-depleted spectra will show negative slope values below and positive slope values above the Q-Pr peak wavelength.

16. The optimal subtraction spectrum is identified by minimizing the residual influence of the Pr component in the Pfr-enriched spectrum. The Q-Pr peak contributes a distinct slope to the derivative spectrum, and as the subtraction progresses, this slope diminishes until the Pr contribution is optimally removed. To achieve this, the first derivative of each subtraction spectrum is calculated, and the wavelengths corresponding to the derivative's minima and maxima near the Q-Pr peak are identified. These points represent the steepest negative and positive slopes on either side of the Q-Pr peak. The optimal subtraction spectrum is determined by minimizing the magnitude of the difference between these derivative values. At this point, the Pr component is appropriately depleted, leaving a pure representation of the Pfr state. The selected spectrum should exhibit the expected smooth slope leading to the Q-Pfr peak without introducing distortions indicative of over- or under-subtraction. This methodology ensures the systematic removal of Pr contributions and facilitates the accurate extraction of the pure-Pfr spectrum for subsequent analyses. This is a heuristic algorithm that has been reliable with our analyses, generating consistent Q-Pfr peaks.

17. This creates a series of spectra with varying levels of Pfr removed.

18. In this region, Pfr dominates the shape of the spectrum, and its removal smooths out variability.

19. The search strategy is based on the fact that pure or nearly pure-Pr spectra do not absorb beyond the Q-Pfr peak, and therefore, no signal is expected apart from the baseline. Consequently, the goal is to find a signal in this region that is as low and invariant as possible. The mean of the derivative reflects the overall slope in this region, while the SD of the derivative measures its variability, both of which decrease as noise and residual Pfr are removed. However, near the point where the Pfr component is fully eliminated, the product of the mean

derivative and SD increases, reaching a local maximum. In this heuristic, the local maximum corresponds to the subtraction spectrum where the Pr signal is most accurately isolated, supporting the preservation of the spectrum shape while minimizing spectral distortions.

20. This calculation is more reliable when working only with the Q bands' range (510–900 nm).
21. For a system with two interconverting species, monoexponential behavior is typically expected. However, a double-exponential model can account for more complex kinetics if necessary. Residual analysis is recommended to evaluate the quality of the monoexponential fit.
22. The signal below 450 nm is typically too noisy and not worth measuring.
23. Free bilin exhibits broad, low-intensity absorption peaks in the visible range. However, when bound to the protein, these peaks become narrower and more intense [21]. Broad peaks from free bilin may overlap and mask the holo-protein signal if the bilin concentration is too high. To optimize the holo-protein signal, minimize the free-bilin signal while maintaining a strong, clear holo-protein signal. To determine the optimal bilin concentration [17]:

    Use a reliable lysate sample of an overexpressed phytochrome variant with proven bilin binding.

    (a) Place a suitable volume of the lysate in a 1-cm pathlength cuvette (a plastic cuvette is acceptable for this purpose).
    (b) Measure an initial spectrum in the range 450–900 nm.
    (c) Add free bilin to a final concentration of 5 μM.
    (d) Measure serial spectra every minute at a controlled temperature of 25 °C until the signal stabilizes (defined as three consecutive overlapping spectra). An increase in the Q-Pr band will be observed.
    (e) Incrementally increase the bilin concentration by 5 μM (e.g., 5 μM, 10 μM, and 15 μM) and repeat the measurements.

    The optimal working bilin concentration is the lower of the two concentrations that produce the same signal.

24. After determining the optimal bilin concentration, establish the time required for complete bilin assembly:

    (a) Place a suitable volume of the lysate in a 1-cm pathlength cuvette.
    (b) Measure an initial spectrum in the range 450–900 nm.
    (c) Add free bilin at the working concentration.

(d) Measure serial spectra every minute at a controlled temperature of 25 °C until the signal stabilizes.

The working incubation time corresponds to the third of three consecutive overlapping spectra.

25. Typically, the highest value corresponds to either the first or last spectrum. However, this may not hold true if bilin assembly is incomplete when dark conversion measurements begin. This would indicate that the incubation time determined in 4.7 was insufficient for complete chromophore assembly in that specific variant.

References

1. Rockwell NC, Lagarias JC (2020) Phytochrome evolution in 3D: deletion, duplication, and diversification. New Phytol 225:2283–2300. https://doi.org/10.1111/nph.16240
2. Rockwell NC, Su Y-S, Lagarias JC (2006) Phytochrome structure and signaling mechanisms. Annu Rev Plant Biol 57:837–858. https://doi.org/10.1146/annurev.arplant.56.032604.144208
3. Hughes J, Winkler A (2024) New insight into phytochromes: connecting structure to function. Annu Rev Plant Biol 75:153–183. https://doi.org/10.1146/annurev-arplant-070623-110636
4. Otero LH, Foscaldi S, Antelo GT et al (2021) Structural basis for the Pr-Pfr long-range signaling mechanism of a full-length bacterial phytochrome at the atomic level. Sci Adv 7: eabh1097. https://doi.org/10.1126/sciadv.abh1097
5. Smith H (2000) Phytochromes and light signal perception by plants—an emerging synthesis. Nature 407:585–591. https://doi.org/10.1038/35036500
6. Nagy F, Schäfer E (2000) Nuclear and cytosolic events of light-induced, phytochrome-regulated signaling in higher plants. EMBO J 19:157–163. https://doi.org/10.1093/emboj/19.2.157
7. Kircher S, Kozma-Bognar L, Kim L et al (1999) Light quality-dependent nuclear import of the plant photoreceptors phytochrome A and B. Plant Cell 11:1445–1456. https://doi.org/10.1105/tpc.11.8.1445
8. Yamaguchi R, Nakamura M, Mochizuki N et al (1999) Light-dependent translocation of a phytochrome B-GFP fusion protein to the nucleus in transgenic Arabidopsis. J Cell Biol 145:437–445. https://doi.org/10.1083/jcb.145.3.437
9. Tarutina M, Ryjenkov DA, Gomelsky M (2006) An unorthodox bacteriophytochrome from Rhodobacter sphaeroides involved in turnover of the second messenger c-di-GMP. J Biol Chem 281:34751–34758. https://doi.org/10.1074/jbc.M604819200
10. Gourinchas G, Etzl S, Göbl C et al (2017) Long-range allosteric signaling in red light-regulated diguanylyl cyclases. Sci Adv 3: e1602498. https://doi.org/10.1126/sciadv.1602498
11. Takala H, Edlund P, Ihalainen JA, Westenhoff S (2020) Tips and turns of bacteriophytochrome photoactivation. Photochem Photobiol Sci 19:1488–1510. https://doi.org/10.1039/d0pp00117a
12. Huber C, Strack M, Schultheiß I et al (2024) Darkness inhibits autokinase activity of bacterial bathy phytochromes. J Biol Chem 300: 107148. https://doi.org/10.1016/j.jbc.2024.107148
13. Conforte V, Otero LH, Toum L et al (2021) Pr-favoured variants of the bacteriophytochrome from the plant pathogen Xanthomonas campestris hint on light regulation of virulence-associated mechanisms. FEBS J 288: 5986–6002. https://doi.org/10.1111/febs.15883
14. Bonomi HR, Toum L, Sycz G et al (2016) Xanthomonas campestris attenuates virulence by sensing light through a bacteriophytochrome photoreceptor. EMBO Rep 17:1565–1577. https://doi.org/10.15252/embr.201541691
15. Hildebrandt P (2023) Vibrational spectroscopy of phytochromes. Biomolecules 13: 1007. https://doi.org/10.3390/biom13061007
16. Otero LH, Klinke S, Rinaldi J et al (2016) Structure of the full-length

bacteriophytochrome from the plant pathogen Xanthomonas campestris provides clues to its long-range signaling mechanism. J Mol Biol 428:3702–3720. https://doi.org/10.1016/j.jmb.2016.04.012

17. Antelo GT, Sánchez-Lamas M, Goldbaum FA et al (2020) A spectroscopy-based methodology for rapid screening and characterization of phytochrome photochemistry in search of Pfr-favored variants. Photochem Photobiol 96:1221–1232. https://doi.org/10.1111/php.13313
18. Assafa TE, Anders K, Linne U et al (2018) Light-driven domain mechanics of a minimal phytochrome photosensory module studied by EPR. Structure 26:1534–1545.e4. https://doi.org/10.1016/j.str.2018.08.003
19. Burgie ES, Bussell AN, Walker JM et al (2014) Crystal structure of the photosensing module from a red/far-red light-absorbing plant phytochrome. Proc Natl Acad Sci USA 111: 10179–10184. https://doi.org/10.1073/pnas.1403096111
20. Gourinchas G, Vide U, Winkler A (2019) Influence of the N-terminal segment and the PHY-tongue element on light-regulation in bacteriophytochromes. J Biol Chem 294: 4498–4510. https://doi.org/10.1074/jbc.RA118.007260
21. Klinke S, Otero LH, Rinaldi J et al (2014) Crystallization and preliminary X-ray characterization of the full-length bacteriophytochrome from the plant pathogen Xanthomonas campestris pv. campestris. Acta Crystallogr Sect F Struct Biol Commun 70: 1636–1639. https://doi.org/10.1107/S2053230X14023243
22. Basalla AJ, Kendrick BS (2023) Correcting ultraviolet-visible spectra for baseline artifacts. J Pharm Sci 112:3240–3247. https://doi.org/10.1016/j.xphs.2023.08.015
23. Winder AF, Gent WL (1971) Correction of light-scattering errors in spectrophotometric protein determinations. Biopolymers 10: 1243–1251. https://doi.org/10.1002/bip.360100713
24. Conforte VP, Rinaldi J, Bonomi HR et al (2025) Xanthomonas campestris pv. Campestris regulates virulence mechanisms by sensing blue light. Photochem Photobiol Sci. https://doi.org/10.1007/s43630-025-00694-y

Chapter 8

Computational Protocols to Study Phytochromes

Probal Nag, Chongyao Wei, Megan J. Mackintosh, and Igor Schapiro

Abstract

Computer simulations allow us to validate the structure, predict properties, and study the reactivity of proteins. In many areas of research, they are well integrated as they are used for structure validation and calculations of spectroscopic properties. In this chapter, we describe a simulation protocol that starts from a protein structure of a cyanobacteriochrome. The procedure to prepare the structure for classical molecular dynamics simulations is explained and broken down into single steps. After the classical simulation, it is shown how spectroscopic properties can be obtained by quantum chemical calculations which are part of a multiscale model. The high dimensionality of the protein environment in the calculation of excitation energies is an issue for the calculation of excitation energies, because a single representative geometry cannot be selected. To alleviate this problem, we explain the conformational sampling and compare it to the optimization procedure.

Key words Photoreceptor, Phytochrome, Cyanobacteriochrome, Bilin, Molecular dynamics simulations, Force field, Hybrid quantum mechanics/molecular mechanics, Excited state

1 Introduction

Computer simulations are an essential area of research that ideally complement experiments in biomolecular sciences. The development of computational methods is tightly linked to the advancement and availability of computer hardware. Protein simulations date back to the 1970s [1]. The first predictive results were obtained for the photoreceptor protein rhodopsin [2], where the photoisomerization was found to be faster than the excited-state proton transfer, ahead of experiments. It took much longer before the first application to phytochromes became available because of the lack of experimental structures and also due to the large size of the bilin chromophore. However, in recent years, multiscale simulations have been efficiently carried out on phytochromes.

Protein simulations require a structure of the phytochrome as a starting point. Typically, this is based on an experimental structure, for example, an X-ray crystal structure. The experimental structure

Andreas Winkler (ed.), *Phytochromes: Methods and Protocols*, Methods in Molecular Biology, vol. 2970,
https://doi.org/10.1007/978-1-0716-4791-2_8,

requires several steps of preparation before the initial simulation. This is due to the fact that crystallographic conditions are often different from physiologically relevant conditions, which concerns the solvent, pH condition, oligomerization, and temperature. This is addressed in several preparatory steps before the actual production run of a molecular dynamics (MD) simulation, as mentioned in Subheadings 3.1, 3.2, and 3.3 below.

More recently, cryogenic-electron microscopy (cryo-EM) has also been successfully applied to solve the full-length phytochrome structure [3, 4]. Structures from cryo-EM usually have poorer resolution compared to X-ray crystallography as it excels in visualizing the structure in the native state because it does not require crystallization. In contrast to X-ray crystallography, it captures both the amplitude and phase information, therefore circumventing the phase problem inherent to X-ray crystallography. Cryo-EM densities can be used as constraints for MD simulations or can be directly refined by simulations to optimize the atomic positions. Besides the conceptual differences between cryo-EM and X-ray crystallography, there is no difference in the computational protocol, as described below.

In the future, the use of predictive models such as AlphaFold [5] or RoseTTAFold [6] might become a viable option to generate a structure for atomistic simulations. However, there are major challenges for the application of phytochromes. The first one is the presence of the bilin chromophore, which is essential for the light response and the subsequent reactivity. The choice of the correct bilin, that is, the type which is bound by the protein and its placement, requires the correct prediction of protein–ligand interaction as well as the covalent bond. Another challenge is that protein structure prediction tools based on machine learning are focused on the secondary structure of the protein backbone. However, the binding of a ligand such as bilin also requires the knowledge of side chain orientation. Recent developments such as RoseTTAFold All-Atom have started to address these shortcomings. In a recent study by Baker and coworkers, the prediction of bilin binding proteins has been demonstrated by using a generative diffusion approach [7]. A further step in the prediction of ligand–protein interactions was reported in AlphaFold3 for wide range of biomolecular systems [8]. However, these recent developments need further testing to demonstrate a generalization of deep learning frameworks and wide range of coverage of biomolecular interactions.

2 Materials

1. Protein structure of TePixJg at 2.00 Å resolution (PDB IDs: 6PRU).

2. Computer hardware: The simulations can be carried out on a standard laptop. This is due to the small size of the protein for classical MD simulations and due to efficient quantum chemical methods in the hybrid quantum mechanics/molecular mechanics (QM/MM) simulation.
3. Software for classical MD simulation: Assisted Model Building and Energy Refinement (AMBER) is available free of charge for academic users.
4. Software for hybrid QM/MM simulation: The ORCA program is used for quantum chemical simulations as part of the QM/MM setup and for excitation energy calculations.

3 Methods

This section serves as a guide to handling protein data bank (PDB) files obtained from such experiments, setting up simulations, and predicting the excited-state properties of a protein–chromophore system.

In the following, we provide a protocol to run protein simulations in three steps:

1. Preparing a computational protein model starting from an experimental protein structure.
2. Setting up classical molecular dynamics and hybrid quantum mechanics/molecular mechanics (QM/MM) simulations.
3. Simulating the UV/Vis absorption spectrum.

Such simulations will provide information about the structural integrity of the protein and an in-depth understanding of the spectral properties of the system, which will complement the experimental findings. As an example, we use the c**G**MP-specific phosphodiesterases, **a**denylyl cyclases and **F**hlA (GAF) domain of TePixJ (TePixJg) which is a cyanobacteriochrome. The latter is a subfamily of phytochromes which only requires a GAF domain to perform the photochemistry and is therefore of compact size. The crystal structure at 2.00 Å resolution has the PDB code 6PRU and was solved by Burgie et al. [9] (Fig. 1). The small size of the protein makes it less demanding in terms of computer resources. Hence, the simulations presented here can be carried out on a standard laptop and without the need for access to a high-performance computing (HPC) facility.

The procedure presented below is performed using the AMBER simulation package [10]. This package is available free of charge for academic users and includes programs for geometry optimization, MD simulation, free energy calculations, and potential of mean force. It is augmented by open-source AmberTools

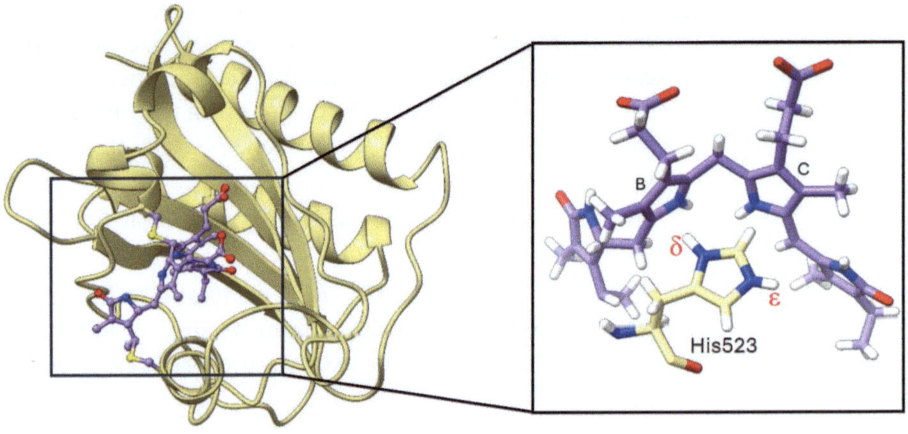

Fig. 1 TePixJg structure based on the crystal structure from Burgie et al. (PDB code 6PRU). The inset shows the PVB chromophore with histidine in position 523. The protonation of this side chain has a direct interaction with the chromophore and might affect the photochemical reaction

[11], which includes auxiliary programs that range from setting up a simulation model to analyzing the results. Another important aspect is that AMBER also has a series of classical force fields for different biomolecules. There are alternative packages like CHARMM, OpenMM, GROMACS, and many more, which go beyond the scope of this book chapter. For quantum chemical simulations, we use the ORCA program package that is available for free for academic users.

3.1 Structure Preparation

Any protein simulation starts with a reference structure or a model. For all purposes, this is usually obtained from X-ray crystallography and/or cryo-EM models. In the initial step, we complete the missing residues, identify the important part of the protein to simulate (e.g., a protomer), and add protons to the selected system. In the selected example, the PDB entry 6PRU contains a dimer of TePixJg. For the simulation, you choose only one protomer because the aim is to prepare the protein for the calculation of spectroscopic data which depends only on one chromophore. The workflow of the structure preparation can be broken down into the following steps:

1. Convert the experimental PDB structure to the AMBER format using the *pdb4amber* tool. This is crucial to fix missing residues, identify disulfide bonds, and address different protonation states of the residues.
2. Identify the correct protonation states from experimental data. Depending on the pH used during the experiments, the protonation state should be adjusted to the AMBER convention for the simulations accordingly. Refer to **Note 1** for the amino

acid sidechains with variable protonation states and their corresponding labels.

3. Process the PDB file using the AMBER tool *LEaP*. This tool generates the topology and coordinate files of the system to start the simulations.

Implementing the correct protonation state is the most crucial part in structure handling because, depending on the protonation state of the residues and the chromophore, the interaction between them will change, thereby affecting the electronic properties of the system.

In the example of TePixJg, the protonation of the pyrrole nitrogen of the bilin chromophore, specifically phycoviolobilin (PVB), and the propionates can change depending on the environment. Using vibrational spectroscopy, it was shown that PVB is completely protonated, and this has also been confirmed by quantum chemical simulations [12]. However, there is a histidine residue in position 523 (His523) which directly interacts with the PVB and might therefore be involved in the photochemistry (Fig. 1). Depending on the experimental conditions, like a change in the pH, this residue can have different protonation forms. Refer to **Note 2** for details of how to determine the protonation state.

Once the PDB file reflects the correct protonation state, use the *LEaP* tool to load the force field parameters of the whole system, which include the amino acids, chromophore, solvent, and ions. While these parameters are available for standard residues and solvents, the PVB parameters need to be prepared in advance. This can be done using the *antechamber* tool or more sophisticated procedures which go beyond the purpose of this book chapter. Once the structure and the parameters are loaded in *LEaP*, the entire protein can be placed in a solvent box with water. In order to make it electrostatically neutral, sodium and chloride ions are added. The procedure is illustrated in Fig. 2. The system is now ready to be used for simulations.

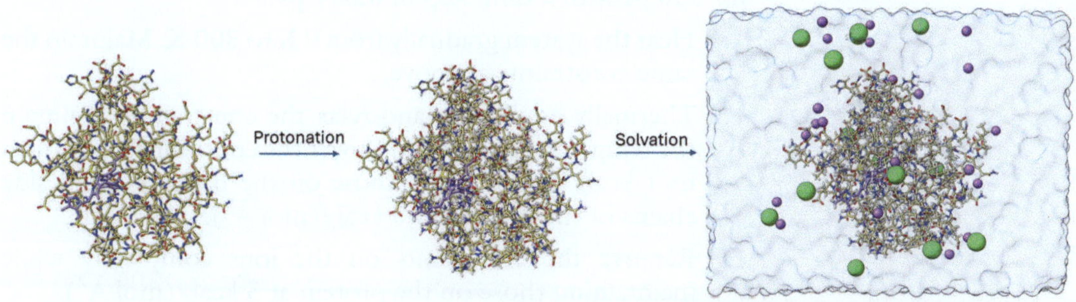

Fig. 2 Preparation of the TePlxJg protein for simulation. In the first step, the protons are added to the entire system. Then, the full protein is solvated in water. Some Na^{+} and Cl^{-} ions are added to make the simulation box electrostatically neutral

3.2 Initial Relaxation

Once the system is correctly protonated, solvated, and neutralized with ions, check if clashes are found between the residues. To remove bad contacts, perform an initial relaxation of the protein structure. Run a classical MD simulation using the *sander* tool of AMBER with the generated force fields (from running *LEaP*) to represent the protein and the PVB chromophore. Using a classical force field, many hundreds of iterations of geometry optimization or MD can be calculated within a few minutes. During this energy minimization, apply constraints on the system (side chains and backbone of the protein, and ions) to prevent overly high potential energies due to rapid movement of the atoms. Set the constraints on the side chains lower than the protein backbone, to allow flexibility for the side chains to reorient themselves to avoid clashes.

In the TePixJg example, apply constraints of 10 kcal/(mol·Å^2) on the protein backbone and ions, and constrain the side chains using 5 kcal/(mol·Å^2). With these constraints, perform an energy minimization for 10,000 iterations. Due to the large size of the simulated system, a convergence is not achieved, but it is sufficient to provide a reasonable structure for the following MD simulations.

3.3 Equilibration

The next step involves the setup of the simulation conditions like temperature, pressure or, pH values. In several steps, the conditions are adjusted to approach the desired values. Start with the same constraints as described above. Over multiple cycles, relax the constraints and allow the system to thermally equilibrate at a temperature of 300 K. Equilibration is performed using NVT and NPT ensembles where the number of molecules (N), volume (V), and temperature (T) or pressure (P) are maintained throughout the system. In the NVT ensemble, the volume is kept constant; hence, the size of the simulation box does not change, while in the NPT ensemble, the pressure is kept constant and the box size can change. In the protocol presented below, all the steps are described; each step includes a MD run to equilibrate the system. These steps can be broken down into the following:

1. Equilibrate the NVT ensemble in three cycles (run each cycle for 500 ps with a time step of 0.001 ps).

   1.1 Heat the system gradually from 0 K to 300 K. Maintain the same constraints as above.

   1.2 Thermally equilibrate and relax the constraints. Maintain the temperature at 300 K. Lower the constraint on the ions to 1 kcal/(mol·Å^2) and those on the backbone and side chains of the protein to 5 kcal/(mol·Å^2).

   1.3 Remove the constraints on the ions completely while maintaining those on the protein at 5 kcal/(mol·Å^2).

2. Equilibrate the NPT ensemble in six cycles (keep the constraints on the protein backbone at 5 kcal/(mol·$Å^2$); run each cycle for 250 ps with a time step of 0.001 ps):
   2.1 Maintain a constraint of 5 kcal/(mol·$Å^2$) on the side chains.
   2.2 Relax to 2.5 kcal/(mol·$Å^2$).
   2.3 Relax to 1 kcal/(mol·$Å^2$).
   2.4 Relax to 0.5 kcal/(mol·$Å^2$).
   2.5 Relax to 0.1 kcal/(mol·$Å^2$).
   2.6 Completely remove the constraint on the protein side chains.
3. Equilibrate the NPT ensemble in five cycles (run each cycle for 1000–3000 ps with a time step of 0.001 ps):
   3.1 Relax the constraint on the protein backbone to 2.5 kcal/(mol·$Å^2$) and run a MD for 1000 ps.
   3.2 Relax to 1 kcal/(mol·$Å^2$) and run a MD for 1000 ps.
   3.3 Relax to 0.5 kcal/(mol·$Å^2$) and run a MD for 1000 ps.
   3.4 Relax to 0.1 kcal/(mol·$Å^2$) and run a MD for 1000 ps.
   3.5 Completely remove all constraints on the system and run a MD for 3000 ps.

3.4 Production Run

Once the equilibration is completed, use the generated coordinates and velocities of the system to propagate TePIxJg for hundreds of nanoseconds or even longer time, depending on the purpose of the simulation. Analyze the results by calculating the root mean square fluctuation (RMSF) or the root mean square deviation (RMSD) (Fig. 3). Properties such as flexibility and side chain orientation can be readily extracted from these classical simulations.

In this example, we performed classical MD simulations for 1500 ns and monitored the structural stability by analyzing the RMSD with respect to the structure at the starting point of the production run. This serves two purposes: (i) understanding the effect of various protonation models on the system and (ii) providing starting points for hybrid QM/MM simulations. Figure 3 shows the RMSD variation of the entire protein, which is further broken down into helices, loops, and the PVB chromophore over the simulation time. The helices show the most structural fluctuations, while the chromophore shows the least. This is attributed to the difference in the degree of flexibility between them. A large fluctuation occurs around 125 ns, which is more pronounced for the protein and helices compared to the other two. This takes place due to a change in the orientation of the C-terminal, which is observed by monitoring the structural distortions along the trajectory. After 125 ns, the protein stabilizes, and

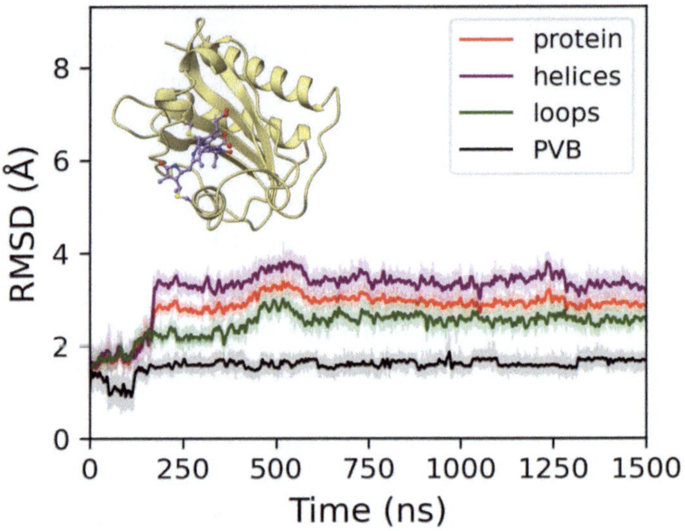

Fig. 3 RMSD of the protein, helices, loops, and the PVB chromophore from the MD simulation

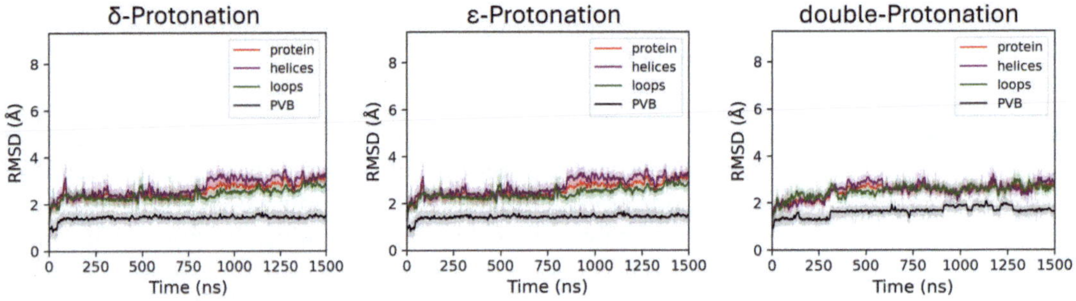

Fig. 4 RMSD of the protein, helices, loops, and the chromophore PVB from the MD simulation for different protonation models of His523

the structural deviations are minimal till the end of the simulation period. The RMSD variation of the PVB chromophore originates mainly from the movement of the propionate groups.

Such an analysis is also useful to understand the effect of protonation of a nearby residue that directly interacts with the chromophore. The side chain of His523 interacts with PVB, and depending on the protonation state of His, the stability of the chromophore can be affected (Fig. 1). Figure 4 shows the RMSD variation under different protonation models of His523. When it is protonated in the ε-position, the PVB shows the highest structural deviations compared to the protonation in the δ-position or the double protonated model. This depicts that the system is unstable compared to the other two models. Hence, the model where

His523 is protonated in the ε-position can be discarded from further calculations. The remaining two protonation models can be used for the calculation of spectroscopic properties using the hybrid QM/MM method as a further critical validation of the model.

3.5 Hybrid Quantum Mechanics/Molecular Mechanics Simulations

In order to consolidate the results from force field simulations or to compute properties in the excited state, a hybrid QM/MM method is used after the production run [13]. The basic idea of the QM/MM method is to divide the simulated system into two regions with different resolutions. The QM region has the highest resolution which includes nuclei and electrons. The description of the electrons allows to compute excited states.

The key step in QM/MM simulations is to partition the total system in two regions (Fig. 5). Typically, the chromophore is chosen to be part of the QM region, while the remaining protein and solvent is part of the MM region (*see* **Note 3**). The QM region can be described using density functional theory (DFT) or wave function-based methods. Refer to **Note 4** for considerations about the choice of the quantum chemical method. The MM region is typically described by a classical force field, but recently, polarizable force fields have also been used. In our example, we use a classical AMBER force field to describe the MM region.

3.6 Calculation of Spectra

Using hybrid QM/MM simulations, it is possible to compute spectral properties such as vibrational or UV/Vis absorption spectra. Vibrational spectra are obtained from the normal mode analysis, which requires the minimum energy of the ground state. However, this is not always accessible and several equivalent minima are present on the complex potential energy surface. An alternative method is to obtain the spectra from an MD trajectory. This requires to run an MD simulation, record the dipole moments

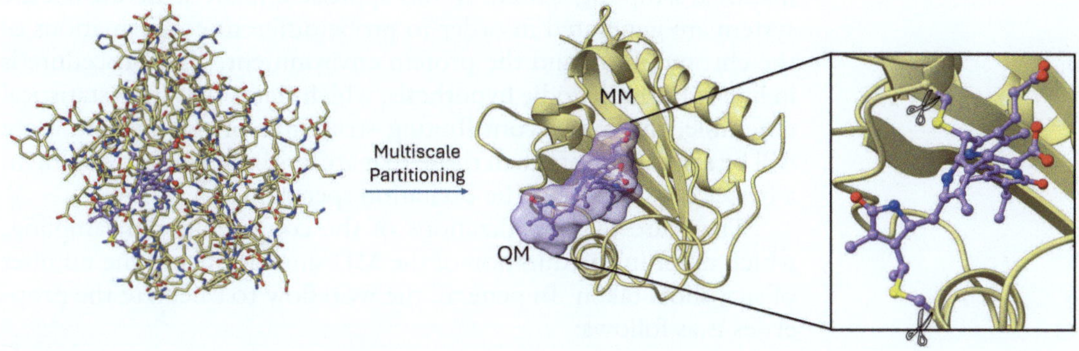

Fig. 5 QM/MM partitioning of the TePlxJg protein. The MM region of the protein is visualized in a cartoon representation. The QM region is shown as electron density. Covalent bonds that are at the boundary between the QM and MM regions are indicated by scissors

along the trajectory, and Fourier-transform them to produce the vibrational spectrum. This protocol includes anharmonic corrections but lacks the information of the normal modes, which makes the assignment cumbersome.

The description of TePIxJg using the hybrid QM/MM method allows the construction of a realistic model in terms of chromophore–protein interactions. However, the calculation of an absorption spectrum for such a high-dimensional system becomes more complicated compared to an isolated chromophore in the gas phase. For an isolated chromophore, a spectrum is determined from a single conformation of the chromophore corresponding to the ground-state equilibrium geometry. A single structure provides one set of excitation energies, which can be visualized in a so-called stick spectrum (Fig. 6a).

To obtain the stick spectrum, perform a calculation of the excitation energies as implemented. In this example, ORCA has been used with the energies computed using the second-order Algebraic Diagrammatic Construction method (ADC[2]) with the correlation-consistent polarized double-zeta(cc-pVDZ) basis set. The height of the stick is determined from the oscillator strength that is related to the experimental extinction coefficient. Often, the resulting stick spectrum can be convoluted by Gaussian to obtain a broadening that resembles to an experimental spectrum.

This approximation only yields reliable results for systems with low-dimensional potential energy surfaces (PES). For small systems, the global minimum of the surface can be obtained during structural relaxation, giving rise to one dominant conformation of the chromophore. However, for systems with high dimensionality, ensuring the localization of a minimum is unviable given the large number of degrees of freedom. Systems in complex environments, such as protein-embedded chromophores or solvated molecules, are often poorly described by a single minimum.

To address the issue of multiple coexisting conformers, conformational sampling is used. In this approach, many structures of the system are generated in order to probe different conformations of the chromophore and the protein environment. This procedure is in line with the ergodic hypothesis, which states that for a statistical ensemble, the major contributing structures of the chromophore will be sampled more than those that are less important, resulting in a better description of the excitation spectrum.

There are many realizations of the conformational sampling, which differ in the duration of the MD simulation and the number of snapshots taken. In general, the workflow to calculate the properties is as follows:

1. Run a ground-state MD simulation to heat and equilibrate the protein for 10 ps with a time step of 0.001 ps.

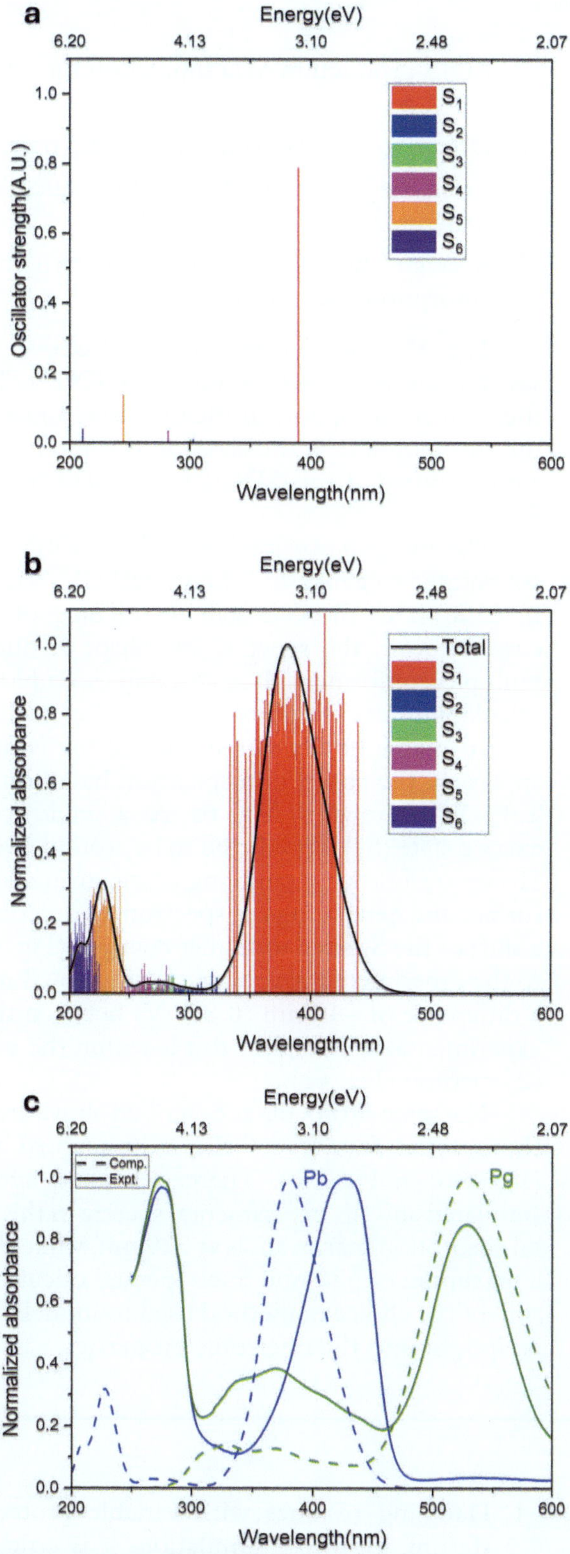

Fig. 6 Calculated UV/Vis absorption spectra of TePlxJg. (**a**) Stick spectrum based on a single geometry of TePlxJg. (**b**) Spectrum computed from 100 snapshots. (**c**) Computational and experimental UV/Vis absorption spectrum of TePixJg in Pb and Pg states. Excitation energies are computed at the ADC(2)/cc-pVDZ level of theory. Experimental UV/Vis spectra from Burgie et al.

2. Run a production MD simulation for 10 ps with a time step of 0.001 ps.
3. Obtain snapshots from the resulting trajectory at 1-ps intervals.
4. Compute the vertical excitation energies for each sampled structure.
5. Average the energies over the structures to generate the absorption spectrum.

The MD simulations described above are run with ORCA software using the semi-empirical GFN2-xTB method by treating the PVB chromophore at the QM level and a region of 5 Å around the chromophore at the classical MM level. The excitation energies are computed at the ADC(2)/cc-pVDZ level of theory (*see* **Note 5**).

The energies obtained from the ensemble of conformers can be convoluted to generate a theoretical UV/Vis absorption spectrum. In contrast to the Gaussian broadening of a stick spectrum discussed above, the spectral line shape is due to the sampling of multiple conformers. Both strategies employed in sampling and calculating the excitations will alter the produced spectra.

To highlight the significance of the conformational sampling approach, the absorption spectrum based on a single geometry of TePIxJg is shown in Fig. 6a. At a single conformation, the first excited state (S_1) is predicted to be around 390 nm for the Pb form. However, following sampling of the ground-state potential energy surface and generating the spectrum as described above, we observe a shift of the S_1 state to higher energies (Fig. 6b). We compare this to the experimental spectrum obtained by Burgie et al. [9]. There is a difference of ~40 nm (0.35 eV) between the computed and the experimental spectra, but this is within the error limit of the ADC (2) method (Fig. 6c).

The same protocols as described above are also used to simulate the spectral features of the light-adapted structure of TePixJg (Pg state in Fig. 6c). There is a good agreement between the simulated and the experimental spectra in this case. The experimental absorption maximum is at 529 nm, while the simulations predict a maximum of 530 nm. Such spectral calculations provide a validation of the choice of method used to model the system and provide an insight into the electronic properties.

4 Notes

1. Handling residues with variable protonation states. Before starting a protein simulation, it is crucial to add protons to the structure. Titratable residues can be protonated differently based on the experimental conditions. It is important to find

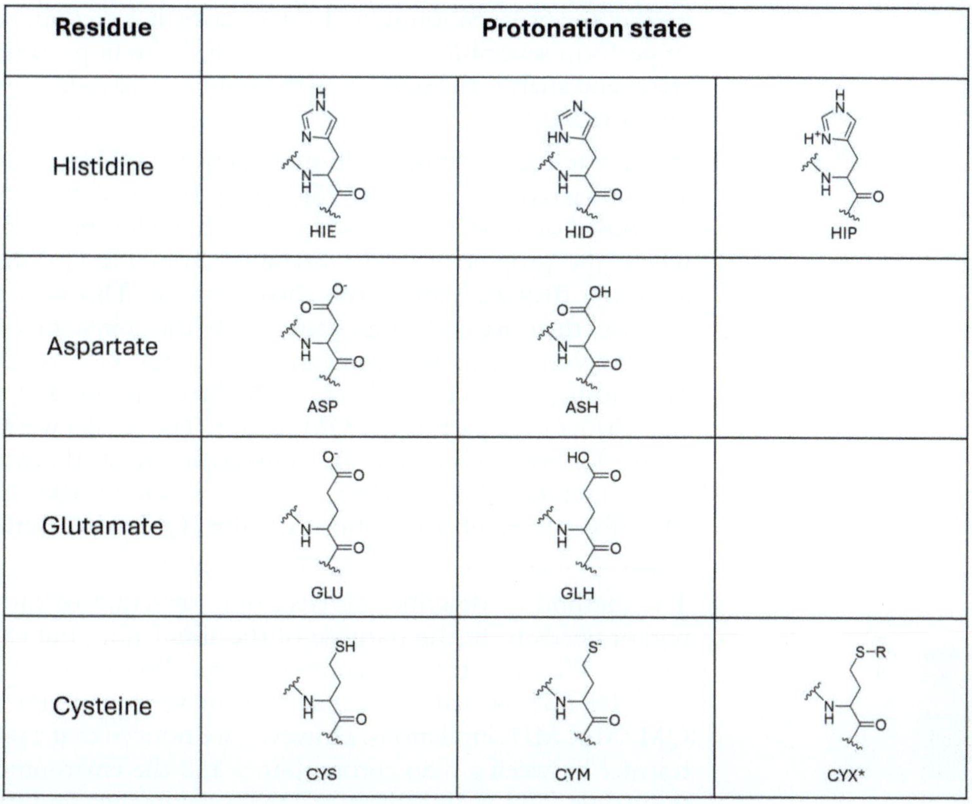

Fig. 7 Residues with variable protonation states and their corresponding AMBER convention. CYX is used to denote disulfide bonds

the right protonation state based on the pH condition used in experiments. For example, histidine can be protonated in three different ways. Such titratable residues have the same residue name in the experimental structure, but different names in the AMBER convention and are associated with different classical force fields. This issue of nomenclature does not arise in crystallographic/cryo-EM structures since the protonation state cannot be easily assigned except for protein structures at ultra-high resolution. The residues with variable protonation states and their corresponding AMBER conventions are shown in Fig. 7. To enable the correct protonation state in the system, the residue name needs to be changed to the corresponding convention. For example, if histidine needs to be protonated at the δ-position, then the residue name should be changed from the three-letter code HIS to HID.

2. Determining the protonation state of titratable groups is generally done using tools like PROPKA [14] or H++ [15], but this requires force field parameters of all molecules. Since the PVB is not a standard residue, these tools will not be able to

predict the protonation state. In such cases, it is recommended to perform several MD simulations with different protonation states and analyze the stability of the system. This is discussed in Subheading 3.4.

3. We have tested different partitioning between QM and MM for a cyanobacteriochrome Slr1393g3 [16]. In a small QM region composed of 66 atoms, we used a truncated bilin chromophore where the propionates were excluded from the QM region although they are part of the chromophore. This is justified because they are not conjugated with the π-system of the chromophore. This was compared to a large QM region of 106 atoms, where the full-size bilin chromophore and three side chains were part of the QM region. The results were very similar for both QM region sizes and did not justify the increase in computational time when going from 66 to 106 atoms. Therefore, careful calibration of the QM/MM setup is required, especially if a new system is studied.
4. The method to describe the electronic structure of the QM region needs to fit the purpose of the simulation, but should also be efficient in terms of computer time. Previously, we have used DFTB2, which is a semiempirical version of DFT for QM/MM MD simulations. However, we noticed that a proton transfer between a bilin chromophore and the environment is occurring. This is an artifact of DFTB parameters and can be avoided if DFT is used instead.
5. When it comes to the calculation of excitation energies, the choice of quantum chemical method is essential. One of the most popular methods for the calculation of excitation energies is time-dependent DFT (TD-DFT). However, it fails in case of bilin chromophore because it can lead to artificial charge transfer states between the carboxylic groups of the propionates and the π-system of the tetrapyrrole. Therefore, a careful benchmark of electronic structure methods for excitation energies is required [17].

Acknowledgments

I.S. acknowledges support by the Deutsche Forschungsgemeinschaft (DFG, German Research Foundation) under Germany's Excellence Strategy EXC-2033-390677874—RESOLV funded by the German Federal Ministry of Education.

References

1. Warshel A, Levitt M (1976) Theoretical studies of enzymic reactions: dielectric, electrostatic and steric stabilization of the carbonium ion in the reaction of lysozyme. J Mol Biol 103(2):227–249
2. Warshel A (1976) Bicycle-pedal model for 1st step in vision process. Nature 260(5553): 679–683
3. Li H, Burgie ES, Gannam ZTK et al (2022) Plant phytochrome B is an asymmetric dimer with unique signalling potential. Nature 604(7904):127–133
4. Bódizs S, Mészáros P, Grunewald L et al (2024) Cryo-EM structures of a bathy phytochrome histidine kinase reveal a unique light-dependent activation mechanism. Structure 32(11):1952–1962.e3
5. Jumper J, Evans R, Pritzel A et al (2021) Highly accurate protein structure prediction with AlphaFold. Nature 596(7873):583–589
6. Baek M, DiMaio F, Anishchenko I et al (2021) Accurate prediction of protein structures and interactions using a three-track neural network. Science 373(6557):871–876
7. Krishna R, Wang J, Ahern W et al (2024) Generalized biomolecular modeling and design with RoseTTAFold All-Atom. Science 384(6693):eadl2528
8. Abramson J, Adler J, Dunger J, Evans R et al (2024) Accurate structure prediction of biomolecular interactions with AlphaFold 3. Nature 630(8016):493–500
9. Burgie ES, Clinger JA, Miller MD et al (2020) Photoreversible interconversion of a phytochrome photosensory module in the crystalline state. Proc Natl Acad Sci 117(1):300–307
10. Case DA, Betz RM, Cerutti DS et al (2016) AMBER 2016. University of California, San Francisco
11. Case DA, Aktulga HM, Belfon K et al (2023) AmberTools. J Chem Inf Model 63(20): 6183–6191
12. Ruf J, Bindschedler F, Buhrke D (2023) The molecular mechanism of light-induced bond formation and breakage in the cyanobacteriochrome TePixJ. Phys Chem Chem Phys 25(8): 6016–6024
13. Senn HM, Thiel W (2009) QM/MM methods for biomolecular systems. Angew Chem Int Ed 48(7):1198–1229
14. Olsson MHM, Søndergaard CR, Rostkowski M, Jensen JH (2011) PROPKA3: consistent treatment of internal and surface residues in empirical pKa predictions. J Chem Theory Comput 7(2):525–537
15. Anandakrishnan R, Aguilar B, Onufriev AV (2012) H++ 3.0: automating pK prediction and the preparation of biomolecular structures for atomistic molecular modeling and simulations. Nucleic Acids Res 40(W1):W537–W541
16. Wiebeler C, Rao AG, Gärtner W, Schapiro I (2019) The effective conjugation length is responsible for the red/green spectral tuning in the cyanobacteriochrome Slr1393g3. Angew Chem Int Ed 58(7):1934–1938
17. Wiebeler C, Schapiro I (2019) QM/MM benchmarking of cyanobacteriochrome Slr1393g3 absorption spectra. Molecules 24(9):1720

Chapter 9

Using Computational Techniques for the Characterization of Structural Changes During the Phytochrome Photocycle

Giacomo Salvadori

Abstract

Phytochromes are red/far-red light-sensing photoreceptor proteins that undergo reversible structural changes when absorbing light. Upon photon absorption, structural changes in the bilin chromophore (an open-chain tetrapyrrole) propagate initially to its binding pocket and subsequently throughout the entire protein, ultimately triggering biological function. This chapter presents a comprehensive multiscale strategy for tracking the entire sequence of events, beginning with the initial electronic excitation in the embedded chromophore up to the large-scale structural changes in the protein. The approach integrates nonadiabatic quantum mechanics/molecular mechanics (QM/MM) molecular dynamics (MD) simulations with fully classical (MM) MD and enhanced sampling techniques. All key steps are provided in a practical, implementation-focused format: from system setup, force-field parametrization, and excited-state simulations, to long-time classical simulations that explore global protein conformational changes. While our examples focus on bacteriophytochromes, the computational pipeline described here can be adapted to any photoactive protein to elucidate its light-driven reaction coordinates.

Key words Photochemistry, Nonadiabatic dynamics, Surface hopping, QM/MM, Molecular dynamics, Enhanced sampling techniques, Phytochromes, Bilin

1 Introduction

Phytochromes are a widespread family of photoreceptor proteins found in plants, bacteria, and fungi, where they regulate numerous light-dependent physiological processes [1–4]. Their remarkable ability to function as biological light switches makes them not only fascinating subjects for basic research but also promising tools for optogenetic applications in synthetic biology [2, 5–11]. Structurally, phytochromes exist as homodimers, with each ~ 125-kDa monomer comprising an N-terminal photosensory module (PSM, typically PAS–GAF–PHY domains) covalently

Andreas Winkler (ed.), *Phytochromes: Methods and Protocols*, Methods in Molecular Biology, vol. 2970,
https://doi.org/10.1007/978-1-0716-4791-2_9,

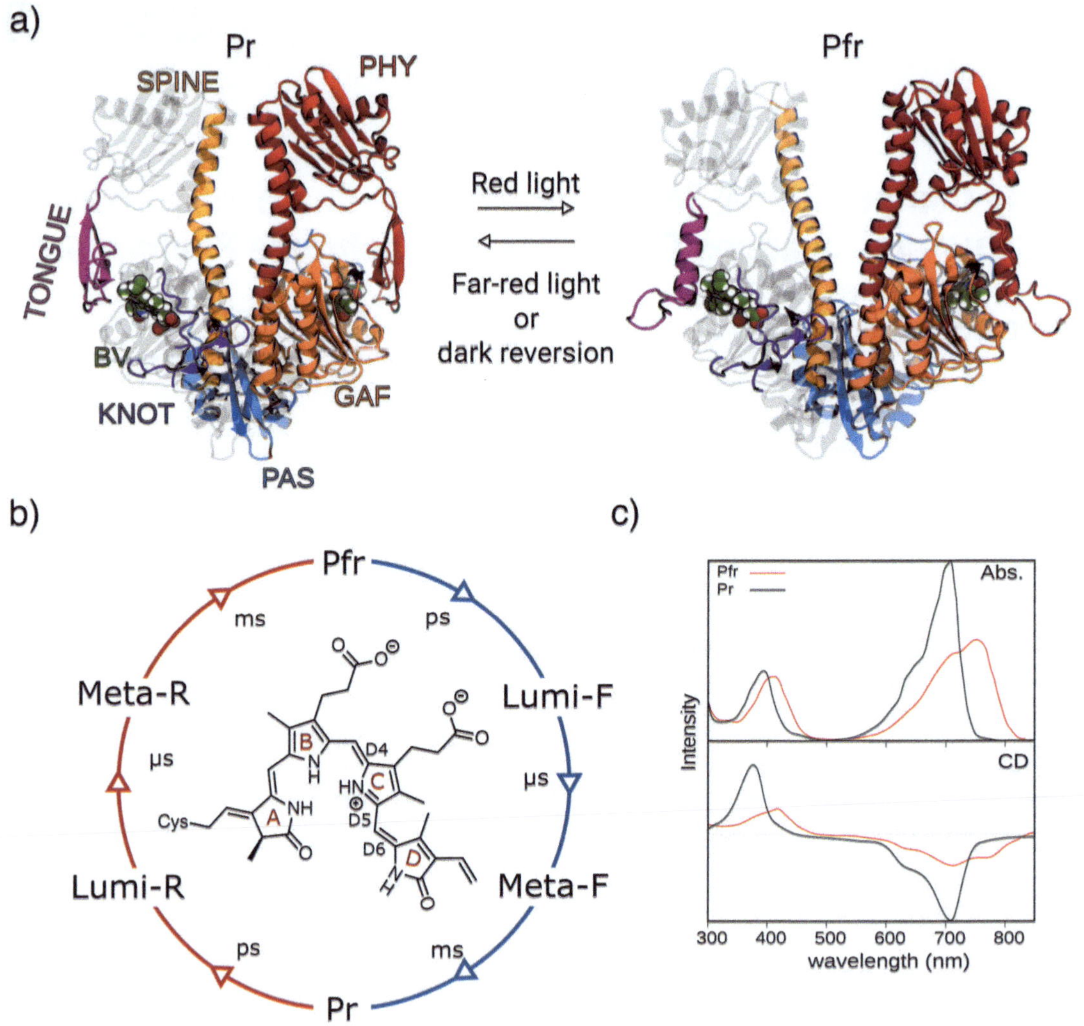

Fig. 1 (**a**) Crystal structure of Pr (PDB ID:4Q0J [23]) and Pfr (PDB ID:5C5K [24]) forms for the *Deinococcus radiodurans* (DrBphP) photosensory module (PSM), highlighting the key domains. The biliverdin IXα chromophore (BV) is represented in green. (**b**) In canonical and bathy phytochromes, Pr and Pfr are the stable dark states, respectively. After light absorption, canonical phytochromes follow the red pathway, while bathy phytochromes follow the blue pathway. The chemical structure and definition of the main dihedral angles for the embedded biliverdin are also shown. The timescales for the different steps are the ones reported in the literature [25]. This panel is adapted from ref. 26. (**c**) Measured absorption and CD spectra of the two forms for *Deinococcus radiodurans* bacteriophytochrome

bound to an open-chain tetrapyrrole (bilin) chromophore, and a C-terminal output module (often a histidine kinase or other signaling domain).

A hallmark of phytochromes is their photoreversible switch between two stable states with distinct spectra and biological activity (Fig. 1a, c): the *Pr* state, which absorbs red light, and the *Pfr* state, which absorbs far-red light [12–15]. These two states differ in both spectroscopic properties and the configuration of the bound bilin chromophore. In the *Pr* state, bilin adopts a ZZZssa configuration, whereas in the *Pfr* state it is ZZEssa (Fig. 1b).

Based on their resting state thermodynamics, phytochromes are classified into two distinct categories. In bathy phytochromes (e.g., *Agrobacterium fabrum*, Agp2), *Pfr* is the thermodynamically favored dark state, while *Pr* is photoinduced [16–18]. By contrast, canonical phytochromes (e.g., *Deinococcus radiodurans*, DrBphP) have *Pr* as the dark state. This difference stems from variations in the chromophore-binding pocket that modulate protonation states of the bilin propionates and local hydrogen bonds [19–22].

Despite these differences in dark-state energetics, both classes share a similar primary photochemistry: Upon photon absorption, the bilin chromophore undergoes an isomerization at the D-ring (specifically at the D6 dihedral, Fig. 1b). This local change leads to early photoproducts (Lumi-R or Lumi-F) and subsequent thermal relaxations (Meta-R or Meta-F). Eventually, large-scale protein rearrangements occur, for example, the PHY "tongue" refolds from a β-hairpin to an α-helix, affecting the dimer interface and generating the activated signaling state (Fig. 1a). The reverse Pfr $\rightarrow$ Pr transition [27] may be driven by far-red light or by slow thermal reversion in darkness.

Despite recent advances in X-ray crystallography and cryo-EM that have captured snapshots of phytochromes in their end states, the transient intermediates and detailed pathways of the Pr-to-Pfr conversion remain only partially understood. Ultrafast spectroscopic studies have revealed intermediate states on pico- to millisecond timescales [2, 25, 28–34], but resolving their structures and the full mechanism requires methods that bridge disparate timescales and resolutions. The phytochrome photoactivation process spans multiple timescales, from femtoseconds (light absorption) to picoseconds (excited-state dynamics), microseconds (protein pocket rearrangements), and milliseconds or longer (larger conformational changes). This range of timescales presents a fundamental challenge: No single computational technique can capture the full photocycle.

To address this challenge, a multiscale computational strategy is needed (Fig. 2). Quantum mechanics (QM) describes the photoexcited chromophore. However, pure QM alone cannot handle the entire protein environment. Classical molecular dynamics (MM MD) can simulate the protein relaxation and large conformational changes, but standard force fields do not model electronically excited states or bond-breaking events. Hybrid QM/MM methods bridge this gap by treating the chromophore (and possibly nearby residues) quantum-mechanically, while the rest of the protein and solvent are treated with molecular mechanics. Furthermore, nonadiabatic QM/MM molecular dynamics techniques are needed to simulate the actual photochemical process. This chapter focuses on

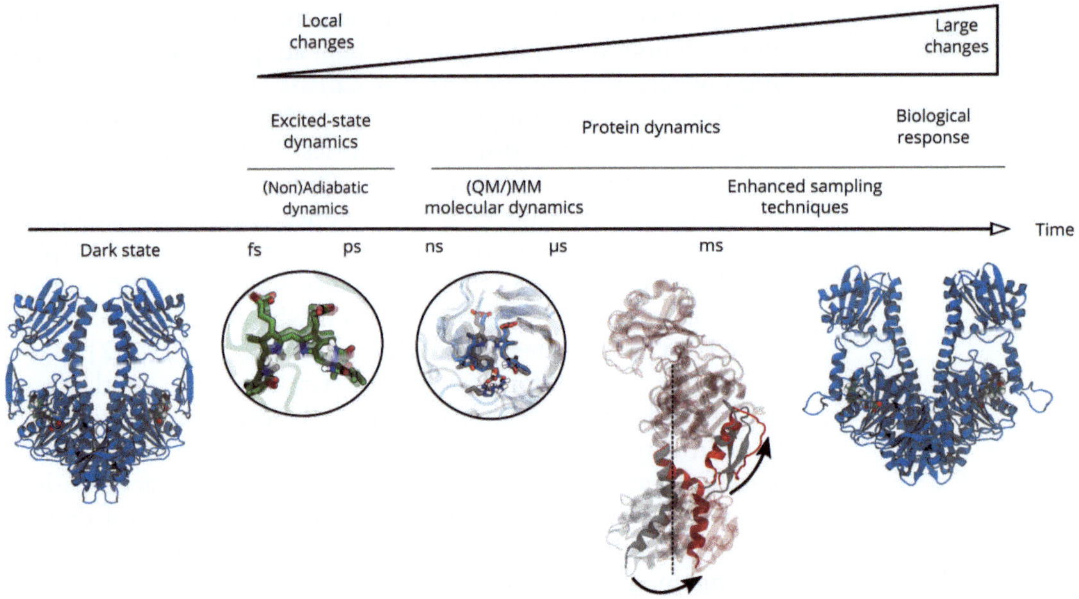

Fig. 2 Multiscale strategy applied for modeling the photoactivation mechanism of phytochromes. The figure illustrates the progression from dark state (left) through excited-state dynamics (QM/MM with nonadiabatic surface hopping, fs-ps timescale), followed by protein conformational dynamics (classical MM MD, ns-μs timescale), and finally large-scale structural changes (enhanced sampling techniques, ms timescale). Each method is mapped to its appropriate timescale and the corresponding phase of the photocycle

the surface hopping algorithm [35, 36] pioneered by Tully, which allows trajectories to stochastically hop between electronic states, capturing the radiationless decay from excited states. Finally, because the later steps of the photocycle (e.g., domain movements, tongue refolding) occur on micro- to millisecond timescales, enhanced sampling techniques are needed to accelerate rare events to characterize intermediate states that would be inaccessible within limited simulation time.

The combination of different computational methods needed to describe such cascade of events covering multiple timescales has as its main drawback the risk of potentially overlooking specific steps, or introducing artifacts, especially in densely packed binding pockets like those in phytochromes (*see* **Note 1**). To minimize these risks, a continuous connection between the experimental and computational investigations is of paramount importance. Computational models provide insights and predictions that guide experimental research. In turn, experimental data validate the computational models and enable their refinement.

This chapter provides a detailed protocol for this multiscale approach, including (i) system setup and force-field parametrization of the bilin, (ii) choice of QM methods for excited states, (iii) nonadiabatic surface hopping simulations, (iv) enhanced sampling for late-stage conformational changes, and (v) validation against experimental data (spectra, crystallographic structures, IR signals, etc.).

Despite the focus on bacteriophytochromes, the methodology is broadly transferable to other photoactive proteins, offering an atomistic window into how light drives large-scale protein conformational changes.

2 Materials

To study structural changes during the phytochrome photocycle, the following materials and resources are required:

1. **Computational resources**: Access to a high-performance computing (HPC) cluster or GPU-equipped nodes is typically necessary (*see* **Note 2**).
2. **Molecular dynamics package**: A simulation program capable of both MM MD and QM/MM. This protocol focuses on AMBER [37], but other engines (GROMACS [38], Cp2K [39], etc.) can be used with appropriate adaptations.
3. **Quantum chemistry software**: Programs for performing ab initio or semiempirical QM calculations, such as Gaussian [40], ORCA [41], or MOPAC [42]. For on-the-fly photodynamics, codes implementing Tully's surface hopping are required. This protocol focuses on MOPAC-PI [43].
4. **Structural data**: Atomic coordinates of the phytochrome in one of its stable states (Pr or Pfr) as a starting model. These coordinates are typically obtained from the Protein Data Bank. Examples include PDB ID 4Q0J [23] (DrBphP photosensory module in Pr state) and 5C5K [24] (DrBphP in Pfr state).
5. **Experimental data for validation**: Published spectral data (UV–Vis absorption peaks, circular dichroism, resonance Raman, or FTIR band positions) and available structural data (distance changes, known intermediate spectra). *See* **Note 3**.

3 Methods

The following protocol delineates the steps involved in simulating the phytochrome photocycle, from preparing the dark state to capturing photochemistry via QM/MM, and finally modeling slow structural relaxations (Fig. 2). While the protocol is presented as a linear sequence, certain steps can be executed iteratively or in parallel. For instance, one might test different QM methods or perform multiple independent MM MD runs concurrently.

3.1 Assemble the Initial Model

1. Begin with a high-resolution crystal structure of the phytochrome in its initial dark state (Pr for canonical or Pfr for bathy). If an intact dimer is biologically relevant, use the full dimeric assembly (*see* **Note 4**). Include any crystallographic waters or cofactors near the chromophore, as they can be functionally important. If no structure is available, construct a homology model or use de novo structural prediction (e.g., AlphaFold [44, 45]).
2. Rebuild any missing loops or residues using homology modeling tools (e.g., MODELLER [46]). Ensure that the bilin linkage to the protein is correctly formed.
3. Parameterize the bilin chromophore using antechamber/GAFF [47] or a similar approach. Derive partial charges via RESP [48] fits to a QM electrostatic potential (e.g., B3LYP/6-31G**). *See* **Note 5**.
4. Assign protonation carefully, especially for the bilin propionates and the histidine interacting with the D-ring. Consult different pKa predictions (e.g., H++ server, MCCE, Epik Schrodinger) and experimental data (*see* **Note 6**).

3.2 Obtain a Proper Sampling

1. Solvate the protein in a suitable water model (e.g., TIP3P, OPC, etc.). Add ions to neutralize or reach physiological ionic strength. Minimize in stages, gradually releasing positional restraints on the backbone and chromophore. Heat from 0 to 300 K under NVT, and then switch to NPT for density equilibration. Finally, run multiple independent MD replicas (each several microseconds long) of the dark-state phytochrome to thoroughly sample its conformational space.
2. Cluster the trajectory based on chromophore-pocket interactions or overall protein conformation and pick representative structures of the dominant clusters. This approach ensures that the subsequent excited-state simulations start from a thermally representative ensemble of protein/chromophore configurations that reflect the equilibrium distribution of the dark state rather than a single arbitrary structure.
3. To confirm the reliability of the equilibrated structures, compute an in silico absorption/CD spectrum of BV (e.g., via TD-DFT on extracted snapshots) and compare it to the known experimental one. A close match suggests that the environment and chromophore parameters are realistic [49, 50]. *See* **Note 7** for details on spectral calculations and validation metrics.

3.3 QM/MM Setup for Photoexcitation

1. Define the QM region (minimally the chromophore) and MM region (rest of the system). A common approach is to cut the bond between the chromophore and the cysteine with a link atom, capping the chromophore.

2. Choose an appropriate QM method for describing the chromophore's excited states. The photoisomerization involves breaking and reforming conjugation, so a single-reference DFT method can be unreliable. Therefore, a multireference method is needed. Define an active space that includes the π and π^* orbitals of the bilin. At minimum, include the ground state (S_0) and first excited singlet state (S_1) for a basic photoisomerization, but it may be necessary to include higher excited states (S_2, S_3, etc.) if they are close in energy or participates in the dynamics. More information about method selection and validation is provided in **Notes 8** and **9**.
3. Perform short (20–50 ps) ground-state QM/MM MD to relax the QM region within the MM environment. This step accounts for any minor geometric adjustments of the chromophore when treated with a higher-level method than the classical force field. It allows the chromophore's electron distribution to adjust and can reveal if the force-field geometry was somewhat inaccurate. This equilibration must be performed at the same level of theory as the subsequent nonadiabatic dynamics to ensure consistency. An example of an equilibration input is reported in **Note 10**.

3.4 Nonadiabatic Surface Hopping Excited-State Dynamics

1. Sample initial conditions (positions and velocities) from the equilibrated ground-state distribution to mimic excitation from a vibrationally "hot" ground-state ensemble. Select structures according to transition dipole moment and initiate independent trajectories with vertical excitation to S_1. If you use MOPAC-PI [43], useful script to sample the initial conditions can be found in the GitLab repository. Initial momenta can also be sampled from a Wigner distribution of the S_0 vibrational ground state, but using classical thermal velocities is a common and reasonable approximation.
2. Launch a swarm of surface hopping trajectories on the excited state (an input file is shown in **Note 14**). The nuclear positions evolve according to forces from the currently occupied electronic state, while the electronic wavefunction is propagated concurrently. A time step of approximately 0.1 fs is typically necessary to accurately integrate fast motions and electronic transitions (*see* **Note 11**). Use Tully's fewest-switches surface hopping to allow transitions between states (*see* **Note 12**). Apply a decoherence correction, such as the energy-based correction. Within this method, we need to define an empirical parameter, whose recommended value is 0.1 hartree. More details are reported in **Note 13**.
3. Run a sufficient number of trajectories (hundreds to thousands) to obtain statistically meaningful results (*see* **Note 15**). Key outcomes to analyze include: (i) photoisomerization

yield: the fraction of trajectories that successfully isomerize to the E configuration; (ii) hop time distribution: fit to an exponential to estimate the excited-state lifetime; (iii) reaction coordinate analysis: track dihedral angles, bond lengths, and other key parameters throughout the isomerization process; (iv) quantum yield validation: compare computational results to experimental quantum yield measurements.

4. Collect final structures from trajectories that end in a "hot" 15E-like configuration (Lumi-R or Lumi-F).

3.5 Ground-State Relaxation to Later Intermediates

1. After the hop to the ground state, it is helpful to continue QM/MM for a few picoseconds/nanoseconds on the ground state to let excess vibrational energy dissipate locally. This ensures the immediate environment around the chromophore is at least partially relaxed at the same level of theory before transitioning to longer timescale methods.
2. For longer μs-timescale MDs, switch to fully MM MD with the chromophore in the 15E geometry. Monitor key interactions that change during the photocycle, such as hydrogen bonds to the D-ring and salt bridges in the binding pocket (*see* **Note 16** for specific interactions that serve as reliable markers of conformational progression).
3. If large-scale changes (tongue refolding, domain bending) do not appear spontaneously within accessible simulation timescales, apply enhanced sampling methods such as metadynamics [51] or Gaussian accelerated MD [52] to accelerate the crossing of energy barriers and explore conformational changes that would otherwise require milliseconds to observe.
4. Validate simulated intermediates against available experimental data, including FTIR frequencies (particularly the ring-D carbonyl IR shift), IR difference spectra between dark and photoactivated states, NMR chemical shifts if available, or structural data from crystal structures of intermediates (e.g., Meta-F state [53]). This validation step is crucial for establishing confidence in your computational model. *See* **Note 17** for specific spectroscopic markers that are most diagnostic of successful intermediate capture.

4 Notes

1. Another significant limitation of combining these computational methods is the extreme computational cost. A single nonadiabatic simulation, even at the semiempirical level, can

require more than a month of computation time. When studying processes with low quantum yields, hundreds (sometimes thousands) of simulations are required to collect adequate statistics and gain atomistic details about both the reaction mechanism and the role of the environment. The same problem applies to enhanced sampling simulations, where you may need to wait several weeks before determining whether a method is effective. Unfortunately, there is no simple solution to these computational challenges. One promising avenue is the utilization of machine learning (ML) strategies. The substantial reduction in computational cost achievable with ML-based approaches could enable tracking of light-induced structural changes within the embedded chromophore, extending beyond the photoproduct to encompass subsequent evolution in the ground state, including the formation of intermediates. However, machine learning heavily relies on the quality and quantity of training data. Such data must be accurate and comprehensive to produce reliable models, which still demands substantial computational resources for data collection and curation. Moreover, the "black-box" nature of many ML-based algorithms makes it challenging to interpret their predictions and understand the underlying physics. While machine learning can be a valuable tool, it is crucial to remain mindful of these critical issues and not rely on it as a silver bullet solution for speeding up molecular dynamics simulations.

2. Nonadiabatic QM/MM simulations are computationally intensive, and multiple trajectories may be required to achieve statistical significance. Microsecond-scale classical MD simulations also benefit greatly from GPU acceleration.

3. The frequency of the D-ring C=O stretch in the IR spectrum serves as a sensitive indicator of hydrogen bond changes. This feature can be used to validate whether the simulation's intermediate has placed the chromophore's carbonyl in an environment similar to experimental observations. If possible, also obtain reference Pr and Pfr spectra or structures to compare against simulation predictions (e.g., compute a difference IR spectrum between simulated Pr and Pfr states and compare it to experimental difference spectra). Critical validation metrics: absorption maxima (λ_{max}) of Pr and Pfr states (typically 650–700 nm vs. 700–750 nm); quantum yield (typical range for phytochromes: 0.10–0.16); D-ring C=O stretch frequency shift from $\sim 1730\ cm^{-1}$ to $\sim 1710\ cm^{-1}$ upon isomerization; CD spectrum shape and sign changes; tongue refolding timescale ($\sim$ milliseconds). Experimental techniques particularly useful for validation: time-resolved IR spectroscopy (can detect

changes in chromophore and protein on ps to ms); transient absorption spectroscopy (provides excited-state lifetimes, ~ 50–800 ps); resonance Raman spectroscopy (probes chromophore configuration).

4. Bacteriophytochromes (and phytochromes in general) naturally function as homodimers, and certain global conformational changes, such as dimer opening or the relative movement of output domains, can only be observed if both subunits are present. However, many simulations focus on the monomeric photosensory core (PAS–GAF–PHY), which is often sufficient to capture local photochemistry (chromophore isomerization, tongue movement, etc.) because the chromophore and immediate responses reside in a single PSM. Still, a monomer simulation may miss cooperative or stabilizing effects provided by the dimer interface or overestimate movements that, in reality, are constrained by the opposing monomer. For example, in DrBphP, the Pfr state involves an increased separation of the PHY domains in the dimer—behavior that would be absent in a monomer-only model. Thus, depending on the experimental question, one may need to include the full dimer or even the output domains to capture later steps such as domain activation or large-scale conformational rearrangements. A common compromise is to simulate only the monomeric PSM for faster calculations when the main interest is local photoisomerization and early structural changes, and then switch to the dimer (or larger constructs) for investigating global events. One practical approach is to run fully MM MD simulations on the dimer to explore its configurational space, and subsequently focus on the monomeric PSM for detailed photochemical studies using more computationally intensive methods.

5. When parameterizing the chromophore, verify that torsional parameters are reasonable. Because photoisomerization will switch from 15Z to 15E, consider deriving parameters for both forms if you plan to do extended ground-state MD of the photoproduct. Pay special attention to: D-ring rotation barriers (often underestimated in generic force fields); planarity constraints in the tetrapyrrole system; proper improper torsions to maintain stereochemistry; propionate side chain flexibility. Sometimes GAFF assigns incorrect atom types for conjugate systems, so it manually corrects atom types or uses specialized parameters from literature.

6. The protonation of the chromophore and nearby residues can be critical. Experimental evidence suggests that in the Pr state,

the bilin chromophore of DrBphP is fully protonated at pH 8 [54], meaning both propionic acid side chains are deprotonated, while all nitrogens remain protonated. However, in the intermediates or in the Pfr state, the protonation pattern can differ significantly. If there is uncertainty, one strategy could be to simulate the protein with different chromophore protonation states to assess stability or compare calculated spectra to experiment. When assigning protonation to histidine residues, consider their role: for instance, a histidine that hydrogen bonds to a chromophore ring nitrogen might be protonated on the interacting nitrogen (HIE vs. HID tautomer) to strengthen the interaction. The H++ server and PROPKA can provide initial suggestions, but manual inspection and literature data should guide final choices for functionally important sites. Alternative approach: Perform constant pH MD simulations for key residues. For the D-ring surrounding histidines, wrong protonation states can lead to premature isomerization or unrealistic stabilization of intermediates.

7. The absorption lineshape is sensitive to the surrounding environment, which can affect both homogeneous and inhomogeneous broadening, through fast and slow dynamics. In this case, it is not enough to compare vertical transition energies with the experimental maximum. Instead, you must simulate the entire band shape, which, through couplings with vibrations and environment, provides critical information about the validity of your sampling. One successful approach is based on the spectral density formalism [55–58]. The spectral density is a function that describes the linear coupling of the excitation $g \rightarrow e$ to an infinite set of harmonic oscillators, with frequencies ω_k,

$$J_{eg}(\omega) = \pi \sum_k S_k^{(eg)} \omega_k^2 (\delta(\omega - \omega_k) - \delta(\omega + \omega_k)), \quad (1)$$

where $S_k^{(eg)}$ is the dimensionless displacement, or Huang-Rhys factor, for excited state e along the normal mode k. These quantities are calculated by projecting the excited-state gradients onto the normal modes. The SD can be obtained from a normal mode analysis of the chromophore in its ground-state optimized geometry. For each mode, compute the reorganization energy, which is used to reconstruct the SD that gives the homogeneous broadening. Additionally, static disorder (inhomogeneous broadening) must be included, due to the fluctuations of the environment. To compute the effect of inhomogeneous disorder on the absorption and CD spectra, the inhomogeneous distribution of both excitation energies and dipole strengths computed along the MD trajectory (after chromophore optimization) must be taken into account.

8. The accuracy of the photoexcitation simulation depends on an appropriate QM method for the chromophore's excited states. For phytochrome chromophores, multireference methods are typically required due to the conjugated system and the need to describe bond isomerization and potential conical intersections. A common choice is to use CASSCF (complete active space self-consistent field) to obtain qualitatively correct potential energy surfaces, possibly followed by second-order perturbation (CASPT2) for improved energetics. However, multireference configuration interaction methods, such as CASPT2 or MRCI, are computationally expensive for on-the-fly dynamics. In this framework, the floating occupation molecular orbitals approach used in a semiempirical context with a CASCI approach (FOMO-CI) [59–62] can be seen as a cheap substitute of state-averaged CASSCF (SA-CASSCF). This method has been successfully used to describe the excite-state dynamics of phytochromes [26, 63]. With respect to SA-CASSCF, FOMO-CI has the advantage that the results do not depend on the subspace of states that are optimized, and therefore is not prone to the root switching problem. A reparameterization of the semiempirical Hamiltonian is needed in the case of FOMO-CI, because most of the parameter sets in use (AM1, MNDO, etc.) have been optimized for restricted HF wavefunctions to represent ground-state properties. However, standard parameter sets provide good starting points and in most cases do not need to be fully reoptimized.
9. Semiempirical methods are orders of magnitude faster than first-principles multireference calculations and can thus support the propagation of dozens or even hundreds of trajectories. However, researchers must exercise caution: Semiempirical methods may not capture all details (such as subtle charge redistribution or state ordering) and should ideally be validated against higher-level calculations. Good validation practices include:
   (a) Computing the absorption spectrum of the chromophore and comparing it to either known experimental spectra or ab initio calculations. A reasonable match of peak positions and relative intensities (e.g., for Pr vs. Pfr states) builds confidence in the chosen method.
   (b) Checking that the method predicts the correct ordering of excited states. For biliverdin, the S1 should be a $\pi - \pi*$ excited state that leads to isomerization; some methods might incorrectly predict a different state ordering that would fundamentally alter the dynamics.

(c) Scanning the excited-state potential energy surface along the D6 dihedral rotation to map the isomerization pathway. Compare the obtained results with either experimental values or higher-level calculations. This approach also helps locate any conical intersection regions, geometries where ground and excited states become degenerate and nonradiative transitions become probable.

10. To run a ground-state QM/MM MD with MOPAC-PI, you need three different files: a .dat file, a .key file, and a .xyz file.

The .dat file contains the QM region, calculation parameters, and simulation settings. For instance:

```
    AM1 VECTORS OPEN(6,6) FLOCC=0.05 PULAY ISEED=1 +
    GEO-OK PREC ITRY=1500 SCFCRT=1.D-8 QMMM=11231 +
    DYNAM NTRAJ=1 TIME=0.50 CYCLES=40000 NCYPR=2 +
    SINGLET ISTAT=1 NSTAT=3 MXROOT=20 +
    EXTERNAL=bestpar ITHERMO=1 TEMP=300 TAU=10 MAXWELL=300 +
    NOHOP CHARGE=-1 MEMORY=2000
Cluster1

C            3.737200  1   5.707300  1   0.747000  1
...
Hx12.011     3.084200  1   6.250500  1  -4.098900  1
...
```

In this file, atoms labeled with modified symbols (like "Hx12.011") represent link atoms, which are typically hydrogen atoms that replace the actual bonds between QM and MM regions. In this example, a carbon atom is replaced by a hydrogen atom at the QM/MM boundary. QMMM=n: specifies that a QM/MM calculation will be performed with "n" MM atoms. The QM atoms are defined in the MOPAC input file, while the MM part is handled by TINKER [64]. ISTAT=n: designates the starting electronic state for the dynamics (typically 1 for ground state). NOHOP: no surface hopping. Nevertheless, the time evolution of the electronic wavefunction is computed anyway. FLOCC=x: sets the Gaussian width for floating occupation in SCF calculations. TIME=x: sets the integration time step in femtoseconds. More information about the parameters is reported in the GitLab repository [43].

The .key file includes force-field parameters and group definitions. For instance:

```
# Output Control
DEBUG

# Force Field Selection
PARAMETERS           parameters.prm

# Potential Function Parameter
CHARGE            158    0.0042  !q_S + q_CH2
CHARGE             57    1.0
CHARGE             79    0.0    !HBA1 (MM) bonded to link atom (CA)
CHARGE             80    0.0    !HBA2 (MM) bonded to CA
CHARGE             85    4.0
CHARGE             87    1.0    !CA
...

# Partial Structure
GROUP                1 -1 75
GROUP                2 -76 11306
GROUP-SELECT         1 1 0.0
GROUP-SELECT         1 2 1.0
GROUP-SELECT         2 2 1.0
```

Important settings in the key file include:

(a) PARAMETERS: Specifies the force-field parameter file.

(b) CHARGE: Modifies atomic charges for specific atom types. For QM/MM simulations: delete inner electrons for QM region atoms (e.g., carbon charge set to 4.0 instead of 6.0); set charges of hydrogen atoms bonded to link atoms to zero; set link atom charges to 1.0 (hydrogen); and rescale the charge of atoms directly bonded to link atoms.

(c) GROUP: defines atom groups. Typically: Group 1: QM atoms; Group 2: MM atoms.

(d) GROUP-SELECT: controls interactions between groups. 1 1 0.0: QM-QM interactions are handled by QM methods (set to 0.0); 1 2 1.0: QM-MM interactions are included (set to 1.0); and 2 2 1.0: MM-MM interactions are included (set to 1.0).

The .xyz file contains the coordinates of all atoms in the system. For instance:

```
11306 xxx
    1 C1A      3.737200     5.707300     0.747000  324     7    38    39
    2 C1B      1.299900     2.895300    -1.761300  323     8    31    40
    ...
    11306 Na+     15.459300    23.709800    -8.046700  466
```

Each line contains the atom index, atom name, x/y/z coordinates, atom type, and connectivity information.

11. The surface hopping implementation in MOPAC-PI relies on the local diabatization algorithm [60, 65]. It is important to note that this approach is used only to integrate the time-dependent Schrödinger equation (TDSE) for electrons, while the time evolution of the system as a whole occurs on an adiabatic basis, not a diabatic one. The locally diabatic basis serves exclusively to integrate the TDSE for electrons from time t to $t + \Delta t$, providing numerical stability during propagation.
12. At each time step in surface hopping simulations, a stochastic decision is made to hop between electronic states (e.g., from S_1 to S_0). When a hop is accepted, it becomes necessary to adjust the kinetic energy of the nuclei to maintain energy conservation, as the new active state will have a different potential energy than the previous one. For backward hops (e.g., $S_0 \rightarrow S_1$), the rescaling is applied only to the component of the velocity vector parallel to the nonadiabatic coupling vector between the old and new active states. In contrast, for forward hops (e.g., $S_1 \rightarrow S_0$), the entire nuclear velocity vector is rescaled while maintaining its direction. The reason for this difference is twofold. First, the QM system typically has sufficient kinetic energy to compensate for backward hops once it reaches the ground state; therefore, it is essential to rescale only the velocity component parallel to the nonadiabatic couplings. Second, computing nonadiabatic couplings represent a significant computational burden that is not justified for forward hops, which are always energetically allowed. Note that the time evolution of the electronic wavefunction is performed through evaluation of wavefunction overlaps, so the nonadiabatic couplings are primarily needed for backward hops.
13. Various decoherence schemes exist for surface hopping simulations, each differing in how they model the decay of non-active state amplitudes [66–69]. A practical and widely used approach to correct the surface hopping method for decoherence effects is the "nonlinear decay of mixing" model, proposed by Truhlar et al. [70]. In this approach, the populations of inactive states are damped with an exponential function when the trajectory moves out of a coupling region with strong nonadiabatic interactions. Here, at each integration time step, the coefficients c_k of the wavefunction are evaluated by solving the time-dependent Schroedinger equation and are used to obtain the transition probabilities, by which the current running state (labeled l) is determined. Therefore:

$$c'_k = c_k e^{-\frac{\Delta t}{\tau_{kl}}} \quad \forall k \neq l \tag{2}$$

$$c'_l = c_l \left[\frac{1 - \sum_{k \neq l} |c'_k|^2}{|c_l|^2}\right] \tag{3}$$

$$\tau_{kl} = \frac{\hbar}{|E_k - E_l|}\left(1 + \frac{\alpha}{E_{kin}}\right), \tag{4}$$

where E_{kin} is the nuclear kinetic energy, Δt is the integration interval, and α is an empirical parameter whose recommended value is $\alpha = 0.1$ hartree. The c'_k are then used to continue the time evolution of the electronic wavefunction. This formulation ensures that states with larger energy gaps from the active state experience faster amplitude decay, a feature that gives this approach its name: the energy-based decoherence correction [67].

14. The following is an example of a dat file for a real surface hopping trajectory. The xyz and key files are the same as the ones shown in **Note 10**.

```
AM1 VECTORS OPEN(6,6) FLOCC=0.05 PULAY ISEED=1 +
GEO-OK PREC ITRY=1500 SCFCRT=1.D-8 QMMM=11231 +
DYNAM NTRAJ=1 TIME=0.10 CYCLES=200000 NCYPR=10 +
SINGLET ISTAT=2 NSTAT=3 MXROOT=20 CHVEL=1 +
EXTERNAL=bestpar RDINI=inicond.0001 +
DECAY=0.1 TOLE=0.005 CHARGE=-1 MEMORY=2000
Cluster1

C            3.737200  1   5.707300  1   0.747000  1
...
Hx12.011     3.084200  1   6.250500  1  -4.098900  1
...
```

In this setup, the system begins in an excited electronic state (ISTAT=2) rather than the ground state, allowing the simulation to model photochemical or photophysical processes. The NOHOP keyword has been removed, enabling transitions between electronic states through surface hopping. The simulation employs a significantly smaller time step (TIME=0.10 fs vs. 0.50 fs) to accurately capture the rapid electronic transitions and nuclear motions during nonadiabatic events. The DECAY=0.1 parameter introduces an energy-based decoherence correction, which addresses a key limitation of standard surface hopping by accounting for quantum decoherence effects. CHVEL=1 ensures that after backward hops (from lower to higher energy states), velocity rescaling occurs along the nonadiabatic coupling vectors rather than maintaining the original direction. Initial conditions are read from an external file (RDINI=inicond.0001). Notably, all thermostat parameters (ITHERMO, TEMP, TAU, MAXWELL) have been removed because photochemical processes are typically simulated in a microcanonical ensemble to properly conserve energy during the electronic relaxation process. TOLE=x allows an hop between two states only when the energy difference between them is smaller than x eV.

15. The surface hopping algorithm can be outlined as follows:
    (a) A trajectory starts from a set of initial conditions consisting of nuclear positions and velocities, the electronic density matrix, and the initial electronic state.
    (b) Newton's equations of motion are numerically integrated using the potential energy surface of the current electronic state, similar to a classical MD simulation.
    (c) The time-dependent Schrödinger equation for electrons is integrated using the local diabatization algorithm to obtain the new density matrix, allowing the use of identical time steps for both electronic and nuclear motion.
    (d) The transition probabilities, $g_{\lambda j}$, from the current state (λ) to all other electronic states are computed based on the evolution of electronic coefficients.
    (e) A random number $\chi \in [0, 1]$ is drawn from a uniform distribution and compared with all hopping probabilities $g_{\lambda j}$: If $g_{\lambda, j-1} < \chi < g_{\lambda j}$, a hop from λ to j occurs.
    (f) When a hop takes place, the system's velocity is modified to ensure energy conservation, such that the nuclear kinetic energy compensates for the sudden change in electronic energy.

    Due to the stochastic nature of surface hopping and the complex, rugged energy landscape of proteins, many independent trajectories (often hundreds) are required to obtain statistically meaningful results. A single trajectory can be highly misleading due to the random nature of the hopping process. Proper ensemble averaging across multiple trajectories is essential for drawing robust conclusions about reaction mechanisms and kinetics.

16. During classical MD of photoproducts, key structural features to monitor include:
    (a) Hydrogen bond rearrangements between the D-ring and surrounding residues
    (b) Salt bridge stability. in the binding pocket (e.g., the Asp-Arg pair in DrBphP)
    (c) Mobility of the PHY tongue region

(d) Water molecule movements in and out of the binding pocket

(e) Domain-domain distances and angles, which reflect the global conformational changes that ultimately lead to signaling

17. The D-ring C=O stretch frequency in IR spectra serves as a particularly valuable marker, as it sensitively reports on hydrogen bonding changes that occur throughout the photocycle. This spectroscopic signature provides one of the most reliable validation points for computational models. Researchers should compare computed frequency shifts between intermediates with experimental difference spectra to verify that their simulations accurately capture the environmental changes around this critical functional group. Protocol for calculating IR difference spectra:

(a) Extract representative structures from equilibrated Pr and Lumi-R states.

(b) Optimize structures with QM/MM (fix protein beyond 5 Å from chromophore).

(c) Calculate normal modes with DFT (B3LYP/6-31G* recommended).

(d) Scale frequencies by standard factor (0.97 for B3LYP/6-31G*).

(e) Apply Gaussian broadening (FWHM $\sim$15 cm^{-1}).

(f) Subtract spectra to obtain difference spectrum.

(g) Compare with experimental difference spectrum.

Focus analysis on the 1600–1800 cm^{-1} region where the most diagnostic peaks appear.

References

1. Quail PH (1998) The phytochrome family: dissection of functional roles and signalling pathways among family members. Phil Trans R Soc Lond B 353(1374):1399–1403. https://doi.org/10.1098/rstb.1998.0294
2. Möglich A, Yang X, Ayers RA, Moffat K (2010) Structure and function of plant photoreceptors. Annu Rev Plant Biol 61(1):21–47. https://doi.org/10.1146/annurev-arplant-042809-112259
3. Anders K, Essen LO (2015) The family of phytochrome-like photoreceptors: diverse, complex and multi-colored, but very useful. Curr Opin Struct Biol 35:7–16. https://doi.org/10.1016/j.sbi.2015.07.005
4. Hughes J, Winkler A (2024) New insight into phytochromes: connecting structure to function. Annu Rev Plant Biol 75(1):153–183. https://doi.org/10.1146/annurev-arplant-070623-110636
5. Chernov KG, Redchuk TA, Omelina ES, Verkhusha VV (2017) Near-infrared fluorescent proteins, biosensors, and optogenetic tools engineered from phytochromes. Chem Rev 117(9):6423–6446. https://doi.org/10.1021/acs.chemrev.6b00700
6. Gourinchas G, Etzl S, Winkler A (2019) Bacteriophytochromes – from informative model systems of phytochrome function to powerful tools in cell biology. Curr Opin Struct Biol 57:72–83. https://doi.org/10.1016/j.sbi.2019.02.005
7. Redchuk TA, Karasev MM, Verkhusha PV, Donnelly SK, Hülsemann M, Virtanen J, Moore HM, Vartiainen MK, Hodgson L, Verkhusha VV (2020) Optogenetic regulation of

endogenous proteins. Nat Commun 11(1): 605. https://doi.org/10.1038/s41467-020-14460-4

8. Tang K, Beyer HM, Zurbriggen MD, Gärtner W (2021) The red edge: bilin-binding photoreceptors as optogenetic tools and fluorescence reporters. Chem Rev 121(24):14906–14956. https://doi.org/10.1021/acs.chemrev.1c00194
9. Kuwasaki Y, Suzuki K, Yu G, Yamamoto S, Otabe T, Kakihara Y, Nishiwaki M, Miyake K, Fushimi K, Bekdash R, Shimizu Y, Narikawa R, Nakajima T, Yazawa M, Sato M (2022) A red light–responsive photoswitch for deep tissue optogenetics. Nat Biotechnol. https://doi.org/10.1038/s41587-022-01351-w
10. Nagano S, Sadeghi M, Balke J, Fleck M, Heckmann N, Psakis G, Alexiev U (2022) Improved fluorescent phytochromes for in situ imaging. Sci Rep 12(1):5587. https://doi.org/10.1038/s41598-022-09169-x
11. Kasatkina LA, Ma C, Matlashov ME, Vu T, Li M, Kaberniuk AA, Yao J, Verkhusha VV (2022) Optogenetic manipulation and photoacoustic imaging using a near-infrared transgenic mouse model. Nat Commun 13(1): 2813. https://doi.org/10.1038/s41467-022-30547-6
12. Rockwell NC, Su YS, Lagarias JC (2006) Phytochrome structure and signaling mechanisms. Annu Rev Plant Biol 57(1):837–858. https://doi.org/10.1146/annurev.arplant.56.032604.144208
13. Takala H, Edlund P, Ihalainen JA, Westenhoff S (2020) Tips and turns of bacteriophytochrome photoactivation. Photochem Photobiol Sci 19:1488–1510. https://doi.org/10.1039/D0PP00117A
14. Rockwell NC, Shang L, Martin SS, Lagarias JC (2009) Distinct classes of red/far-red photochemistry within the phytochrome superfamily. Proc Natl Acad Sci U S A 106(15):6123–6127. https://doi.org/10.1073/pnas.0902370106
15. Burgie ES, Vierstra RD (2015) Phytochromes: an atomic perspective on photoactivation and signaling. Plant Cell 26(12):4568–4583. https://doi.org/10.1105/tpc.114.131623
16. Rottwinkel G, Oberpichler I, Lamparter T (2010) Bathy phytochromes in rhizobial soil bacteria. J Bacteriol 192(19):5124–5133. https://doi.org/10.1128/jb.00672-10
17. Karniol B, Vierstra RD (2003) The pair of bacteriophytochromes from agrobacterium tumefaciens are histidine kinases with opposing photobiological properties. Proc Natl Acad Sci USA 100(5):2807–2812. https://doi.org/10.1073/pnas.0437914100
18. Krieger A, Molina I, Oberpichler I, Michael N, Lamparter T (2008) Spectral properties of phytochrome agp2 from agrobacterium tumefaciens are specifically modified by a compound of the cell extract. J Photochem Photobiol B, Biol 93(1):16–22. https://doi.org/10.1016/j.jphotobiol.2008.07.001
19. Zienicke B, Molina I, Glenz R, Singer P, Ehmer D, Escobar FV, Hildebrandt P, Diller R, Lamparter T (2013) Unusual spectral properties of bacteriophytochrome agp2 result from a deprotonation of the chromophore in the red-absorbing form pr. J Biol Chem 288(44):31738–31751. https://doi.org/10.1074/jbc.m113.479535
20. Escobar FV, Piwowarski P, Salewski J, Michael N, Lopez MF, Rupp A, Qureshi BM, Scheerer P, Bartl F, Frankenberg-Dinkel N, Siebert F, Mroginski MA, Hildebrandt P (2015) A protonation-coupled feedback mechanism controls the signalling process in bathy phytochromes. Nat Chem 7(5):423–430. https://doi.org/10.1038/nchem.2225
21. Escobar FV, Buhrke D, Michael N, Sauthof L, Wilkening S, Tavraz NN, Salewski J, Frankenberg-Dinkel N, Mroginski MA, Scheerer P, Friedrich T, Siebert F, Hildebrandt P (2017) Common structural elements in the chromophore binding pocket of the pfr state of bathy phytochromes. Photochem Photobiol 93(3):724–732. https://doi.org/10.1111/php.12742
22. Fernandez Lopez M, Nguyen AD, Velazquez Escobar F, González R, Michael N, Nogacz Z, Piwowarski P, Bartl F, Siebert F, Heise I, Scheerer P, Gärtner W, Mroginski MA, Hildebrandt P (2019) Role of the propionic side chains for the photoconversion of bacterial phytochromes. Biochemistry 58(33): 3504–3519. https://doi.org/10.1021/acs.biochem.9b00526
23. Burgie ES, Wang T, Bussell AN, Walker JM, Li H, Vierstra RD (2014) Crystallographic and electron microscopic analyses of a bacterial phytochrome reveal local and global rearrangements during photoconversion. J Biol Chem 289(35):24,573–24,587. https://doi.org/10.1074/jbc.m114.571661
24. Burgie ES, Zhang J, Vierstra RD (2016) Crystal structure of deinococcus phytochrome in the photoactivated state reveals a cascade of structural rearrangements during photoconversion. Structure 24(3):448–457. https://doi.org/10.1016/j.str.2016.01.001
25. Kübel J, Chenchiliyan M, Ooi SA, Gustavsson E, Isaksson L, Kuznetsova V, Ihalainen JA, Westenhoff S, Maj M (2020) Transient IR spectroscopy identifies key interactions and unravels new intermediates in the

photocycle of a bacterial phytochrome. Phys Chem Chem Phys 22(17):9195–9203. https://doi.org/10.1039/c9cp06995j

26. Salvadori G, Mennucci B (2024) Analogies and differences in the photoactivation mechanism of bathy and canonical bacteriophytochromes revealed by multiscale modeling. J Phys Chem Lett 15(31):8078–8084. https://doi.org/10.1021/acs.jpclett.4c01823

27. Klose C, Nagy F, Schäfer E (2020) Thermal reversion of plant phytochromes. Mol Plant 13(3):386–397. https://doi.org/10.1016/j.molp.2019.12.004

28. Ihalainen JA, Gustavsson E, Schroeder L, Donnini S, Lehtivuori H, Isaksson L, Thöing C, Modi V, Berntsson O, Stucki-Buchli B, Liukkonen A, Häkkänen H, Kalenius E, Westenhoff S, Kottke T (2018) Chromophore–protein interplay during the phytochrome photocycle revealed by step-scan FTIR spectroscopy. J Am Chem Soc 140(39): 12396–12404. https://doi.org/10.1021/jacs.8b04659

29. Claesson E, Wahlgren WY, Takala H, Pandey S, Castillon L, Kuznetsova V, Henry L, Panman M, Carrillo M, Kübel J, Nanekar R, Isaksson L, Nimmrich A, Cellini A, Morozov D, Maj M, Kurttila M, Bosman R, Nango E, Tanaka R, Tanaka T, Fangjia L, Iwata S, Owada S, Moffat K, Groenhof G, Stojković EA, Ihalainen JA, Schmidt M, Westenhoff S (2020) The primary structural photoresponse of phytochrome proteins captured by a femtosecond X-ray laser. eLife 9:1–18. https://doi.org/10.7554/eLife.53514

30. Heyne K, Herbst J, Stehlik D, Esteban B, Lamparter T, Hughes J, Diller R (2002) Ultrafast dynamics of phytochrome from the cyanobacterium synechocystis, reconstituted with phycocyanobilin and phycoerythrobilin. Biophys J 82(2):1004–1016. https://doi.org/10.1016/s0006-3495(02)75460-x

31. Isaksson L, Gustavsson E, Persson C, Brath U, Vrhovac L, Karlsson G, Orekhov V, Westenhoff S (2021) Signaling mechanism of phytochromes in solution. Structure 29(2): 151–160.e3. https://doi.org/10.1016/j.str.2020.08.009

32. Chenchiliyan M, Kübel J, Ooi SA, Salvadori G, Mennucci B, Westenhoff S, Maj M (2023) Ground-state heterogeneity and vibrational energy redistribution in bacterial phytochrome observed with femtosecond 2D IR spectroscopy. J Chem Phys 158(8):085,103. https://doi.org/10.1063/5.0135268

33. Do TN, Menendez D, Bizhga D, Stojković EA, Kennis JT (2024) Two-photon absorption and photoionization of a bacterial phytochrome. J Mol Biol 436(5):168357. https://doi.org/10.1016/j.jmb.2023.168357

34. Ihalainen JA, Takala H, Lehtivuori H (2015) Fast photochemistry of prototypical phytochromes—a species vs. subunit specific comparison. Front Mol Biosci 2:75. https://doi.org/10.3389/fmolb.2015.00075

35. Tully JC (1990) Molecular dynamics with electronic transitions. J Chem Phys 93(2): 1061–1071. https://doi.org/10.1063/1.459170

36. Tully JC (2012) Perspective: nonadiabatic dynamics theory. J Chem Phys 137(22): 22A301–7. https://doi.org/10.1063/1.4757762

37. Case D, Aktulga H, Belfon K, Ben-Shalom I, Berryman J, Brozell S, Cerutti D, Cheatham T, III, Cisneros G, Cruzeiro V, Darden T, Forouzesh N, Giambaşu G, Giese T, Gilson M, Gohlke H, Goetz A, Harris J, Izadi S, Izmailov S, Kasavajhala K, Kaymak M, King E, Kovalenko A, Kurtzman T, Lee T, Li P, Lin C, Liu J, Luchko T, Luo R, Machado M, Man V, Manathunga M, Merz K, Miao Y, Mikhailovskii O, Monard G, Nguyen H, O'Hearn K, Onufriev A, Pan F, Pantano S, Qi R, Rahnamoun A, Roe D, Roitberg A, Sagui C, Schott-Verdugo S, Shajan A, Shen J, Simmerling C, Skrynnikov N, Smith J, Swails J, Walker R, Wang J, Wang J, Wei H, Wu X, Wu Y, Xiong Y, Xue Y, York D, Zhao S, Zhu Q, Kollman P (2023) Amber 2023. University of California, San Francisco

38. Abraham MJ, Murtola T, Schulz R, Páll S, Smith JC, Hess B, Lindahl E (2015) GROMACS: high performance molecular simulations through multi-level parallelism from laptops to supercomputers. SoftwareX 1–2: 19–25. https://doi.org/10.1016/j.softx.2015.06.001

39. Kühne TD, Iannuzzi M, Del Ben M, Rybkin VV, Seewald P, Stein F, Laino T, Khaliullin RZ, Schütt O, Schiffmann F, Golze D, Wilhelm J, Chulkov S, Bani-Hashemian MH, Weber V, Borštnik U, Taillefumier M, Jakobovits AS, Lazzaro A, Pabst H, Müller T, Schade R, Guidon M, Andermatt S, Holmberg N, Schenter GK, Hehn A, Bussy A, Belleflamme F, Tabacchi G, Glöß A, Lass M, Bethune I, Mundy CJ, Plessl C, Watkins M, VandeVondele J, Krack M, Hutter J (2020) Cp2k: an electronic structure and molecular dynamics software package—quickstep: efficient and accurate electronic structure calculations. J Chem Phys 152(19). https://doi.org/10.1063/5.0007045

40. Frisch MJ, Trucks GW, Schlegel HB, Scuseria GE, Robb MA, Cheeseman JR, Scalmani G,

Barone V, Petersson GA, Nakatsuji H, Li X, Caricato M, Marenich AV, Bloino J, Janesko BG, Gomperts R, Mennucci B, Hratchian HP, Ortiz JV, Izmaylov AF, Sonnenberg JL, Williams-Young D, Ding F, Lipparini F, Egidi F, Goings J, Peng B, Petrone A, Henderson T, Ranasinghe D, Zakrzewski VG, Gao J, Rega N, Zheng G, Liang W, Hada M, Ehara M, Toyota K, Fukuda R, Hasegawa J, Ishida M, Nakajima T, Honda Y, Kitao O, Nakai H, Vreven T, Throssell K, Montgomery JA Jr, Peralta JE, Ogliaro F, Bearpark MJ, Heyd JJ, Brothers EN, Kudin KN, Staroverov VN, Keith TA, Kobayashi R, Normand J, Raghavachari K, Rendell AP, Burant JC, Iyengar SS, Tomasi J, Cossi M, Millam JM, Klene M, Adamo C, Cammi R, Ochterski JW, Martin RL, Morokuma K, Farkas O, Foresman JB, Fox DJ (2016) Gaussian 16 revision A.03. Gaussian, Wallingford

41. Neese F (2012) The ORCA program system. WIRES Comput Molec Sci 2(1):73–78. https://doi.org/10.1002/wcms.81
42. Stewart JJ (2002) Mopac2002. Fujitsu Limited, Tokyo
43. Granucci G, Persico M, Accomasso D, Sangiogo Gil E, Corni S, Fregoni J, Laino T, Botticelli S, Tesi M, Toniolo A (2024) MOPAC-PI: a program for excited state dynamics simulations based on nonadiabatic trajectories and semiempirical electronic structure calculations. https://gitlab.com/granucci/mopacpi.git
44. Akdel M, Pires DEV, Pardo EP, Jänes J, Zalevsky AO, Mészáros B, Bryant P, Good LL, Laskowski RA, Pozzati G, Shenoy A, Zhu W, Kundrotas P, Serra VR, Rodrigues CHM, Dunham AS, Burke D, Borkakoti N, Velankar S, Frost A, Basquin J, Lindorff-Larsen K, Bateman A, Kajava AV, Valencia A, Ovchinnikov S, Durairaj J, Ascher DB, Thornton JM, Davey NE, Stein A, Elofsson A, Croll TI, Beltrao P (2022) A structural biology community assessment of AlphaFold2 applications. Nat Struct Mol Biol 29(11):1056–1067. https://doi.org/10.1038/s41594-022-00849-w
45. Jumper J, Evans R, Pritzel A, Green T, Figurnov M, Ronneberger O, Tunyasuvunakool K, Bates R, Žídek A, Potapenko A, Bridgland A, Meyer C, Kohl SAA, Ballard AJ, Cowie A, Romera-Paredes B, Nikolov S, Jain R, Adler J, Back T, Petersen S, Reiman D, Clancy E, Zielinski M, Steinegger M, Pacholska M, Berghammer T, Bodenstein S, Silver D, Vinyals O, Senior AW, Kavukcuoglu K, Kohli P, Hassabis D (2021) Highly accurate protein structure prediction with AlphaFold. Nature 596(7873):583–589. https://doi.org/10.1038/s41586-021-03819-2
46. Sánchez R, Šali A (1997) Evaluation of comparative protein structure modeling by MODELLER-3. Proteins Struct Funct Genet 29(SUPPL. 1):50–58. https://doi.org/10.1002/(SICI)1097-0134(1997)1+%3C50::AID-PROT8%3E3.0.CO;2-S
47. Wang J, Wolf RM, Caldwell JW, Kollman PA, Case DA (2004) Development and testing of a general amber force field. J Comput Chem 25(9):1157–1174. https://doi.org/10.1002/jcc.20035
48. Bayly CI, Cieplak P, Cornell WD, Kollman PA (1993) A well-behaved electrostatic potential based method using charge restraints for deriving atomic charges: the RESP model. J Phys Chem 97(40):10269–10280. https://doi.org/10.1021/j100142a004
49. Macaluso V, Cupellini L, Salvadori G, Lipparini F, Mennucci B (2020) Elucidating the role of structural fluctuations, and intermolecular and vibronic interactions in the spectroscopic response of a bacteriophytochrome. Phys Chem Chem Phys 22(16):8585–8594. https://doi.org/10.1039/d0cp00372g
50. Macaluso V, Salvadori G, Cupellini L, Mennucci B (2021) The structural changes in the signaling mechanism of bacteriophytochromes in solution revealed by a multiscale computational investigation. Chem Sci 12(15):5555–5565. https://doi.org/10.1039/D1SC00186H
51. Barducci A, Bonomi M, Parrinello M (2011) Metadynamics. WIREs Comput Mol Sci 1(5):826–843. https://doi.org/10.1002/wcms.31
52. Miao Y, Feher VA, McCammon JA (2015) Gaussian accelerated molecular dynamics: unconstrained enhanced sampling and free energy calculation. J Chem Theory Comput 11(8):3584–3595. https://doi.org/10.1021/acs.jctc.5b00436
53. Schmidt A, Sauthof L, Szczepek M, Lopez MF, Escobar FV, Qureshi BM, Michael N, Buhrke D, Stevens T, Kwiatkowski D, von Stetten D, Mroginski MA, Krauß N, Lamparter T, Hildebrandt P, Scheerer P (2018) Structural snapshot of a bacterial phytochrome in its functional intermediate state. Nat Commun 9(1):4912. https://doi.org/10.1038/s41467-018-07392-7
54. Rumfeldt JA, Takala H, Liukkonen A, Ihalainen JA (2019) UV-Vis spectroscopy reveals a correlation between y263 and BV protonation

states in bacteriophytochromes. Photochem Photobiol 95(4):969–979. https://doi.org/10.1111/php.13095

55. Mukamel S (1995) Principles of nonlinear optical spectroscopy. Oxford University Press, New York
56. Mukamel S (1985) Fluorescence and absorption of large anharmonic molecules-spectroscopy without eigenstates. J Phys Chem 89(7):1077–1087. https://doi.org/10.1021/j100253a008
57. Loco D, Cupellini L (2018) Modeling the absorption lineshape of embedded systems from molecular dynamics: a tutorial review. Int J Quantum Chem 119(1). https://doi.org/10.1002/qua.25726
58. Georgievskii Y, Hsu CP, Marcus RA (1999) Linear response in theory of electron transfer reactions as an alternative to the molecular harmonic oscillator model. J Phys Chem 110(11): 5307–5317. https://doi.org/10.1063/1.478425
59. Granucci G, Toniolo A (2000) Molecular gradients for semiempirical CI wavefunctions with floating occupation molecular orbitals. Chem Phys Lett 325(1–3):79–85. https://doi.org/10.1016/s0009-2614(00)00691-6
60. Granucci G, Persico M, Toniolo A (2001) Direct semiclassical simulation of photochemical processes with semiempirical wave functions. J Chem Phys 114(24):10608–10615. https://doi.org/10.1063/1.1376633
61. Toniolo A, Ben-Nun M, Martínez TJ (2002) Optimization of conical intersections with floating occupation semiempirical configuration interaction wave functions. J Phys Chem A 106(18):4679–4689. https://doi.org/10.1021/jp014289y
62. Toniolo A, Granucci G, Martínez TJ (2003) Conical intersections in solution: a QM/MM study using floating occupation semiempirical configuration interaction wave functions. J Phys Chem A 107(19):3822–3830. https://doi.org/10.1021/jp022468p
63. Salvadori G, Macaluso V, Pellicci G, Cupellini L, Granucci G, Mennucci B (2022) Protein control of photochemistry and transient intermediates in phytochromes. Nat Commun 13:6838. https://doi.org/10.1038/s41467-022-34640-8
64. Rackers JA, Wang Z, Lu C, Laury ML, Lagardère L, Schnieders MJ, Piquemal JP, Ren P, Ponder JW (2018) Tinker 8: software tools for molecular design. J Chem Theory Comput 14(10):5273–5289. https://doi.org/10.1021/acs.jctc.8b00529
65. Persico M, Granucci G (2014) An overview of nonadiabatic dynamics simulations methods, with focus on the direct approach versus the fitting of potential energy surfaces. Theor Chem Acc 133(9):1526. https://doi.org/10.1007/s00214-014-1526-1
66. Subotnik JE, Shenvi N (2011) A new approach to decoherence and momentum rescaling in the surface hopping algorithm. J Chem Phys 134(2). https://doi.org/10.1063/1.3506779
67. Granucci G, Persico M, Zoccante A (2010) Including quantum decoherence in surface hopping. J Chem Phys 133(13). https://doi.org/10.1063/1.3489004
68. Cheng SC, Zhu C, Liang KK, Lin SH, Truhlar DG (2008) Algorithmic decoherence time for decay-of-mixing non–born–oppenheimer dynamics. J Chem Phys 129(2). https://doi.org/10.1063/1.2948395
69. Min SK, Agostini F, Tavernelli I, Gross EKU (2017) Ab initio nonadiabatic dynamics with coupled trajectories: a rigorous approach to quantum (de)coherence. J Phys Chem Lett 8(13):3048–3055. https://doi.org/10.1021/acs.jpclett.7b01249
70. Zhu C, Jasper AW, Truhlar DG (2005) Non-born-Oppenheimer Liouville-von Neumann dynamics. evolution of a subsystem controlled by linear and population-driven decay of mixing with decoherent and coherent switching. J Chem Theory Comput 1(4):527–540. https://doi.org/10.1021/ct050021p

Chapter 10

FTIR Spectroscopy of Red and Far-Red-Light Sensing Phytochromes In vitro

Moona Kurttila and Janne A. Ihalainen

Abstract

The action of phytochromes is controlled by a tetrapyrrole bilin chromophore that isomerizes upon red or far-red-light illumination. Photoactivation results in a multitude of structural rearrangements of the protein, including secondary structural changes and changes in the interaction patterns of the chromophore and the protein scaffold. The IR signals of phytochromes originate from molecular vibrations of both the chromophore and the protein. Being sensitive to molecular interactions, the vibrational frequencies reveal sub-angstrom-level information about structural changes of a phytochrome protein in solution. Phytochromes are relatively easy to produce and modify, they can be concentrated into high concentrations, and their photocycle is reversible, both thermally and by light. These qualities make phytochromes excellent molecules for FTIR spectroscopic studies. Furthermore, the effect of pH, temperature, or buffer conditions on photoactivation or intramolecular interactions can be studied by means of FTIR spectroscopy.

Key words Molecular vibrations, Photoactive protein, Amide I vibration, Carbonyl vibration, Molecular interactions

1 Introduction

Fourier transform infrared (FTIR) spectroscopy utilizes absorption of infrared (IR) light to detect molecular vibrations that arise from covalent bonds between atoms [1, 2]. A vibrational frequency describes the vibrational energy of the bond, which in classical terms depends on the bond strength (single, double, triple) and on the mass of the bonding atoms, which is tuned by weak molecular interactions with neighboring atoms as well as affected by coupling with near-by vibrations [3]. The vibrational spectrum of a molecule comprises, in principle, all the information about the molecular structure and intramolecular interactions, and therefore, FTIR is a sensitive and powerful tool to study the properties of photoactive proteins in vivo and in vitro [4, 5].

As nonlinear molecules contain 3N-6 normal modes, where N is the number of atoms in the molecule, IR absorption spectra of

Andreas Winkler (ed.), *Phytochromes: Methods and Protocols*, Methods in Molecular Biology, vol. 2970, https://doi.org/10.1007/978-1-0716-4791-2_10,

proteins are often saturated by the number of signals arising from the multitude of bonds in the system. Therefore, for phytochromes, and for photoactive proteins in general, the most effective approach is to directly measure the changes in the vibrational spectra, which substantially decreases the number of observed signals. In phytochromes, this is done between the red (Pr) and far-red (Pfr) absorbing states, and the spectral changes reflect the changes of molecular interactions among the chromophore and protein scaffold during the photoconversion between the two states. The short-lived intermediate photoproducts can be measured by means of time-resolved IR spectroscopic techniques, such as step-scan FTIR spectroscopy [6], time-resolved IR spectroscopy with quantum cascade laser detection [7, 8], or ultra-fast pump-probe IR spectroscopy [9–11], but they are out of the scope of this contribution.

One of the advantages and at the same time challenges of FTIR spectroscopy is that it collects all structural information of a molecule in solution. Due to a large amount of overlapping vibrations, assignment of the signals to single functional groups or intramolecular interactions, even in difference spectra, is very demanding. To some extent, successful interpretations have been achieved by means of isotope labeling of the chromophore [12] and protein [6, 13, 14], truncations [15], point mutations [16], and computations [9, 17]. IR probes, such as nitrile and azide labels that shift the frequency of the labeled residue to a window free from other protein vibrations, have been utilized in phytochromes to extract information from a single residue [18, 19]. All these techniques are useful; however, labeling or other modifications are done in the protein production phase. Therefore, this chapter does not provide methods to perform modifications. Yet, the methods for measuring the FTIR difference spectra of these samples are the same as for any unmodified sample and can be performed as advised here. Utilizing two-dimensional (2D) IR spectroscopy, Buhrke et al. revealed intramolecular coupling between the protein scaffold and the chromophore [20].

Another challenge in protein IR measurements is that the frequency of the bending motion of the water molecule (between 1700 cm^{-1} and 1500 cm^{-1}) overlaps with the protein fingerprint region (Fig. 3) suppressing essentially all information of the protein vibrations. To overcome this challenge, one can either concentrate the sample to very high concentration of tens, even hundreds of mg/ml, and use of a thin measurement cuvette (7–10 μ m). Together, high concentration and short path length are to minimize the absorption of water that overlaps with the amide I vibrations at around 1650 cm^{-1}. Alternatively, diluting the sample with D_2O results in a shift of the bending motion of water to lower frequency, out from the protein fingerprint region. In D_2O, relatively high concentration and path lengths such as 30–100 μm are needed for proper amount of light and high-quality spectrum to be obtained.

We note that the methods described below are described for canonical bacteriophytochrome holoprotein samples and optimized for *Deinococcus radiodurans* bacteriophytochrome (*Dr*BphP) which is also used as an example system here. The absorption maxima of *Dr*BphP, like in many bacterial phytochromes, are at 700 nm for the Pr state and at 750 nm in the Pfr state [21]. In case of bathy phytochrome samples, where the Pfr state is the resting state [22–24], analogous information from the protein can be obtained with the same sample preparation methods, but we suggest to start the difference spectral measurements with the far-red illumination, as noted in the methods section.

2 Materials

2.1 Sample Preparation

1. Purified phytochrome sample: ideally 1–3 mg of protein (*see* **Note 1**)
2. Centrifugal concentrator with a cutoff of 10 kDa: Vivaspin 500 (Sartorius) (*see* **Note 2**)
3. Centrifugal filters: Amicon Ultra 0.22 μm (Merck-Millipore)
4. A microcentrifuge: with capacity up to 12,000 rcf
5. Sample buffer: 30 mM Tris-HCl, pH 8.0 (*see* **Note 3**)

2.1.1 For Buffer Exchange into D_2O

1. Buffered D_2O, which should match the used H_2O buffer. We use 30 mM Tris-HCl in D_2O, pD 8.0 (adjusted with concentrated HCl, (*see* **Note 4**).
2. Desalting columns with 40K MWCO and 50–150 μl sample volume: for buffer exchange.

2.2 FTIR Sample Cell

1. Two round CaF_2 windows (Fig. 1a, 2.). Our windows have 1 inch diameter and are 2 mm thick.
2. Kimtech wipes and ethanol or isopropanol: for cleaning the windows.
3. CaF_2 sample cell (*see* **Note 5**). We show two different custom-made sample cell options in Fig. 1a (1 and 3) and closing accessories such as screw driver or tightening block (Fig. 1a, 5).
4. Silicone-free vacuum grease: for measuring sample in H_2O to seal the sample between the windows (Fig. 1a, 4.). At least Glisseal N grease from Borer Chemie is known to work.
5. 30–100 μm-thick Teflon spacer ring: for measuring samples in D_2O. An equal circumference to the CaF_2 windows and about 3 mm ring width leaving about 20 mm diameter for the sample.

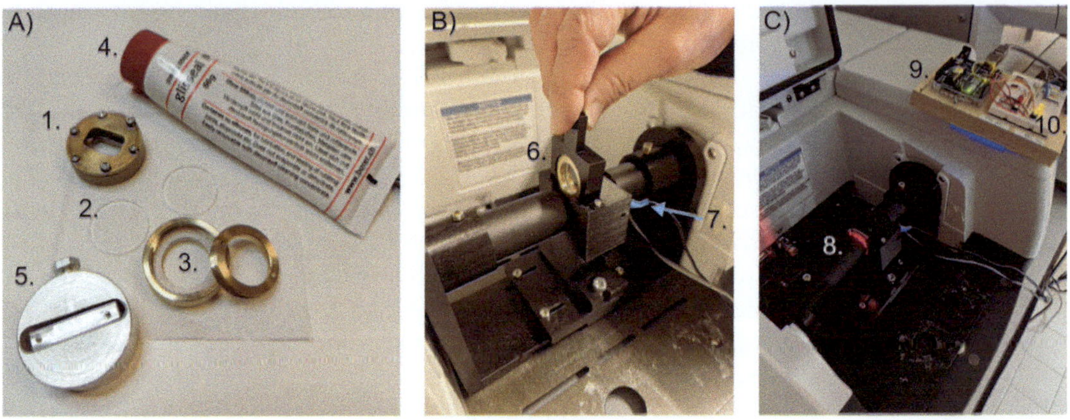

Fig. 1 Materials and FTIR layout. (**a**) Materials for loading sample cell: 1. Sample cell with screws, 2. CaF_2 windows, 3. Sample cell with grooves, Teflon insulator ring inside, 4. Sealing grease, 5. Tightening block with pins matching the holes in the top of 3. (**b**) FTIR setup. 6. Sample cell mount that fits inside the vertical sample holder. 7. LEDs (one shown) are mounted in both sides of the sample holder, pointing at the sample. The IR-beam path is insulated with tubing. (**c**) 8. The red LED in action. 9. Arduino computer to control the illumination LEDs. 10. Push buttons to manually switch the LEDs on and off, and indicator LEDs on the breadboard

2.3 FTIR Setup

1. FTIR spectrophotometer, preferably with liquid-nitrogen-cooled MCT detector for better signal-to-noise ratio (*see* **Note 6**). The measurements are performed in transmission mode with Nicolet iS50 FTIR spectrophotometer with a liquid-nitrogen-cooled MCT detector. Equally good spectrometers are available for example from Bruker.
2. Vertical sample holder (*see* **Note 7**) with LED holders and a matching sample cell mount to move the sample in and out (Fig. 1b, 6. and Fig. 2).
3. Red (660 nm) and far-red (780 nm) LEDs (*see* **Note 8**) mounted into the sample holder (Fig. 1b, c). We have used unmounted LEDs from Thorlabs (LED660L and LED780L).
4. Power source for the LEDs, in our case 5V, is required.
5. Arduino computer and a breadboard for controlling the LEDs (Fig. 1c, 9. and Fig. 2).

3 Methods

3.1 Sample Preparation

The sample can be either prepared in aqueous solution or changed into D_2O. If the sample is to be measured in aqueous buffer, move directly to Subheading 3.1.2.

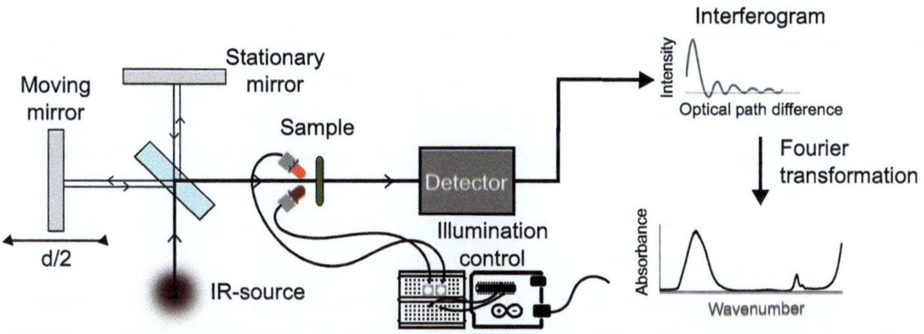

Fig. 2 Materials and FTIR layout. (**a**) Materials for loading sample cell. 1. Sample cell with screws, 2. CaF_2 windows, 3. Sample cell with grooves, Teflon insulator ring inside, 4. Sealing grease, 5. Tightening block with pins matching the holes in the top of 3. (**b**) FTIR setup. 6. Sample cell mount that fits inside the vertical sample holder. 7. LEDs (one shown) are mounted in both sides of the sample holder, pointing at the sample. The IR-beam path is insulated with tubing. (**c**) 8. The red LED in action. 9. Arduino computer to control the illumination LEDs. 10. Push buttons to manually switch the LEDs on and off, and indicator LEDs on the breadboard

3.1.1 Buffer Exchange for Samples in D_2O

1. Remove the cap from the bottom of the Zeba Spin Desalting column, loosen the cap, and place the column in an empty microtube. Centrifuge at 1500 × g for 1 minute to remove the storage solution.
2. The compacted resin is slanted upward during centrifugation. Place a mark on the slanted side of the column with a marker. Perform the subsequent centrifugations in the same position, the mark toward the outside of the centrifuge.
3. Empty the collection tube, and load 300 μl of the buffered D_2O on top of the resin. Close the cap loosely and centrifuge at 1500 × g for 1 minute.
4. Repeat the step 3 for three additional times to equilibrate the column with the D_2O buffer.
5. Thaw your phytochrome sample(s) rapidly up to room temperature (*see* **Note 9**).
6. Place column in a clean collection tube and apply 100 μl of purified phytochrome sample (20–30 mg/ml concentration) to the top of the compact resin bed (*see* **Note 10**).
7. Centrifuge at 1500 × g for 2 minutes to collect the sample.
8. Save the sample on ice.
9. Re-equilibrate the column with the D_2O buffer following steps 3 and 4.
10. (In parallel with step 8) Illuminate the recovered protein from step 6 with 660 nm LED for 5 minutes to drive the phytochrome molecules to light-activated Pfr state. Repeat but with 750 nm LED to drive it back. Repeat illumination of the sample 3–4 times with both LEDs (*see* **Note 11**).

3.1.2 Concentrating the Sample

1. If not yet done, thaw your phytochrome sample(s) rapidly to room temperature (*see* **Note 9**).
2. Prerinse the Vivaspin concentrator to remove excess glycerin. Pipette at least 500 μl of the sample buffer where the sample is placed in the concentrator, and wash the buffer by centrifuging at 12,000 rcf for 5 min. Empty the collection tube and concentrator from excess buffer.
3. Pipette the sample inside the concentrator and centrifuge the sample at 12,000 rcf in 10–15 min intervals to as small volume as possible, often around the 25 μl line (*see* **Note 12**). The sample can be mixed by pipetting up and down in between cycles.
4. Prerinse the centrifugal filter by pipetting 500 μl of the sample buffer on top of the filter membrane, and wash the buffer through by centrifuging at 12,000 rcf for 2 min. Empty the collection tube carefully by pipetting.
5. Filter the concentrated sample by pipetting it in the middle of the membrane. Centrifuge at 12,000 rcf for 2 min (*see* **Note 13**).
6. Dilute a small fraction of the sample sufficiently to measure the UV-vis absorption spectrum of the protein (*see* **Notes 14**). Calculate the concentration using the extinction coefficient of the protein at 280 nm (*see* **Note 15**) and the respective dilution factor. The concentration should be about 100 mg/ml (which corresponds to 1–3 mM depending on the molecular weight of the construct) for H_2O measurements and about a quarter of this to D_2O measurements.
7. The concentrated and filtered sample is best used fresh (*see* **Note 16**).

3.2 Sample Assembling

3.2.1 Sample in H_2O

1. Take two CaF_2 windows and make sure they are clean. Wipe them with Kimtech wipes using pure water and ethanol/isopropanol to clean them (*see* **Note 17**).
2. Create a very thin (about 1 mm), narrow (1–2 mm wide), and continuous ring of sealing grease on the top edge of one window (*see* **Note 18**). Leave no gaps, and keep the amount of grease as similar as possible throughout.
3. Load 2.0 μl of the sample in the middle of the greased window by pipetting (*see* **Note 19**). Avoid bubbles.

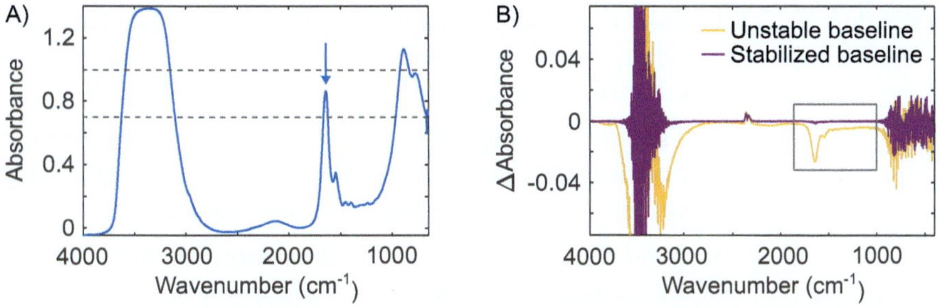

Fig. 3 Checking sample quality by FTIR. (**a**) Absorption spectrum of the sample in H_2O using air spectrum as a background. The 1645 cm^{-1} peak should be less than 1 but ideally above 0.7 for the best difference signal quality. (**b**) The baseline should be checked and let to stabilize prior to starting the difference measurements. The background is taken from the sample, and without any modifications, a baseline check spectrum is measured after 2 min (time of a spectral acquisition). In theory, a completely stable sample would result in a flat spectrum at 0. If the sample has not set, the baseline with often display bleaches around 1500–1700 cm^{-1} (yellow line). In a stabilized sample, which is ready to be measured, almost no difference is observed in that region (purple line)

4. Carefully lay the second window on top of the sample, take the window sandwich in two hands, and gently squeeze the windows together with slight rotation to opposite directions for about one quarter to half of a round (*see* **Note 20**).
5. Place an insulator ring (from Teflon) inside the sample cell, carefully lay the window sandwich on top, place another insulator ring on top of the second glass, and close the sample cell with the top part of the sample cell.
6. Screw the sample cell together by turning until you can feel the contact with the windows.
7. Optimize the sample thickness by utilizing the FTIR spectrophotometer (*see* **Note 21**). Measure a background spectrum (air) without any sample in the compartment. Place the sample in, and measure an absorption spectrum of the sample. The absorbance of the 1650 cm^{-1} (OH-stretching band and amide I) band should be at 0.7–1.0 (Fig. 3a) (*see* **Note 22**).
8. If the absorbance value is below 0.7, start over from Step 1 (*see* **Note 23**).
9. If the absorbance value is above the target, continue to screw the sample cell further together and repeat the step 7.
10. Repeat the step 9 until the target absorbance value of the sample is reached (*see* **Note 24**).
11. Once a proper absorption value is obtained, switch the sample three times back and forth between Pr and Pfr states with illumination LEDs (*see* **Note 25**).

3.2.2 Sample in D_2O

1. Take two CaF_2 windows (*see* **Note 26**) and make sure they are clean. Wipe them with Kimtech wipes using pure water and ethanol/isopropanol to clean them (*see* **Note 17**).
2. Take the sample holder with screws, put insulation ring in the bottom part, lay one window inside the cuvette, and a 50 μm spacer on top of the window (*see* **Note 27**).
3. Load 8 μl of the sample in the center of the window (*see* **Note 19**). Avoid bubbles.
4. Carefully lay the second window on top of the sample. Place a second insulation Teflon ring and the top part of the cuvette on top. Match the screw holes of the cuvette, but do not apply any extra pressure (*see* **Note 28**).
5. Carefully place the screws in the holes and start to tighten them. Make sure that the tightening happens equally in all sides, tighten only a bit at a time, and always move from opposite side after tightening one screw. Let the screws slowly squeeze the sample together. Tighten the screws all the way.
6. Measure a background spectrum (air) without any sample in the compartment. Place the sample in, and measure an absorbance spectrum of the sample. The absorbance of the 1650 cm^{-1} (amide I) band should be at 0.4–0.8 (*see* **Note 29**).
7. Once a proper absorbance value is obtained, switch the sample three times back and forth between Pr and Pfr states with illumination LEDs (*see* **Note 25**).

3.3 Data Collection

3.3.1 Sample Quality Checking

1. Set the measurement parameters to the FTIR program. When measured with the sample concentrations described above and normal steady-state difference spectra, 32 scans per spectrum with a frequency resolution of 2 cm^{-1} can be used (*see* **Note 30**). The spectra are measured and displayed in absorbance.
2. Take a baseline measurement of the sample. To do this, place the sample in the FTIR sample compartment, measure a background, wait for 2 min (the time that illumination and measuring a spectrum would take), and measure a sample spectrum without any changes in the sample conditions.
3. Based on the acquired baseline measurement, ensure that the sample is settled between the windows. Ideally, the resulting spectrum should be a flat line with absorption value of zero which indicates that the sample has not changed between the background and the spectral collection (*see* **Note 31**). If the sample has not settled, negative (or rarely positive) spectral features will be observed at 1650 and 1550 cm^{-1} (Fig. 3b, *see* **Note 32**). If the sample has settled, only symmetrical noise will be observed.

4. Wait until the baseline of the sample stays constant. Once the baseline is settled, the sample is ready for difference spectra measurements.
5. Take an absorption spectrum (air as a background) of your sample before and after your difference spectral measurement (*see* **Note 33**).

3.3.2 Measuring Difference Spectra

1. The difference spectra express the spectral differences between the Pr and Pfr states. Ideally, the sample should remain still in the sample compartment for the whole time, and use of the integrated illumination LEDs is encouraged to minimize other changes in the sample environment (*see* **Note 34**).
2. First, drive the sample to the photoactivated (Pfr) state by illuminating with the 660 nm LED (*see* **Note 35**). Then, turn off the LED and measure the background (*see* **Note 36**).
3. Right after the background is measured, drive the sample to the resting (Pr) state by illuminating with the 780 nm LED (*see* **Note 37**). Switch off the LED and measure a sample spectrum in absorption mode, which will give you the difference spectrum Pr *minus* Pfr.
4. Once the sample spectrum collection is finished, measure the background before switching on any LEDs (Pr becomes the new background).
5. After a new background is measured, drive the sample back to the photoactivated state with LED illumination (Pfr by 660 nm). Switch the LED off and measure a sample spectrum in absorption mode. This is the second difference spectrum Pfr *minus* Pr. After this, measure a new background from the sample in the photoactivated state.
6. Repeat the steps 3–5 for multiple times in order to improve the signal-to-noise ratio (Fig. 4). The number of individual spectra needed depends on the signal size of the studied protein sample and the quality of the FTIR instrument (*see* **Note 38**).

3.4 Data Analysis

1. Start by visually inspecting the spectra. Two opposite difference spectra (i.e., Pr *minus* Pfr and Pfr *minus* Pr) should be mirror images (*see* **Note 39**) of each other, and the two spectra should cross each other where ΔA is zero (Fig. 4a, b) (*see* **Note 40**).
2. Average all the two blocks of difference spectra (from red to far-red, and from far-red to red) together (Fig. 4). These spectra should again be mirror images of each other (*see* **Note 41**).
3. If the two spectra are mirror images, combine the two mirror-imaged spectra into a single spectrum by subtracting Pr *minus* Pfr from Pfr *minus* Pr (Fig. 4b, c). If the signal-to-noise ratio

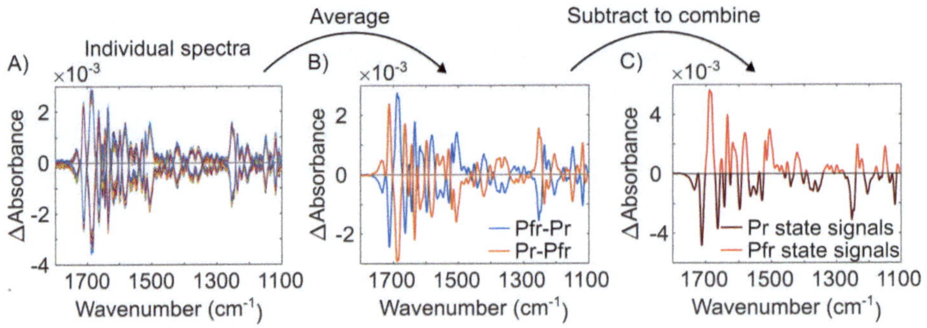

Fig. 4 Analysis of collected data using *Dr*BphP CBD-PHY data as an example. (**a**) A bunch of raw difference spectra, Pfr *minus* Pr and *vice versa.* (**b**) All Pfr *minus* Pr (blue) and all Pr *minus* Pfr spectra (orange) are averaged. The two spectra should be mirror images of each other. (**c**) The mirror images can be brought together by subtracting Pr *minus* Pfr from Pfr *minus* Pr. Note the signal intensity is doubled, but this allows utilizing all collected data in one spectrum. The bleaches (dark red) are present in Pr state but missing in Pfr, and the positive contributions appear in Pfr but are absent in Pr (red)

does not seem adequate, measure and average more data and/or try to improve the signal size by optimizing your sample concentration and the optical path length.

4. The data is now processed, and you have an IR difference spectrum of the sample. In summary, the bleaches are vibrational features that are present in Pr state but disappear upon photoactivation, and the positive signals appear upon illumination (Fig. 4c). The spectral data can be compared to other measured spectra from similar type of phytochromes or variants of the same sample. Information about the protonation states of the side chains, interaction patterns between the chromophore and the protein environment, secondary structural changes along the illumination of the sample is at your hand. Enjoy!

4 Notes

1. We often store our samples in −80 °C in aliquots with about 15–30 mg/ml concentration. For an FTIR experiment, we would take one 100 μl aliquot. The concentration and filtering process often results in about 10–20 % sample loss, and this amount also allows few trials for the assembling of the sample.
2. Other brands' concentrators work as well, albeit in our hands their dead volume tends to be higher.
3. Other buffers can work equally well, but this buffer is typically used for *Deinococcus radiodurans* phytochrome (*Dr*BphP) samples.

4. The buffer pD can be measured with a normal pH probe, but note that pD = pH reading + 0.4.
5. We use a sample cell with grooves for samples in H_2O and holder with screws for samples in D_2O, but these can possibly work crosswise as well. Commercial options for liquid sample cells are also available, for example, from Bruker, Harrick, and Specac.
6. Be sure that no external light gets to the sample space once the experiment is ongoing. We have used, e.g., plastic mats to cover sample compartment windows.
7. We have 3D printed our sample holder, which is attached to the FTIR sample chamber plate with screws. The sample cell mount has a handle for putting the sample in or taking it out. The y-position of the sample relative to the IR beam can be adjusted by loosening or tightening a screw inside the sample holder. The beam path is covered with aluminum tubing to minimize the changes in the air gas concentrations in the beam path. The LEDs are placed in the holder holes toward the sample, in an angle, which directs them away from the detector and thus allows LEDs to be on during measurement. This is beneficial especially for quickly dark-reverting samples.
8. The LED wavelengths should be selected based on the visible light absorption properties of the sample. Look at the absorption spectra of Pr and Pfr spectra, and for red-light illumination, choose a wavelength where the Pfr state does not substantially absorb light (blue edge of Pr spectrum). For far-red-light illumination, choose a wavelength where the Pr absorption has decreased (red edge of Pfr spectrum).
9. Samples are best thawed for example holding the microtube in hand until thawed. To our experience, at least *Dr*BphP samples do not suffer from storage in room temperature during working day. The sample durability can be increased by storing the sample at +4 °C or on ice, but only after the sample has been properly thawed. Storing the sample at − 20 °C or thawing the sample slowly (e.g., on ice) from − 80 °C will destroy the photoswitchability of the system.
10. Aim to load 1–3 mg of protein in the desalting column. Use a volume of 50–150 μl for best result with the desalting column. For sample volumes < 70 μl, apply a 15 μl stacker of the D_2O buffer to the top of the gel bed after the sample has fully absorbed to ensure maximal protein recovery.
11. The H/D exchange is sped up (or even completed) by switching the sample between the two states. The illumination time or power does not need to be accurate as long as the sample is converted and not destroyed by too high power.

12. Samples in D_2O can have a lower concentration due to longer path length and for those 5 to 15 min concentration should be sufficient. For samples in H_2O, 15 to 30 min in total is usually enough. If concentrated for too long, the sample can become viscous. The sample can still be potentially recovered with filtration, but this is not encouraged and extreme concentrations should be avoided.
13. Because the sample volume is very small, the sample is best recovered by centrifuging for 1 min, then turning the tube in the centrifuge by 180 °, and repeating the centrifugation.
14. The absorption spectrum should also be used to estimate the viability of the sample. The $A_{700\ nm}/A_{280\ nm}$ describes the binding of the bilin chromophore (the exact value depends on the sample), and the ratio should remain the same as it was before the quite extreme concentration process.
15. We estimate the extinction coefficient ε_{280} of biliverdin to be around 11800 $M^{-1}\ cm^{-1}$. The biliverdin contribution at 280 nm equals the absorbance value at 350 nm, and a direct subtraction can be made to calculate the concentration based on the protein's extinction coefficient.
16. The sample is best used fresh. If required, the sample can be stored overnight at +4 °C and well sealed, e.g., with Parafilm. In D_2O samples, the risk is exposing the sample to air for a long time resulting in partial D_2O to H_2O exchange. In H_2O samples, the volume is often very small, which poses the risk of drying out for the sample.
17. The final touch of cleaning is best done by wiping, only once to avoid static interactions that attract dust, with a Kimtech wipe wet a bit with ethanol or methanol. Prior to that, the windows are best cleaned with water and distilled water. The wipes can be used to remove any excess sample. If the sealing grease has been used with the window, that is best removed using water and detergent.
18. The grease is best applied by applying a thin layer of it on the top of your clean index finger and then by pressing the side of the window against the finger and rotating the finger along the side of the window allowing the grease to create "a barrier" on the top side of the glass. Make sure to have a continuous ring of grease to make sure the sample is properly sealed. However, keep it thin enough; if too much grease is applied, the sample will become too thick, and not enough light will be passed through for a spectroscopic measurement.
19. Use reverse pipetting or simply do not press the pipette plunger all the way to the bottom to minimize the risk of forming bubbles. If using smaller or larger CaF_2 windows, the sample volume should be adjusted accordingly.

20. When rotating the windows, you should observe spreading of the sample droplet to a circle with about 1 cm diameter. Do not apply force at this point to make the sample area any larger and path length thinner at this point.
21. The sample can be only made thinner (less absorption), not to thicker, at this point so squeeze the sample gently, only one quarter of a round at a time. After tightening the sample, the sample will take a while to set. Therefore, take a (coffee) break and measure the absorbance spectrum of your sample only after about 15 min as the absorbance might change extensively, while the sample is stabilizing between the windows. Also, make sure the FTIR purged and the detector is cooled down if needed by this step.
22. The intensity target of the amide I is based on a) aiming to maximize the signal intensity throughout the difference spectra b) without compromising the signal in the amide I region by allowing enough light (10–20%) to pass through. The thickness of samples in H_2O is about 10 μm. However, it will depend on how tightly the windows are squeezed together. The optimal sample thickness is dictated by the 1650 cm^{-1} vibrational band where both amide I vibrations of the protein and bending of the water are present.
23. Ensure that the sample concentration is high enough (from the UV-vis measurement). You can also visually check the IR absorbance spectrum (Fig. 3); you should see the amide II band at 1580–1480 cm^{-1}, and it should be about half to one third of the amide I band. If this is not the case, the protein concentration is too low, and the sample needs to be concentrated more. If the concentration appears to be good enough, consider slightly increasing the amount of sealing grease, and most importantly, do not squeeze the sample too tightly too early.
24. If the absorbance value stays above 1.0 despite squeezing the sample as tight as possible, the amount of grease is too high, and the sample assembly needs to be started from step 1. In this case, substantially decrease the amount of sealing grease loaded between the windows.
25. This is done to prepare the sample for the difference measurement and stabilize the photoswitching.
26. Alternatively, you can use a CaF_2 window that contains two holes, connected to a syringe system with Luer lock connectors. Such flow-cell sample holder is described in more detail in [25]. In this case, you can pull the sample from a microtube into the sample space with a syringe in Step 3. However, in this case, the sample consumption is considerably higher.

27. Depending on the sample concentration, the spacer can be thicker or thinner. For example, 70 μm or 100 μm spacer can be used for a more diluted sample, or 30 μm if high concentration is achieved. Check that the absorption in the finger print region does not exceed 1.0 (Step 6).
28. During this step, the sample drop should not yet spread.
29. Because in the D_2O sample, the water absorption does not overlap with the amide band, the target absorbance can be lower and a good difference signal still be achieved. If the amide I band absorption is too high or too low, adjust the sample concentration accordingly either by diluting the sample with the D_2O-based buffer (too high absorbance) or by concentrating the sample further (too low absorbance).
30. These settings allow us to retain an acceptable signal-to-noise ratio without compromising the baseline due to long time difference between the background and sample spectra. If required, the settings can be adjusted, for example, with a DTGS detector, and more scans (64 or even 100 scans) should be acquired.
31. Samples in D_2O, assembled with the spacer, rarely have issues with baseline. However, the stabilization of H_2O-based samples can take time from minutes to hours. Take time, and take breaks. The sample cuvette can be turned upside down in the sample holder, which sometimes helps. If the sample has not settled within few hours, to our experience, it probably will not. Therefore, it might be better to give up, and remake the sample. Upon disassembly, leakage of the sample can be observed in cases where the baseline never settled, which is often due to incomplete grease sealing.
32. If signal-like bleaches and positive signals are observed in the same spectrum, it is possible that the light conditions have changed between the background and sample collection. Illuminate the sample to the ground state to minimize the effect of dark reversion. Make sure no external light gets in the sample chamber. Repeat the baseline measurement.
33. Having a "before" and an "after" spectrum of the sample will help to troubleshoot changes in the sample absorption during the measurement which can be hours long. For example, the signal size can drop during the measurement due to slow leakage of the sample, which is not necessarily visible in the baseline check.
34. Integrated LEDs in the sample holder allow keeping the conditions in the sample chamber and with the sample as steady as possible. Opening and closing the sample chamber of the instrument can change the air composition from the

background. Moving the sample has a risk of changing the measurement position that can also result in spectral artifacts.

35. This step is described for a canonical phytochrome sample. With bathy samples, we would also advise to start the measurements by driving the sample to the photoactivated state, i.e., the Pr state. This is because the LED illumination most probably will not result in 100% Pr or Pfr population which might be the case in the thermally reverted sample. In the following steps, perform the illumination with the opposite LED/to the opposite photostate than what is described for canonical samples.

36. The LED can, in principle, be left on when measuring the sample. We suggest to do it only for quickly (within minutes) dark-reverting samples. Make sure that the LED does not affect the measurement; do this without the sample in by measuring a background with the LED off and then a spectrum with the LED on. LED light getting to the detector during a measurement can potentially damage it, but we are not sure about this.

37. The illumination durations depend on the illumination fluxes of the LEDs, on the sample, and the conversion rate of the sample. To test proper illumination parameters, aim to have maximum difference signal without cost of the quality of the baseline, which typically remains the same about 5 min. To test for the maximal signal, one can measure consecutive spectra with the same illumination LED and see if the signal increases between two consecutive scans.

38. The number of spectra needed depends on the instrument sensitivity, and the sample concentration, the achieved path length, and the conversion rate of the sample in relation to the instrument sensitivity/noise. A certain amount of noise is created by the instrument, and that can be affected only by changing the instrument. The amount of signal depends on the sample, and the more there is signal, the better the signal-to-noise ratio. For a sample that is difficult to concentrate to a high concentration, or if the path length is small (e.g., there are less switchable proteins in the light path), or the population difference between Pr and Pfr is small, more spectral repeats are needed to make up for that. Even a triplicate can sometimes be enough. Other times, 50 spectra to both directions (i.e., Pr *minus* Pfr and *vice versa*) are required for adequate signal to noise.

39. The lack of mirror image of the spectra means a lack of two reversible states, i.e., the sample may degrade, aggregate, or something else might have happened during the experiment.

40. If the crossing point is above or below zero, there might be a baseline shift, i.e., the sample is not stabilized in the sample cell. You might want to leave these spectra out of the analysis. The baseline issue possibly does not continue through the dataset.
41. You can multiply one of the spectra by − 1, which brings the spectra to the same side. Ideally, the two spectra should now be identical.

Acknowledgements

We thank Heikki Häkkänen for instrument development, in terms of 3D printing, sample holder design, and LED illumination implementation with Arduino. We thank Brigitte Stücki-Buchli and Nils Lenngren who provided the initial Matlab code for the data analysis. We thank Tilman Kottke and Lea Schröder for sharing and showing their tips and tricks for measuring H_2O diluted samples.

References

1. Siebert F, Hildebrant P (2008) Vibrational spectroscopy in life science tutorials in biophysics Wiley
2. Creighton TE (2010) The physical and chemical basis of molecular biology, 1st edn. Helvetian Press
3. Barth A (2007) Infrared spectroscopy of proteins. Biochimica et Biophysica Acta - Bioenergetics 1767:1073–1101
4. Kottke T, Lórenz-Fonfría VA, Heberle J (2017) The grateful infrared: sequential protein structural changes resolved by infrared difference spectroscopy. J Phys Chem A 121:335–350
5. Goett-Zink L, Klocke JL, Bögeholz LA et al (2020) In-cell infrared difference spectroscopy of LOV photoreceptors reveals structural responses to light altered in living cells. J Biol Chem 295(33):11729–11741
6. Ihalainen JA, Gustavsson E, Schroeder L et al (2018) Chromophore-protein interplay during the phytochrome photocycle revealed by step-scan FTIR spectroscopy. J Am Chem Soc 140: 12396–12404
7. Klocke JL, Mangold M, Allmendinger P et al (2018) Single-shot sub-microsecond mid-infrared spectroscopy on protein reactions with quantum cascade laser frequency combs. Anal Chem 90(17):10494–10500. PMID: 30080964
8. Stritt P, Jawurek M, Hauser K (2023) Mid-IR quantum cascade laser spectroscopy to resolve lipid dynamics during the photocycle of bacteriorhodopsin. J Chem Phys 158(15):154202
9. Van Thor JJ, Ronayne KL, Towrie M (2007) Formation of the early photoproduct Lumi-R of cyanobacterial phytochrome Cph1 observed by ultrafast mid-infrared spectroscopy. J Am Chem Soc 129:126–132
10. Toh KC, Stojković EA, Rupenyan AB et al (2011) Primary reactions of bacteriophytochrome observed with ultrafast mid-infrared spectroscopy. J Phys Chem A 115:3778–3786
11. Kübel J, Chenchiliyan M, Ooi SA et al (2020) Transient IR spectroscopy identifies key interactions and unravels new intermediates in the photocycle of a bacterial phytochrome. Phys Chem Chem Phys 22:9195–9203
12. Foerstendorf H, Benda C, Gärtner W et al (2001) FTIR studies of phytochrome photoreactions reveal the c=o bands of the chromophore: consequences for its protonation states, conformation, and protein interaction. Biochemistry 40:14952–14959
13. Van Thor JJ, Fisher N, Rich PR (2005) Assignments of the Pfr - Pr FTIR difference spectrum of cyanobacterial phytochrome Cph1 using ^{15}N and ^{13}C isotopically labeled phycocyanobilin chromophore. J Phys Chem B 109:20597–20604
14. Takala H, Lehtivuori H, Berntsson O et al (2018) On the (un)coupling of the chromophore, tongue interactions, and overall

conformation in a bacterial phytochrome. J Biol Chem 293:8161–8172

15. Takala H, Niebling S, Berntsson O et al (2016) Light-induced structural changes in a monomeric bacteriophytochrome. Struct Dyn 3:12
16. Lenngren N, Edlund P, Takala H et al (2018) Coordination of the biliverdin d-ring in bacteriophytochromes. Phys Chem Chem Phys 20: 18216
17. Takiden A, Velazquez-Escobar F, Dragelj J et al (2017) Structural and vibrational characterization of the chromophore binding site of bacterial phytochrome agp1. Photochem Photobiol 93:713–723
18. Kurttila M, Stucki-Buchli B, Rumfeldt J et al (2021) Site-by-site tracking of signal transduction in an azidophenylalanine-labeled bacteriophytochrome with step-scan FTIR spectroscopy. Phys Chem Chem Phys 23: 5615–5628
19. Kraskov A, von Sass J, Nguyen AD,et al (2021) Local electric field changes during the photoconversion of the bathy phytochrome agp2. Biochemistry 60(40):2967–2977
20. Buhrke D, Michael N, Hamm P (2022) Vibrational couplings between protein and cofactor in bacterial phytochrome agp1 revealed by 2d-IR spectroscopy. Proc Natl Acad Sci U S A 119:e2206400119
21. Takala H, Björling A, Berntsson O et al (2014) Signal amplification and transduction in phytochrome photosensors. Nature 509:245
22. Giraud E, Fardoux J, Fourrier N et al (2002) Bacteriophytochrome controls photosystem synthesis in anoxygenic bacteria. Nature 417: 202–205
23. Karniol B, Vierstra RD (2003) The pair of bacteriophytochromes from *Agrobacterium tumefaciens* are histidine kinases with opposing photobiological properties. Proc Natl Acad Sci U S A 100:2807–2812
24. Kraskov A, Nguyen AD, Goerling J et al (2020) Intramolecular proton transfer controls protein structural changes in phytochrome. Biochemistry 59(9):1023–1037. PMID: 32073262
25. Bredenbeck J, Hamm P (2003) Versatile small volume closed-cycle flow cell system for transient spectroscopy at high repetition rates. Rev Sci Instrum 74:3188–3189

Chapter 11

Spectroscopic Analyses of Photoconversion in Phytochromes

Chengwei Yi, Stefanie S. M. Meier, Maria Kehr, and Andreas Möglich

Abstract

Phytochrome photoreceptors integrate red and far-red light to inform vital adaptations of physiology, development, and lifestyle in manifold plants, bacteria, and fungi. Once covalently assembled with bilin chromophores, phytochromes traverse their metastable Pr and Pfr states depending on light quality and intensity. This chapter provides protocols for the analyses of phytochromes by absorption spectroscopy to determine their molar extinction coefficients, their photostationary state assumed under constant illumination, and their quantum yields for the light-driven Pr ⇄ Pfr interconversion. Knowledge of these parameters pertains to the natural role of phytochromes and their deployment in optogenetics and biotechnology.

Key words Bilin, Chromophore, Optogenetics, Photobiology, Phytochrome, Photoreceptor, Quantum yield, Spectroscopy

1 Introduction

In nature, organisms must perceive environmental cues to adapt to changing living conditions. Light is one of the most important stimuli, and organisms have evolved numerous sensory photoreceptors to react to it. As one major class, phytochromes (Phy) [1, 2] recur in plants, algae, fungi, and bacteria where they underlie diverse physiological responses to red and far-red light, for example, morphogenesis [3], germination [4], and photoprotection [5]. As different as these processes and the underlying photoreceptors are, the so-called photosensory core module (PCM), responsible for light absorption, is conserved throughout the Phy family. A bilin chromophore nestles within the PCM and is covalently attached to a cysteine residue, thereby granting sensitivity to red and far-red light [1, 2]. Plant Phys use phytochromobilin (PΦB) as their

Chengwei Yi, Stefanie S. M. Meier and Maria Kehr contributed equally with all other contributors.

Andreas Winkler (ed.), *Phytochromes: Methods and Protocols*, Methods in Molecular Biology, vol. 2970,
https://doi.org/10.1007/978-1-0716-4791-2_11,

natural chromophore but can be functionally reconstituted with the cyanobacterial phycocyanobilin (PCB) [6, 7]. By contrast, bacterial Phys incorporate biliverdin (BV) [5, 8], a more oxidized bilin that serves as the precursor to both PΦB and PCB biosynthesis. Bilins exhibit characteristic absorbance bands within the visible portion of the electromagnetic spectrum, namely the Q band in the red/far-red region [9] and the Soret band in the blue region [10]. Not least owing to differing extents of the conjugated π-electron systems in the respective bilins, bacterial Phys display a bathochromic shift by around 50 nm of their Q-band absorbance compared to plant Phys [1, 2, 11]. That notwithstanding, the principal photochemistry playing out upon light absorption is conserved across the Phy superfamily. Phys can adopt two metastable states, called Pr and Pfr (pigment red-absorbing and pigment far-red absorbing [12]), which differ in the isomerization of the C15=C16 double bond of the bilin [1, 2]. In the red-light-absorbing Pr state, the chromophore adopts its 15*Z* isomer, and in the far-red-light-absorbing Pfr state, it is the 15*E* configuration. In plant and prototypical bacterial Phys, the Pr state prevails in darkness, whereas a group of bacterial Phys adopts the Pfr state in darkness and are denoted as bathy-phytochromes. Red and far-red light promotes the photoconversion in the Pr → Pfr and Pfr → Pr directions, respectively. The underlying transition rates between the two states depend on the applied light quality and intensity [13], as well as on the bilin absorption cross section at the excitation wavelength multiplied by the quantum yield (*see* **Note 1**) for photoconversion. Returned to darkness, phytochromes generally revert to their original dark-adapted state in a thermal reaction denoted dark recovery that can be protracted and last up to several days.

This chapter treats the spectroscopic analyses of absorbance and Pr ⇄ Pfr photoconversion in plant and bacterial phytochromes. With the PCMs of *Arabidopsis thaliana* phytochrome B (*At*PhyB) [14], the prototypical Phy from *Deinococcus maricopensis* (*Dm*PCM) [15, 16], and the bathy-phytochrome from *Agrobacterium vitis* (*Av*PCM) [15, 17, 18] as paradigms, we discuss in turn the calculation of molar extinction coefficients for Phy holoproteins (*see* Subheading 3.1), the quantification of the Pr/Pfr ratio assumed at photostationary state under constant red light (*see* Subheading 3.2), and the determination of quantum yields for photoconversion between the Pr and Pfr states (*see* Subheading 3.3). Collectively, the methods collated in this chapter are suited for characterizing photochemical traits in Phys that equally pertain to their natural roles and their applications for biotechnology and optogenetics [6, 16, 19, 20].

2 Materials

1. Purified proteins: *At*PhyB PCM [7, 14], *Dm*PCM [15, 16], and *Av*PCM [15, 18], stored at −80 °C.
2. Native buffer: 50 mM Tris-HCl pH 8.0, 20 mM NaCl. This buffer is used herein, but other buffers can be used as demanded by the protein of interest.
3. Denaturing buffer containing 6 M guanidine hydrochloride. As mentioned earlier, different buffers may be used as dictated by the specific protein under study.
4. UV/vis absorption spectrophotometer: for example, Cary60 (Agilent Technologies), equipped with Peltier temperature control.
5. Quartz glass microcuvette: 1 cm light path, 100 μL volume.
6. Quartz glass fluorescence microcuvette: 0.5 cm light path, 850 μL volume.
7. Light-emitting diodes (LED): for example, all obtained from Roithner Lasertechnik: (650 ± 10) nm (no. H2A1-H650), (658 ± 10) nm (no. JET-660-12), (733 ± 12) nm (no. JET-730-10), (757 ± 12) nm (no. SMB1N-760D-02), and (800 ± 18) nm (no. JET-800-10) (*see* **Note 2**).
8. Three-dimensional (3D)-printed adapters for Cary60 spectrophotometer: available from https://doi.org/10.5281/zenodo.14752680.
9. Power meter: for example, Newport model 842-PE equipped with a 918D-UV-OD3 silicon detector.

3 Methods

3.1 Determination of Phytochrome Extinction Coefficients

The concentration determination of phytochromes and, consequently, their use in any functional in vitro assays hinge on accurate molar extinction coefficients for the bilin-bound holoproteins. Owing to its strong absorbance, the bilin Q band is well suited for pertinent absorption-spectroscopic analyses, with the caveat that its extinction coefficient strongly depends on the surrounding protein scaffold and protonation status. To address this challenge, two separate UV/vis absorbance spectra are recorded at the same phytochrome concentration, but once in native and once in denaturing buffer. As within the denatured state, the extinction coefficients of the protein-bound bilins are to first approximation independent of the protein context, the bilin absorbance properties within the native holo-phytochrome can be inferred by comparison of the two spectra. Using this approach, we routinely determine molar extinction coefficients at the isosbestic point for the Pr ⇄ Pfr transition which affords the advantage that concentrations can be calculated irrespective of the illumination status of the phytochrome sample.

1. All experiments should be performed under dim green safe light to minimize inadvertent photoactivation of the phytochrome samples.
2. Equilibrate the UV/vis absorption spectrophotometer at a temperature of 22 °C and record a blank spectrum for the native buffer.
3. Dilute phytochromes into native buffer (*see* **Note 3**) and load 120 μL into the quartz microcuvette with an optical path length of 1 cm. Record absorbance spectra in darkness and after saturating illumination with LEDs emitting red and far-red light, respectively. For plant phytochromes, 660-nm and 730-nm LEDs can for instance be used to drive the Pr → Pfr and Pfr → Pr transitions, respectively. In the case of bacterial phytochromes, a far-red LED emitting around 800 nm can be deployed instead of the 730-nm LED.
4. Record a blank spectrum for the denaturing buffer. Dilute phytochromes into the denatured buffer by the same factor as above for the native buffer (*see* **Note 4**). Incubate for at least 5 min to allow for denaturation. Record spectra as described above.
5. As required, correct all spectra for potential baseline shifts by considering the absorbance around 800 nm (for plant Phys) or around 900 nm (for bacterial Phys). Plot the spectra, for example, with the freely available Fit-o-mat software [21] (*see* Fig. 1).
6. Calculate the chromophore concentration in the Phy proteins based on the denatured spectra and the molar extinction coefficients of the protein-bound bilins. For PΦB, PCB, and BV, the relevant extinction coefficients within the Soret band are 36,000 M^{-1} cm^{-1} at 380 nm [22], 34,700 M^{-1} cm^{-1} at 352 nm [22], and 39,900 M^{-1} cm^{-1} at 388 nm [23], respectively.
7. With the knowledge of the bilin concentration, calculate the molar extinction coefficients of the native holoproteins at the isosbestic point (or, at other desired wavelengths) via the Beer-Lambert law [24, 25] (*see* **Note 5**, Table 1).

3.2 Determination of the Pr/Pfr Ratio at Photostationary State

Due to the considerable overlap between the Pr and Pfr action spectra, illumination with red light generally populates a Pr/Pfr mixture rather than the pure Pfr state (*see* Fig. 1). By contrast, exposure to far-red light can in principle drive the phytochrome to its Pr state essentially in entirety. Not least for quantitative

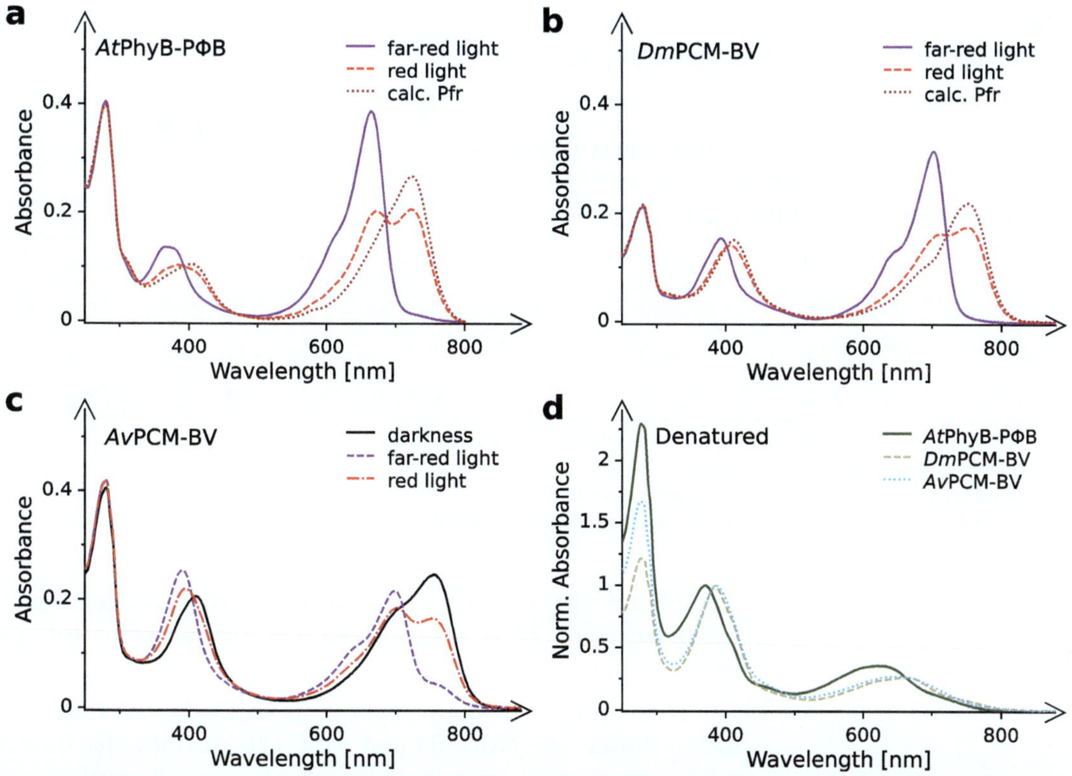

Fig. 1 Spectroscopic analyses of phytochromes. (**a**) UV/vis absorbance spectra of phytochromobilin (PΦB)-bound photosensory core module (PCM) of the *Arabidopsis thaliana* phytochrome B (*At*PhyB) in its Pr state (purple solid line) after far-red light illumination and in the mixed Pr/Pfr state (red dashed line) following red-light illumination. The dotted line represents the calculated absorbance spectrum for the pure Pfr state as described in Subheading 3.2. (**b**) As in (**a**), but for the biliverdin (BV)-bound PCM from the *Deinococcus maricopensis* phytochrome (*Dm*PCM). (**c**) As in (**a**), but for the BV-bound PCM of the *Agrobacterium vitis* phytochrome (*Av*PCM) in its dark-adapted Pfr state (black solid line), in its Pr state (purple dashed line) after far-red light illumination, and in the mixed Pr/Pfr state (red dash-dotted line) under red light. (**d**) Absorbance spectra of the denatured *At*PhyB (dark green solid line), *Dm*PCM (light green dashed line), and *Av*PCM (blue dotted line) samples from panels (**a**–**c**) normalized to their Soret absorbance maxima

analyses, for example, Yi et al. [13], it is often imperative to know the precise Pr and Pfr proportions assumed at photostationary state under red light. To this end, Butler et al. [26] developed a formalism to calculate the Pr/Pfr ratio at the photostationary state and hence also the pure Pfr absorbance spectrum. We describe in the following the application of the Butler method to the *At*PhyB and *Dm*PCM samples as representatives for plant and bacterial phytochromes.

1. All experiments should be performed under dim green safe light to minimize inadvertent photoactivation of the phytochrome samples.

Table 1
Collated parameters of the investigated phytochromes

Phytochrome	Bilin	Isosbestic point [nm]	Holoprotein extinction coefficient[a] $[M^{-1}\ cm^{-1}]$	Pr/Pfr ratio	Quantum yield[b]		
						λ [nm]	Φ_l
*At*PhyB	PΦB	686	41,300	0.24:0.76[c]	Pr → Pfr	658	0.20 ± 0.02
*At*PhyB	PCB	674	48,200	0.27:0.73[c]	Pr → Pfr	658	0.24 ± 0.01
*Dm*PCM	BV	721	36,400	0.21:0.79[c]	Pr → Pfr	660	0.23 ± 0.01
*Av*PCM	BV	710	38,900	0.21:0.79[d]	Pfr → Pr	757	0.12 ± 0.01

[a]The extinction coefficients are reported for the isosbestic points listed in the table, *see* Subheading 3.1
[b]Determined as described in Subheading 3.3
[c]Calculated according to Butler et al. [26], *see* Subheading 3.2
[d]Determined as a linear combination of the near-pure Pr and Pfr spectra

2. Equilibrate the UV/vis absorption spectrophotometer at a temperature of 22 °C and record a blank spectrum for the native buffer.
3. Dilute the phytochrome protein in its dark-adapted state into native buffer (*see* **Note 3**) and load 120 μL into the quartz microcuvette with an optical length of 1 cm. Record absorbance spectra in darkness and after saturating exposure to red (emission around 650–660 nm) and far-red LEDs (emission around 730 nm for plant Phys and around 800 nm for bacterial ones), respectively.
4. Populate the Pr state by exposing the sample to saturating far-red light.
5. Expose the Pr-state sample to red light at an intensity of around 1 mW cm^{-2} while tracking the resultant Pr → Pfr conversion kinetics by recording absorbance at 725 nm (*At*PhyB) or 760 nm (*Dm*PCM) (*see* Fig. 2, **Note 6**).
6. Partially populate the Pfr state by exposing the sample to saturating red light.
7. Expose sample to far-red light at an intensity of around 1–2 mW cm^{-2} and track the resultant Pfr → Pr conversion kinetics by recording absorbance at 725 nm (*At*PhyB) or 760 nm (*Dm*PCM) (*see* Fig. 2).
8. Evaluate the Pr ⇄ Pfr photoconversion kinetics [26] by fitting the time-resolved absorbance data to a single-exponential function Eq. (1), for example, using the Fit-o-mat software [21]:

$$A(t) = A_0 + A_1 \times \exp(-k_i \times t) \tag{1}$$

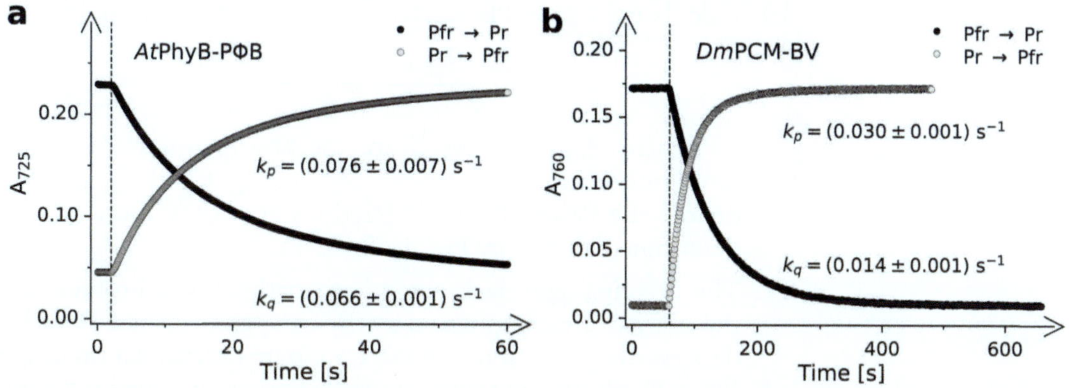

Fig. 2 Analyses of the Pr ⇄ Pfr photoconversion kinetics in phytochrome photosensory core modules (PCM) to determine the Pr/Pfr ratio adopted under red light according to Butler [26]. (**a**) Pr ⟶ Pfr (greycircles) and Pfr ⟶ Pr photoconversion kinetics (black circles) for the phytochromobilin (PΦB)-bound PCM of *Arabidopsis thaliana* phytochrome B (*At*PhyB) under red (660 nm, 1 mW cm^{-2}) and far-red light (730 nm, 2 mW cm^{-2}), respectively. The dashed vertical line marks the onset of irradiation, and the ensuing kinetics were monitored by the absorbance at 725 nm. The values for k_p and k_q in the graph represent mean ± SD of three technical replicates. (**b**) As for (**a**), but for the biliverdin (BV)-bound PCM from the *Deinococcus maricopensis* phytochrome (*Dm*PCM). The Pr ⟶ Pfr and Pfr ⟶ Pr transitions were driven by exposure to 650-nm (1 mW cm^{-2}) and 800-nm light (2.5 mW cm^{-2}), respectively, and followed by the absorbance at 760 nm

where A_0 is the final absorbance reading, $A_0 + A_1$ is the initial absorbance value, and k_i represents the rate constant of the Pr ⇄ Pfr photoconversion kinetics for a given illumination regime. For the Pr ⟶ Pfr and Pfr ⟶ Pr conversion kinetics, we denote the underlying rate constants as k_p and k_q, respectively (*see* Fig. 2).

9. Calculate the mole fractions of P_R and P_{FR} at red-light induced photostationary state according to Eqs. (2, 3) adapted from Butler et al. [26]:

$$\frac{C_R}{C_{FR}} = \frac{E_{FR}}{E_R} \times \frac{A_{light}^{FR}}{A_{dark}^{R}} \times \frac{k_p}{k_q} \tag{2}$$

$$P_R = 1 - P_{FR} = \frac{1}{1 + C_R/C_{FR}} \times \frac{A_{light}^{R}}{A_{dark}^{R}} \tag{3}$$

where C_R and C_{FR} are the photoconversion amplitudes for red and far-red light with photon fluxes of E_R and E_{FR}, respectively (*see* **Note 7**). A_{dark}^{R} and A_{light}^{R} refer to the absorbance within the dark-adapted and the red-light-induced photostationary states at the emission maximum of the red-light source. (Note that the dark-adapted Pr spectrum can be obtained upon saturating application of far-red light.) Likewise, A_{light}^{FR} denotes the absorbance of the photostationary state at the peak wavelength of the far-red-light source.

10. Calculate the pure Pfr spectrum A_{Pfr} according to Eq. (4):

$$A_{\text{Pfr}}(\lambda) = \frac{A_{\text{light}}(\lambda) - A_{\text{dark}}(\lambda) \times P_{\text{R}}}{P_{\text{FR}}} \tag{4}$$

where A_{dark} and A_{light} are the absorbance spectra after saturating exposure to far-red and red light and P_{R} and P_{FR} denote the Pr and Pfr mole fractions at photostationary state determined above (*see* Fig. 1, Table 1).

11. The principal approach extends to bathy-phytochromes, but the analyses are complicated by the comparatively fast Pr → Pfr dark recovery which may hinder accurate determination of the Pfr ⇄ Pr photoconversion kinetics. Rather, an alternative and simpler approach is feasible as the pure Pr-state and Pfr-state spectra can be directly recorded and need not be calculated. That is because bathy-phytochromes adopt their Pfr state in darkness, and far-red light can prompt the complete photoconversion to the Pr state. With the knowledge of the pure states, Pr/Pfr mixtures assumed at photostationary state can be analyzed in terms of a linear combination of the pure Pr and Pfr spectra.

3.3 Determination of Photoconversion Quantum Yields

This section regards the Pr ⇄ Pfr interconversion efficiency and the associated photoconversion quantum yields (Φ_i). As described in Yi et al. [14] (*see* **Note 8**) and similar to Stadler et al. [27], the quantum yield Φ_p for the Pr → Pfr photoconversion of prototypical phytochromes like *At*PhyB and *Dm*PCM can be inferred from absorption spectroscopy (*see* **Note 9**). Likewise, the quantum yield Φ_q for the Pfr → Pr photoconversion can be determined for bathy-phytochromes, for example, *Av*PCM.

1. All experiments should be performed under dim green safe light to minimize inadvertent photoactivation of the phytochrome samples.
2. Equip UV/vis spectrophotometer with a cuvette holder for the fluorescence microcuvette with 0.5 cm optical path length, for example, using the template file for a 3D-printed adapter for the Cary60 instrument, as supplied in the Subheading 2 (*see* **Note 10**).
3. Equilibrate the UV/vis absorption spectrophotometer at ambient temperature and record a blank spectrum for the native buffer.
4. Dilute the dark-adapted phytochrome sample into native buffer (*see* **Note 1**) and load 400 μL into the 0.5-cm fluorescence cuvette. Record absorbance spectra after saturating exposure to red (emission around 650–660 nm) and far-red LEDs (emission around 730 nm for plant Phys, and around 800 nm for bacterial ones), respectively.

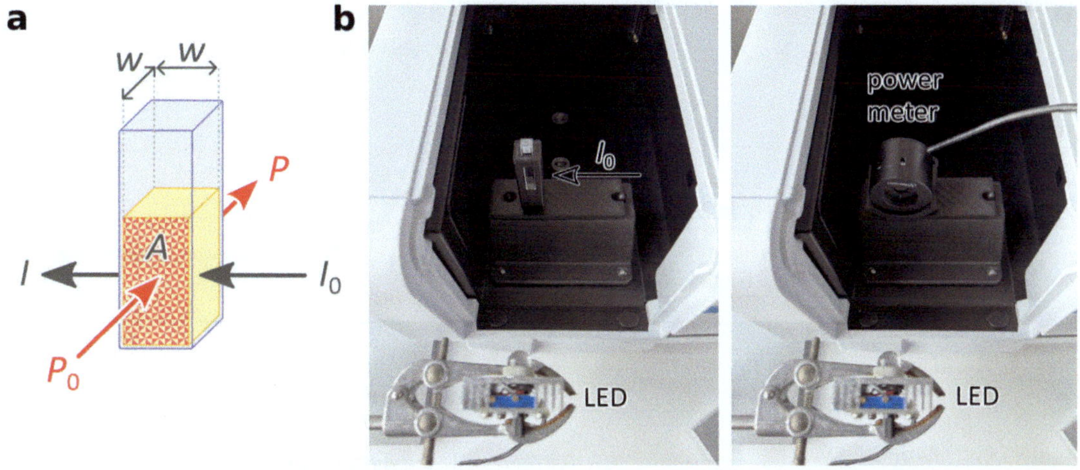

Fig. 3 Experimental setup for the photoconversion quantum-yield measurements. (**a**) Schematic of the experiment. Protein samples are loaded into a quartz fluorescence microcuvette of path length w and are exposed to actinic light perpendicular to the probe beam of the UV/vis spectrophotometer at a light power of P_0 (red arrow). The light intensity of the probe beam I_0 and the light intensity I passing through the sample are indicated by black arrows. (**b**) Photograph of the experimental setup schematized in panel (**a**) (left). Setup for determination of the power P_0 of the actinic light source using a light power meter (right)

5. Calculate the holoprotein concentration c_0 using the Beer–Lambert law [25, 26] with the extinction coefficients determined in Subheading 3.1 (*see* Table 1).
6. Remove the cuvette and position an actinic light source in front of the instrument perpendicular to the UV/vis detection beam of the spectrophotometer (*see* Fig. 3). To drive the Pr → Pfr transition of the prototypical *At*PhyB and *Dm*PCM, LEDs emitting around 650 to 660 nm are used. In the case of the bathy-phytochrome *Av*PCM, a 760-nm LED is employed to promote the Pfr → Pr photoconversion. Adjust the light intensity P_0 of the actinic LED to around 0.5 mW cm^{-2} using a light power meter.
7. Switch off the actinic LED and place the cuvette back into the instrument. Convert the sample to its dark-adapted state. To this end, *At*PhyB and *Dm*PCM are illuminated with LEDs emitting around 730 nm and 800 nm, respectively, to drive the Pfr → Pr photoconversion. Owing to its fast dark recovery [17, 18], the *Av*PCM can be incubated inside the spectrophotometer for around 5 minutes to passively return to its dark-adapted Pfr state.
8. Switch on the actinic light and follow the Pr → Pfr photoconversion (for *At*PhyB and *Dm*PCM) or the Pfr → Pr photoconversion (for *Av*PCM), respectively, by tracking the absorbance at the peak wavelength λ of the actinic light source (*see* **Note 9**, Fig. 4).

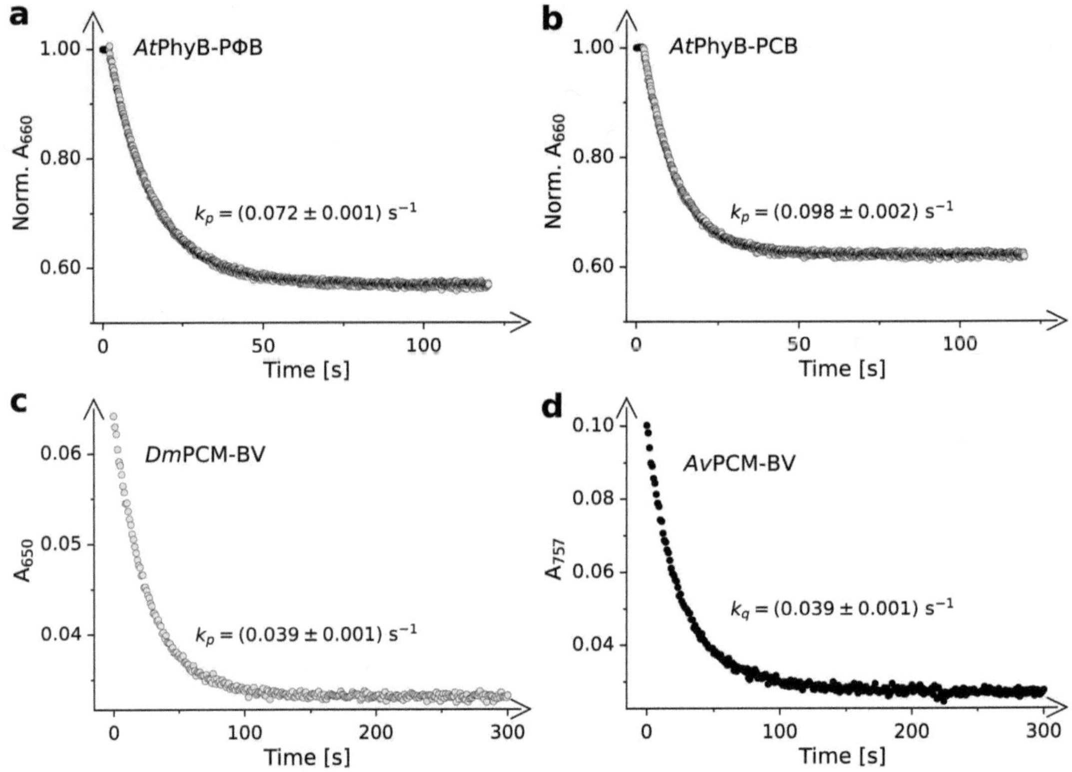

Fig. 4 Photoconversion kinetics of phytochrome photosensory core modules (PCM). (**a**) Pr → Pfr conversion kinetics of phytochromobilin (PΦB)-bound PCM of *Arabidopsis thaliana* phytochrome B (*At*PhyB) during irradiation with 660-nm light (0.45 mW cm^{-2}) as monitored by the absorbance at 660 nm. The photoconversion rate constant stated in the graph refers to the mean ± SD of three technical replicates. (**b**) As in (**a**), but for the *At*PhyB PCM bound to phycocyanobilin (PCB). (**c**) As in (**a**), but for the biliverdin (BV)-bound PCM from the *Deinococcus maricopensis* phytochrome (*Dm*PCM) exposed to 650-nm light (0.5 mW cm^{-2}) and monitored at 650 nm. (**d**) Pfr → Pr conversion kinetics of the BV-bound PCM of the *Agrobacterium vitis* phytochrome (*Av*PCM) exposed to 757-nm light (0.5 mW cm^{-2}) as tracked by the absorbance at 757 nm.

9. Evaluate the resultant absorbance kinetics by fitting to a single-exponential function Eq. (1), for example, using the Fit-o-mat software [21], and thereby determine the photoconversion rate constant k_i (*see* Fig. 4).
10. Calculate the photoconversion quantum yield Φ_i according to Eq. (5) (Table 1):

$$\Phi_i = \frac{k_i \times c_0 \times w \times h \times c \times N_A}{\lambda \times P_0 \times \left(1 - 10^{-A_0}\right)} \tag{5}$$

where k_i is the photoconversion rate constant, c_0 is the sample concentration inside the cuvette, w is the cuvette width, λ is the peak wavelength of the actinic LED source, h is Planck's constant (6.63×10^{-34} J s), c denotes the speed of light (3×10^8 m s^{-1}), N_A is Avogadro's constant

(6.022×10^{23} mol^{-1}), P_0 denotes the applied light intensity (in J m^{-2} s^{-1}), and A_0 is the initial absorbance at wavelength λ of the dark-adapted sample prior to illumination. For more details, *see* **Note 8** and Yi et al. [14].

11. In the case of prototypical phytochromes, the knowledge of the Pr → Pfr photoconversion quantum yield Φ_p enables the calculation of the quantum yield Φ_q for the reverse Pfr → Pr photoconversion at the same excitation wavelength [26] (*see* **Note 11**). The pure Pr and Pfr spectra determined in Subheading 3.2 yield the absorbance values for the two states, A_{Pr} and A_{Pfr}, at a given excitation wavelength λ. Then, the quotient $(1 - 10^{-APr})/(1 - 10^{-APfr})$ is proportional to the number of photons absorbed by the Pr versus the Pfr state. Upon saturating illumination at the excitation wavelength λ, a photostationary state with Pr and Pfr proportions of P_R and P_{FR} is assumed (*see* Subheading 3.2). The ratio of the quantum yields Φ_p and Φ_q for the bidirectional Pr ⇄ Pfr interconversion is then just given by Eq. (6):

$$\Phi_q/\Phi_p = P_R/P_{FR} \times \left(1 - 10^{-A_{Pr}}\right)/\left(1 - 10^{-A_{Pfr}}\right) \quad (6)$$

Similarly, for bathy-phytochromes, Φ_p can be derived from the experimentally determined value of Φ_q.

4 Notes

1. The quantum yield denotes the number of defined events, here Pr ⇄ Pfr photoconversion, occurring per photon absorbed.
2. Other lamp, LED, or laser light sources emitting at similar wavelengths and with sufficient output power can be used as well. Generally, lasers bring the advantage of narrower emission spectra compared to LEDs. As required, LED and lamp emission spectra can be narrowed by using band-pass interference filters.
3. Adjust concentration such that the peak absorbance within the Q band ranges between 0.1 and 0.2.
4. It is imperative that the dilution into the native and denaturing buffers be prepared in the same manner, that is, using a joint dilution factor. Any inaccuracy in preparing the dilutions will directly factor into the subsequent analyses. Ideally, repeat the experiment several times and average the results.
5. The isosbestic point within the Q band is determined for the native samples and refers to the intersection of the absorbance traces obtained upon illumination with red and far-red light. Note that this approach implicitly assumes a two-state transition between the Pr and Pfr forms of the phytochrome. If

additional states are populated throughout the photocycle, isosbestic points where the absorbance spectra of all these states would intersect may not exist.

6. The subsequent analyses require a stable LED output. Therefore, turn on the LEDs 30 min before measurement to reduce intensity fluctuations.

7. Light intensities are measured with a power meter and assumed to be in units $W \times cm^{-2}$. For use in Eq. (2), the intensity can be converted into units of $photons \times m^{-2} \times s^{-1}$ with Eq. (7):

$$E_\lambda = \frac{I_\lambda \times \lambda}{h \times c \times N_A} \quad (7)$$

where E_λ is photon flux per unit time and area, I_λ is the measured light intensity per area at the excitation wavelength λ, h is Planck's constant, N_A is Avogadro's constant, and c is the speed of light.

8. The quantum yield Φ_i can be calculated as described [14] by correlating the absorbed light power per unit time with the initial drop in the Pr-state population (or, Pfr for *Av*PCM). According to the Beer–Lambert law [24, 25], the light intensity $d\Delta P/dt$ absorbed by the sample per unit time and area amounts to:

$$d\Delta P/dt = P_0 \times \left(1 - 10^{-A_\lambda(t)}\right) \quad (8)$$

where $A_\lambda(t)$ denotes the sample absorbance at the peak wavelength λ of the actinic light. Converting the light intensity to photon flux yields the number of absorbed photons per unit time and area, dn'/dt:

$$dn'/dt = d\Delta P/dt \times \lambda \times (h \times c \times N_A)^{-1} \quad (9)$$

where h is the Planck constant, c is the speed of light, and N_A is Avogadro's number. Considering the cross-sectional area of the cuvette (A), one obtains the number of absorbed photons within the cuvette per unit of time:

$$dn/dt = A \times dn'/dt = A \times d\Delta P/dt \times \lambda \times (h \times c \times N_A)^{-1} \quad (10)$$

The absorbance kinetics provide the decrease of Phy concentration c_i in the Pr state (for *At*PhyB and *Dm*PCM) or the Pfr state (for *Av*PCM) per unit time:

$$dc_i/dt = -k_i \times c_i(t) \quad (11)$$

Multiplication by the reaction volume V yields the decrease in the number of Pr-state (for *At*PhyB and *Dm*PCM) or Pfr-state (for *Av*PCM) molecules over time:

$$dn_i/dt = dc_i/dt \times V = -k_i \times c_i(t) \times A \times w \quad (12)$$

where w is the width of the cuvette (*see* Fig. 3).

Accordingly, the quantum yield Φ_i corresponds to the ratio of reacting molecules per unit time, -dn_i/dt, over the number of absorbed photons per time, dn/dt:

$$\Phi_i = \frac{-\mathrm{d}n_i/\mathrm{d}t}{\mathrm{d}n/\mathrm{d}t} = k_i \times c_i(t) \times \frac{w}{\lambda \times (h \times c \times N_A)^{-1} \times P_0 \times \left(1 - 10^{-A_\lambda(t)}\right)} \quad (13)$$

At the beginning of the experiment, that is, $t \rightarrow 0$, the quantum yield Φ_i can be thus calculated from the measurable quantities k_i, c_0, P_0, and A_0 as described in Eq. (5).

9. The photoconversion quantum yield Φ_i is determined for a specific temperature T and excitation wavelength λ. The absorption of light of other wavelengths may drive different electronic and vibronic transitions within the bilin, and hence, the resultant quantum yield for photoconversion may differ in principle. However, as further elaborated in **Note 11**, evidence exists that the photoconversion quantum yield remains largely invariant across the entire Q band at least.
10. A cuvette with 0.5 cm path length is chosen to lower sample consumption and to reduce the inner-filter effect. Other cuvettes (and, if required, sample adapters) may be used provided that the actinic light illuminates the entire sample surface A and passes through the cuvette without being reflected back (*see* Fig. 3).
11. Apart from it being straightforward, the calculation of the quantum yield for the reverse photoconversion is mandated by at least two principal considerations. First, owing to the substantial spectral overlap between the Pr and Pfr states (*see* Fig. 1), many light qualities, for example, red light, will drive both the Pr → Pfr and the Pfr → Pr photoconversion. The flux in either direction is proportional to the respective molar extinction coefficients, that is, for the Pr and Pfr states, at the excitation wavelength multiplied by the quantum yields Φ_p and Φ_q, respectively. In turn, these quantities govern the photostationary state and the speed of bidirectional Pr ⇄ Pfr interconversion under constant illumination. As we recently elucidated for plant phytochromes [13], the knowledge of these parameters is of paramount importance when assessing and rationalizing the downstream signaling responses. Second, to the extent that action spectra of phytochromes have been determined, they often closely resemble their absorbance spectra, at least within the Q band. This immediately argues for largely uniform photoconversion quantum yields across the entire Q band, as indeed experimentally borne out for plant

phytochromes [26]. Therefore, it is expected that the determination of Φ_p and Φ_q at a discrete wavelength provides a good estimator for other wavelengths within the Q band, too.

Acknowledgments

We thank our coworkers for the discussion. This work was supported by the European Commission (FET Open NEUROPA grant 863214 to A.M.) and the German Research Foundation (grant MO2192/4 2 to A.M.).

References

1. Rockwell NC, Lagarias JC (2010) A brief history of phytochromes. ChemPhysChem 11: 1172–1180. https://doi.org/10.1002/cphc.200900894
2. Rockwell NC, Lagarias JC (2020) Phytochrome evolution in 3D: deletion, duplication, and diversification. New Phytol 225:2283–2300. https://doi.org/10.1111/nph.16240
3. Legris M, Ince YÇ, Fankhauser C (2019) Molecular mechanisms underlying phytochrome-controlled morphogenesis in plants. Nat Commun 10:1–15. https://doi.org/10.1038/s41467-019-13045-0
4. Borthwick HA, Hendricks SB, Parker MW et al (1952) A reversible photoreaction controlling seed germination. PNAS 38:662–666
5. Davis SJ, Vener AV, Vierstra RD (1999) Bacteriophytochromes: phytochrome-like photoreceptors from nonphotosynthetic eubacteria. Science 286:2517–2520. https://doi.org/10.1126/science.286.5449.2517
6. Shimizu-Sato S, Huq E, Tepperman JM, Quail PH (2002) A light-switchable gene promoter system. Nat Biotechnol 20:1041–1044. https://doi.org/10.1038/nbt734
7. Golonka D, Fischbach P, Jena SG et al (2019) Deconstructing and repurposing the light-regulated interplay between Arabidopsis phytochromes and interacting factors. Commun Biol 2:448. https://doi.org/10.1038/s42003-019-0687-9
8. Hughes J, Lamparter T, Mittmann F et al (1997) A prokaryotic phytochrome. Nature 386:663. https://doi.org/10.1038/386663a0
9. Scheidt WR (1978) Porphyrin stereochemistry. In: The porphyrins. Academic Press, Cambridge, MA, USA. pp 436–511
10. Soret J-L (1883) Analyse spectrale: Sur le spectre d'absorption du sang dans la partie violette et ultra-violette. Compt Rend 97:1269–1270
11. Hughes J, Winkler A (2024) New insight into phytochromes: connecting structure to function. Annu Rev Plant Biol 75:153–183. https://doi.org/10.1146/annurev-arplant-070623-110636
12. Butler WL, Norris KH, Siegelman HW, Hendricks SB (1959) Detection, assay, and preliminary purification of the pigment controlling photoresponsive development of plants. PNAS 45:1703–1708
13. Yi C, Gerken U, Tang K et al (2024) Plant phytochrome interactions decode light and temperature signals. Plant Cell 36(12): 4819–4839. https://doi.org/10.1093/plcell/koae249
14. Yi C, Golonka D, Meier SSM et al (2025) Commutability of bilin chromophores in plant phytochromes. Under review.
15. Xu Q, Vogt A, Frechen F et al (2024) Engineering bacteriophytochrome-coupled photoactivated adenylyl cyclases for enhanced optogenetic cAMP modulation. J Mol Biol 168257. https://doi.org/10.1016/j.jmb.2023.168257
16. Meier SSM, Multamäki E, Ranzani AT et al (2024) Leveraging the histidine kinase-phosphatase duality to sculpt two-component signaling. Nat Commun 15:4876. https://doi.org/10.1038/s41467-024-49251-8
17. Rottwinkel G, Oberpichler I, Lamparter T (2010) Bathy phytochromes in rhizobial soil bacteria. J Bacteriol 192:5124–5133. https://doi.org/10.1128/JB.00672-10
18. Meier SSM, Hörzing M, Böhm C, Düthorn ELR, Takala H, Uebe R, Möglich A (2025) Engineering NIR-Sighted Bacteria. eLife. https://doi.org/10.7554/eLife.107069.1

19. Deisseroth K, Feng G, Majewska AK et al (2006) Next-generation optical technologies for illuminating genetically targeted brain circuits. J Neurosci 26:10380–10386. https://doi.org/10.1523/JNEUROSCI.3863-06.2006
20. Tang K, Beyer HM, Zurbriggen MD, Gärtner W (2021) The red edge: bilin-binding photoreceptors as optogenetic tools and fluorescence reporters. Chem Rev 121:14906–14956. https://doi.org/10.1021/acs.chemrev.1c00194
21. Möglich A (2018) An open-source, cross-platform resource for nonlinear least-squares curve fitting. J Chem Educ 95:2273–2278. https://doi.org/10.1021/acs.jchemed.8b00649
22. Brandlmeier T, Scheer H, Rüdiger W (1981) Chromophore content and molar absorptivity of phytochrome in the Pr form. Zeitschrift für Naturforschung C 36:431–439. https://doi.org/10.1515/znc-1981-5-615
23. Shu X, Royant A, Lin MZ et al (2009) Mammalian expression of infrared fluorescent proteins engineered from a bacterial phytochrome. Science 324:804–807. https://doi.org/10.1126/science.1168683
24. Beer (1852) Bestimmung der Absorption des rothen Lichts in farbigen Flüssigkeiten. Ann Phys 162:78–88. https://doi.org/10.1002/andp.18521620505
25. Lambert JH (1760) Photometria, sive de mensura et gradibus luminis, colorum et umbrae. Eberhard Klett, Augsburg
26. Butler WL, Hendricks SB, Siegelman HW (1964) Action spectra of phytochrome in vitro. Photochem Photobiol 3:521–528. https://doi.org/10.1111/j.1751-1097.1964.tb08171.x
27. Stadler E, Eibel A, Fast D et al (2018) A versatile method for the determination of photochemical quantum yields via online UV-Vis spectroscopy. Photochem Photobiol Sci 17:660–669. https://doi.org/10.1039/C7PP00401J

Part III

Structural Analysis Approaches

Chapter 12

Small-Angle X-Ray Scattering Studies on Phytochromes

Mao Oide and Masayoshi Nakasako

Abstract

Small-angle X-ray scattering (SAXS) is one of the techniques to investigate the molecular structures of proteins in solution. Here, we describe how SAXS of proteins are measured using synchrotron X-rays and what kind of structural information are obtainable on the low-resolution structures of phytochromes in solution in referring to our previous SAXS studies on phytochrome A and phytochrome B.

Key words Small-angle X-ray scattering, Phytochrome, Photoconversion, Low-resolution structure, Synchrotron X-ray, Size-exclusion chromatography

1 Introduction

Phytochrome is a photoreceptor protein displaying the photoreversible conversion between the red-light absorbing form (Pr form) and the far-red light absorbing form (Pfr form) [1]. The atomic structures of phytochromes are necessary to understand the physicochemical mechanisms underlying red-light-induced biological events such as photomorphogenesis in higher plants. X-ray crystallography [2] and cryogenic electron microscopy (cryo-EM) [3–5] are the major techniques to yield the atomic structures of proteins in crystals and frozen-hydrated solution, respectively. Especially, X-ray crystallography has contributed to determine the structures of various bacteriophytochromes [6–11] and a truncated module of plant phytochrome [12]. In addition, cryogenic electron microscopy (cryo-EM) made great contributions to determine the structures of intact phytochromes A [13] and B [14] from *Arabidopsis thaliana*, which were never crystallized. However, even in the cryo-EM structures of those plant phytochromes, flexibly fluctuating domains are missing in the electrostatic potential maps. Furthermore, it is still difficult to visualize the structures of the Pfr forms of

Andreas Winkler (ed.), *Phytochromes: Methods and Protocols*, Methods in Molecular Biology, vol. 2970,
https://doi.org/10.1007/978-1-0716-4791-2_12,

the full-length phytochromes possibly due to the photostationary equilibrium of the Pr and Pfr forms. Hence, only several partial structures of the Pfr form have been solved [15–19].

In contrast to the two techniques, small-angle X-ray scattering (SAXS) is a technique to investigate low-resolution structures of proteins in solution [20–22]. As SAXS measurements are applied to solution samples, it is easy to study conformational responses of proteins to external physicochemical stimuli, such as light irradiation [23–29], changes in chemical environment [30, 31], and conformational changes under enzymatic reaction [32]. In addition, by combining molecular dynamics simulation, SAXS data may provide the information on protein dynamics in solution including hydration structures [33], which are not accessible with frozen-hydrated structures in X-ray crystallography and cryo-EM [2, 34].

If we have an amount of purified phytochrome solution sufficient to perform SAXS measurements, SAXS studies may provide the structural and physicochemical information of phytochromes in solution [35–41]. In this chapter, we first describe briefly the theoretical background of SAXS and structural information obtainable from SAXS, such as the molecular weight, radius of gyration, maximum dimension, and low-resolution molecular shape of a protein. Next, we introduce how we conduct SAXS measurements on a specific phytochrome system to reveal low-resolution molecular structures and conformational changes in the photoconversion including examples of our previous SAXS studies on phytochrome A (phyA) [36, 41] and phytochrome B (phyB) [40].

1.1 Theoretical Background of Small-Angle X-Ray Scattering

A typical SAXS measurement is very simple as illustrated in Fig. 1. A protein solution at a concentration of a few mg mL^{-1} is irradiated by X-rays with a wavelength of approximately 1 Å, and X-rays scattered by proteins are recorded by an area detector. When using synchrotron as a light source, the exposure time is shorter than 60 s. In solution, protein molecules are randomly oriented against the direction of the incident X-ray beam and under equilibrium of conformations possible in solution. Therefore, we obtain an isotropic scattering pattern averaged over the possible conformations and orientations. After subtracting the scattering from the buffer solution, the SAXS profile, the variation of the scattering intensity depending on the scattering angle, is obtained by circularly averaging the SAXS pattern.

1.2 Structural Information Obtainable from Small-Angle X-Ray Scattering

When a monodisperse protein solution is sufficiently diluted to ignore intermolecular interactions between protein molecules, the net scattering intensity profile from a protein solution $I(S)$ at scattering vector length S is proportional to that from a single protein as [20–22, 42].

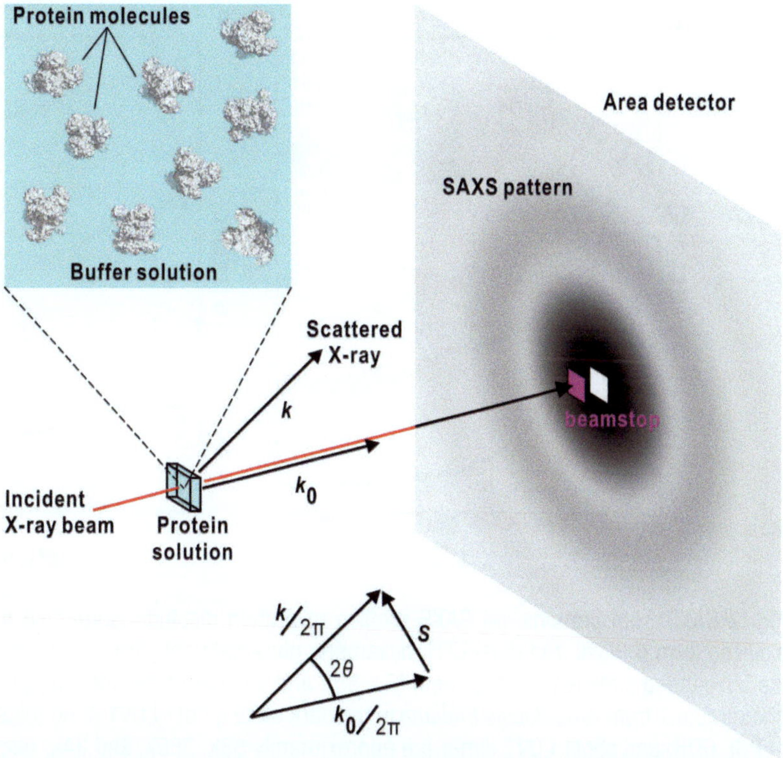

Fig. 1 Schematic of SAXS. Protein solution is irradiated by X-rays, and the scattering pattern is recorded by an area detector. $\boldsymbol{k}$ and $\boldsymbol{k}_0$ are the propagation vectors of the incident and scattered X-rays, respectively. The scattering vector $\boldsymbol{S}$ is defined as the difference between the two propagation vectors as $(\boldsymbol{k} - \boldsymbol{k}_0)/2\pi$. The length of the scattering vector is defined using the scattering angle 2θ and the X-ray wavelength λ as $|\boldsymbol{S}| = 2\sin\theta/\lambda$. (Published with kind permission of Mao Oide and Masayoshi Nakasako, 2025. All Rights Reserved)

$$
\begin{aligned}
I(S) &= \frac{K}{4\pi}\int_{\Omega}\langle I(\boldsymbol{S})\rangle \mathrm{d}\Omega \\
F(\boldsymbol{S}) &= \int_{r}\rho(\mathbf{r})\exp[2\pi i\boldsymbol{S}\cdot\boldsymbol{r}]\mathrm{d}^3 r = |F(\mathbf{S})|\exp(i\alpha(\boldsymbol{S})) \quad , \\
I(\boldsymbol{S}) &= |F(\mathbf{S})|\exp(i\alpha(\boldsymbol{S}))|F(\mathbf{S})|\exp(-i\alpha(\boldsymbol{S})) = |F(\boldsymbol{S})|^2
\end{aligned}
\qquad (1)
$$

where $\rho(\boldsymbol{r})$ is the electron density distribution of the protein in a single conformation, and $\boldsymbol{S}$ is the scattering vector as defined in Fig. 1. Ω is the solid angle against the direction of the incident X-ray. Each integration is taken over all values of the variable. The Fourier transform of $\rho(\boldsymbol{r})$ is the structure factor, $F(\boldsymbol{S})$, which is expressed by the structure amplitude $|F(\boldsymbol{S})|$ and the phase $\alpha(\boldsymbol{S})$. The phase term necessary to reconstruct $\rho(\boldsymbol{r})$ by the inverse Fourier transform of $F(\boldsymbol{S})$ is completely lost in X-ray detection. $< >$ denotes the average over the possible conformations of the protein in

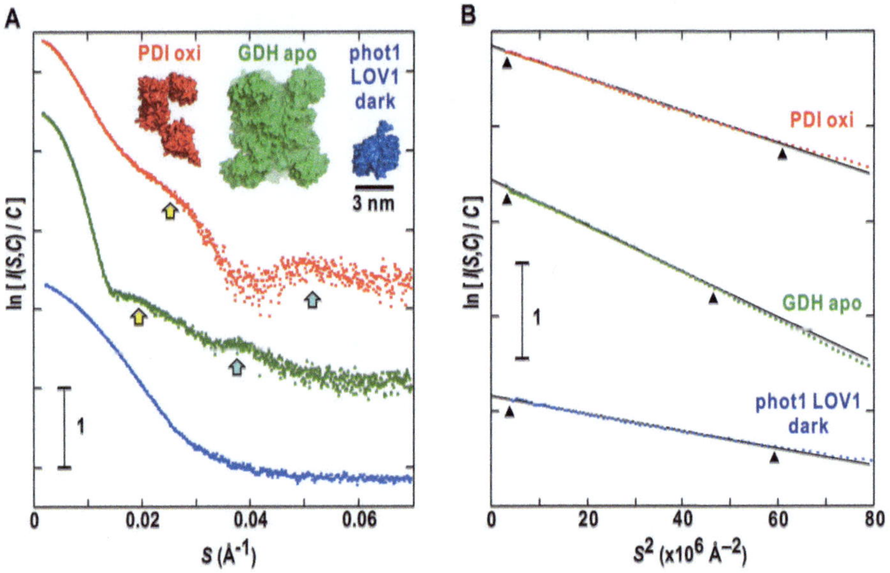

Fig. 2 SAXS from proteins. (**a**) SAXS profiles of protein disulfide isomerase from *Humicola insolens* in the oxidized form (PDIoxi) (red dots) [31], hexameric glutamate dehydrogenase from *Thermococcus profundus* in the unliganded state (GDHapo) (green) [54, 55], and a dimer of the light-oxygen-voltage-sensing domain 1 of phototropin 1 from *Arabidopsis thaliana* in the dark state (phot1 LOV1 dark) (blue) [24]. The molecular weights of PDI, GDH, and phot1 LOV1 dimer are approximately 53k, 280k, and 34k, respectively. The inset shows the surface-rendered crystal structures of yeast PDI (PDB ID: 2B5E) [56], GDH apo (PDB ID: 1EUZ) [54], and phot1 LOV1 dark (PDB ID: 2Z6C) [57] drawn using *PyMOL* [58]. The yellow arrow indicates the enhancement predominantly due to the molecular shape, while the enhancement indicated by the cyan arrow arises from the arrangement of subunits/domains. (**b**) Guinier plots for the SAXS profiles in panel (**a**). The Guinier approximation was applied to the region indicated by the pair of arrows, where the criteria for the Guinier approximation ($SRg < (2\pi)^{-1}$) are satisfied. The *R*g values of PDI, GDH, and phot1 LOV1 dimer are 33.4, 43.9, and 23.5 Å, respectively. (Published with kind permission of Mao Oide and Masayoshi Nakasako, 2025. All Rights Reserved)

solution. Furthermore, as protein molecules in solution are randomly oriented against the direction of the incident X-ray beam, $<I(S)>$ is averaged over Ω. K is a constant reflecting the experimental geometry and the concentration of proteins.

Figure 2a shows examples of SAXS profiles of proteins with different molecular sizes and structures. Each profile in the region of $S < 0.005$ Å^{-1} is likely approximated by Gaussian. The profiles beyond $S > 0.01$ Å^{-1} depend on the sizes, shapes, and the distribution of high electron density regions. When a protein comprises domains and/or subunits, a broad enhancement appears around specific scattering vector length, which is the reciprocal of the interdomain/subunit distance (*see* arrows in Fig. 2a).

In the small-angle region of $S < 0.01$ Å^{-1}, the SAXS profile from a single protein is described by Guinier approximation [42] as

$$I(S) = KI(0)\exp\left[-\frac{4\pi^2}{3}S^2 Rg^2\right], \tag{2}$$

where $I(0)$ is the forward scattering intensity (or zero-angle scattering intensity), which is the square of the total electrons of the protein as

$$I(0) = \int_r \int_{r'} \rho(\boldsymbol{r}')\,\rho(\boldsymbol{r})\mathrm{d}^3\boldsymbol{r}\mathrm{d}^3\boldsymbol{r}'. \tag{3}$$

Rg is the radius of gyration of the protein written as

$$Rg^2 = \frac{1}{2}\frac{\int_r \int_{r'} \rho(\boldsymbol{r}')\,\rho(\boldsymbol{r})\,|\boldsymbol{r}-\boldsymbol{r}'|^2\mathrm{d}^3\boldsymbol{r}\mathrm{d}^3\boldsymbol{r}'}{\int_r \int_{r'} \rho(\boldsymbol{r}')\,\rho(\boldsymbol{r})\mathrm{d}^3\boldsymbol{r}\mathrm{d}^3\boldsymbol{r}'}. \tag{4}$$

Therefore, Rg reflects the variance of the electron density distribution inside the protein. For the monodisperse protein solution, the data points in the Guinier plot, where the logarithm of the scattering intensity is plotted against the square of the scattering vector length, are approximated by a straight line (*see* Fig. 2b). Then, $I(0)$ and Rg are determined from the abscissa and the slope of the line in the region satisfying $SRg < (2\pi)^{-1}$, respectively.

If the data points are out of the Guinier approximation, the concentration of the protein may be too high and/or protein molecules may form nonspecific aggregation. In the former case, it is best to dilute the protein solution. In the latter case, sample preparation and handling must be modified.

If protein molecules interact with each other under diluted conditions, both $I(0)$ and Rg depend on the concentration of proteins, C, [42] as

$$\frac{1}{I(S=0,C)} = D\left(\frac{1}{M_W} + 2A_2C\right), \tag{5}$$

$$Rg(C)^2 = Rg(C=0)^2 - B_{if}C, \tag{6}$$

where D is an experimental constant. M_w is the molecular weight of the protein, and A_2 is the second virial coefficient. Attractive inter-protein interactions result in a negative A_2, while repulsive interactions give a positive A_2. B_{if} also reflects the mode of intermolecular interactions, and the sign is the same as A_2. Additionally, the M_w of a protein is directly estimated from the scattering profiles of the protein and buffer solutions using the program *MoW* [43]. The concentration dependence of Rg is under the influences of the physicochemical properties and the conformational variety of protein surfaces (*see* Fig. 3a).

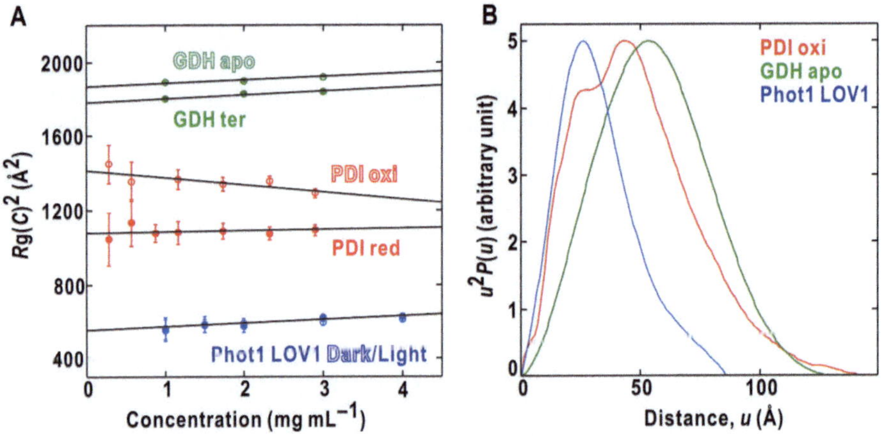

Fig. 3 Radii of gyration and maximum dimensions of proteins obtainable from SAXS. (**a**) Concentration dependences of Rg^2 values. GDH under enzymatic reaction (GDHter, filled green circles) [55] became more compact than in the unliganded state (GDHapo, open green circles) [54], and both were in attractive interactions as suggested from the sign of B_{If}. PDI in the oxidized state (PDIoxi, open red circles) is in repulsive interaction, while little intermolecular interactions are expected for PDI in the reduced state (PDIred, filled red circles), which has a more compact molecular size than that in the oxidized state [31]. In the case of phot1 LOV1, the molecular interactions are unchanged before (Dark, open blue circles hidden by filled blue circles) and after the blue-light irradiation (Light, filled blue circles) [24]. (**b**) Distance distribution functions of proteins in Fig. 2a calculated using the *GNOM* program [44]. The estimated maximum distances of PDI in the reduced state, GDH in the unliganded state, and phot1 LOV1 in the dark are approximately 127 [31], 130 [54], and 100 Å [24], respectively. (Published with kind permission of Mao Oide and Masayoshi Nakasako, 2025. All Rights Reserved)

Since the scattering intensity $I(\mathbf{S})$ is the square of $F(\mathbf{S})$, the Fourier transform of $I(\mathbf{S})$ gives $P(\mathbf{u})$, an autocorrelation function of $\rho(\mathbf{r})$, according to the convolution theorem [20]. The function is defined as

$$P(\boldsymbol{u}) = \int_r \rho(\boldsymbol{r})\rho(\boldsymbol{r} - \boldsymbol{u})\mathrm{d}^3 r = \int_S I(\boldsymbol{S}) \exp[-2\pi i \boldsymbol{S}\cdot\boldsymbol{u}]\, \mathrm{d}^3 S. \quad (7)$$

We can only measure orientationally averaged $I(S)$ in the SAXS experiment (*see* Eq. 1), and consequently we will obtain $u^2P(u)$, the distance distribution function, instead of $P(\boldsymbol{u})$:

$$u^2 P(u) = \frac{1}{\pi} \int_S I(S)\, 2\pi S r\, \sin[2\pi S r]\, \mathrm{d}S. \quad (8)$$

In general, $u^2P(u)$ has a bell-shaped curve, and therefore we can estimate the maximum dimension of the protein from the distance where the curve approaches zero [44] (*see* Fig. 3b).

The *ATSAS* program package for SAXS structure analysis (download site of *ATSAS*: https://www.embl-hamburg.de/biosaxs/download.html) is useful for calculating Rg, $I(0)$ and $u^2\mathrm{P}(u)$ function during experiments [44–46].

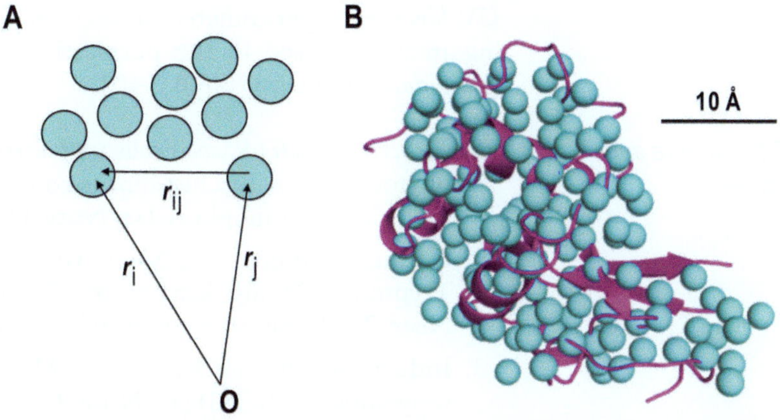

Fig. 4 Ab initio calculation to restore low-resolution structures of proteins in solution. (**a**) Schematic of the distribution of identical spheres approximating a protein structure. (**b**) An assembly of 130 spheres predicted using GASBOR [48] approximates the crystal structure of hen egg white lysozyme. Each sphere has a radius of 3.8 Å and represents an amino acid residue. (Published with kind permission of Mao Oide and Masayoshi Nakasako, 2025. All Rights Reserved)

1.3 Low-Resolution Molecular Structures of Proteins

Low-resolution molecular structures of proteins in solution are restored from the SAXS profiles using only ab initio calculation methods [47, 48]. In the methods, the molecular structure of a protein is modeled as an assembly of identical spheres with a constant radius and electron density (*see* Fig. 4a). Then, the SAXS profile of the model, $I_{\mathrm{cal}}(S)$, is described using Debye formula [49] as

$$I_{\mathrm{cal}}(S) = I_{\mathrm{sph}}(S) \sum_{i=1}^{M} \sum_{j=1}^{M} \frac{\sin Sr_{ij}}{Sr_{ij}}, \tag{9}$$

where r_{ij} is the distance between the ith and jth spheres (*see* Fig. 4a), and M is the number of spheres. $I_{\mathrm{sph}}(S)$ is the scattering intensity from a single sphere. The molecular shape is modeled as an arrangement of the spheres by minimizing the sum of the intensity differences between the model and observation and the penalty function with respect to the packing modes of the spheres [47, 48]. As an example, Fig. 4b demonstrates that an assembly of dummy spheres restored from the SAXS profile of lysozyme nicely approximated the molecular shape of the crystal structure.

2 Materials

For purification of phytochromes, a fast protein liquid chromatography (FPLC) equipped with an ultraviolet–visible (UV-Vis) monitor to measure the absorption of eluent at 280 and 660 nm and a

UV-Vis spectrophotometer are necessary. For instance, the following items are used in the purification of phytochrome B from *Arabidopsis thaliana* (phyB) [40].

2.1 Culture and Harvest

1. *Escherichia coli* (*E. coli*) cells transformed by vectors to express phytochrome with a histidine tag and other proteins for synthesizing the chromophore (*see* **Note 1**).
2. Culture medium: 5 g NaCl, 10 g yeast extract, 16 g Bacto Tryptone, 70 mg kanamycin, 23 mg chloramphenicol (*see* **Note 2**), 23 mg streptomycin in 1 L water.
3. Induction solution: sterilized 1 M isopropyl-β-D-1-thiogalactopyranoside (IPTG) (*see* **Note 3**).
4. Sonication buffer: 50 mM NaCl, 20 mM 4-(2-hydroxyethyl)-1-piperazineethanesulfonic acid (HEPES)–NaOH, pH 7.5, one tablet of protease inhibitor cocktail/100 mL, 1% (w/v) *n*-octyl-β-D-glucopyranoside, lysozyme at 1 mg mL^{-1}.
5. Filter membrane with 0.45-μm pore.

2.2 Nickel-Affinity Chromatography

1. A nickel-affinity chromatography (NAC) column.
2. 2 M imidazole solution.
3. NAC-equilibration buffer: 20 mM HEPES–NaOH, pH 7.5, 30 mM imidazole, 50 mM NaCl, 1 mM benzamidine.
4. NAC-elution buffer: 20 mM HEPES–NaOH, pH 7.5, 1 M imidazole, 50 mM NaCl, 1 mM benzamidine.

2.3 Size-Exclusion Chromatography

1. A size-exclusion chromatography (SEC) column suitable to separate proteins with M_w greater than 200k.
2. SEC buffer: 20 mM Tris–HCl, pH 7.5, 150 mM NaCl.

2.4 Anion Exchange Chromatography

1. An anion exchange chromatography (AEC) column.
2. AEC-equilibration buffer: 20 mM Tris–HCl, pH 7.5, 150 mM NaCl.
3. AEC-elution buffer: 20 mM Tris–HCl, pH 7.5, 1 M NaCl, 1 mM benzamidine.

2.5 Concentration

1. Sample buffer: 20 mM HEPES–NaOH, pH 7.5, 200 mM NaCl.
2. Centrifugal filter.

3 Methods

3.1 Phytochrome Purification

Prepare 100 μL solution in a concentration range of 1–5 mg mL^{-1} necessary to conduct SAXS experiments for phytochromes [36, 40]. Prepare all solutions using analytical-grade reagent and ultrapure water. Perform the following procedure based on our phyB purification [40] under dim green light and at 277 K.

1. Prepare 2-L tryptone yeast extract (TY) medium, and separate the medium into two Erlenmeyer flasks of 2 L. Sterilize the TY medium.
2. Add *E. coli* cells expressing phytochrome to the sterilized medium at 310 K.
3. Shake the flask at 140 rpm at 310 K for 18 h.
4. After the 18-h incubation, the absorption at 600 nm will exceed 0.3 optical density (OD). Then, add 100 μL of the induction solution to 1-L culture medium. The final concentration of IPTG is 0.2 mM.
5. Shake the culture at 140 rpm at 289 K for 24 h.
6. Harvest cells by centrifugation at 3900× *g* for 10 min.
7. Suspend freeze-thawed cells in the sonication buffer.
8. Lyse the harvested cells by sonication on ice.
9. Centrifuge the lysed cell solution at 38,000× *g* and 277 K for 40 min.
10. Add 2 M imidazole solution to the crude phytochrome solution at a final concentration of 30 mM.
11. Remove debris from the supernatant using the filter membrane with 0.45-μm pores.
12. Apply the filtered supernatant to an NAC column already equilibrated using the NAC-equilibration buffer.
13. Elute crude phytochrome solution by a linear gradient (0%–100%) of the NAC-elution buffer (*see* Fig. 5a and **Note 4**). Monitor the absorption of the eluent at 280 nm and 660 nm.
14. After each chromatography step, examine the purity of phytochrome in the eluted fractions by Coomassie Brilliant Blue (CBB) staining after sodium dodecyl sulfate–polyacrylamide gel electrophoresis (SDS-PAGE) (*see* right panels of Fig. 5a–c).
15. Apply the phytochrome fractions to an SEC column equilibrated using the SEC buffer, and collect fractions containing phytochrome (*see* Fig. 5b and **Note 5**).
16. Apply the fractions containing phytochrome to an AEC column.

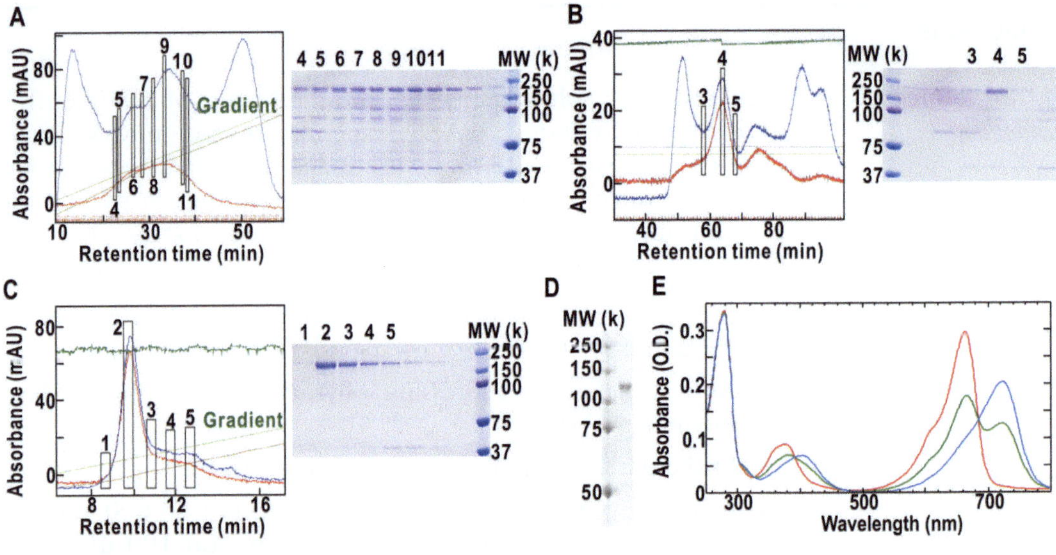

Fig. 5 Purification of phyB using three-step chromatography [40]. Raw elution profiles monitored at 280 nm (blue curve) and 660 nm (red curve) are shown for NAC (**a**), SEC (**b**), and AEC (**c**) with CBB-stained SDS-PAGE patterns of several fractions. The purity of phyB solution supplied to SAXS measurements was examined by the CBB-stained SDS-PAGE pattern (**d**) and the UV-VIS absorption spectra (**e**). The spectra were measured in the dark (the Pr form: red curve) and under red-light irradiation (the Pr–Pfr photostationary state: green). The spectrum of 100% Pfr (blue) is calculated from the spectra of the Pr form and photostationary state. (Published with kind permission of Mao Oide and Masayoshi Nakasako, 2025. All Rights Reserved)

17. Elute phytochrome solution using a linear gradient (0%–100%) of the AEC-elution buffer (*see* Fig. 5c and **Note 6**).
18. Exchange the buffer of purified phytochrome solution to the sample buffer and simultaneously concentrate the phytochrome solution to the concentration necessary in SAXS measurements using an ultrafiltration membrane (*see* **Note** 7).
19. Check the purity of the final phytochrome solution by SDS-PAGE and the absorption spectrum (*see* Fig. 5e). For instance, highly purified solution of phyB in the Pr form will display the absorption at 666 nm as high as that at 280 nm (*see* Fig. 5e).

3.2 Static Measurement and Analysis

Perform static measurement and analysis on phytochrome at a beamline of a synchrotron facility (*see* Fig. 6 and **Note 8**) according to the following procedure.

1. Prior to SAXS measurements, make the experimental hutch of the beamline as a darkroom using black curtains.
2. Set dim green light as safelight for sample handling, and far-red and red light for the photoconversion near the sample position. Then, measure the fluence rate at the sample position using a photometer. For convenience in SAXS experiment, extend the switch of each light outside the experimental hutch.

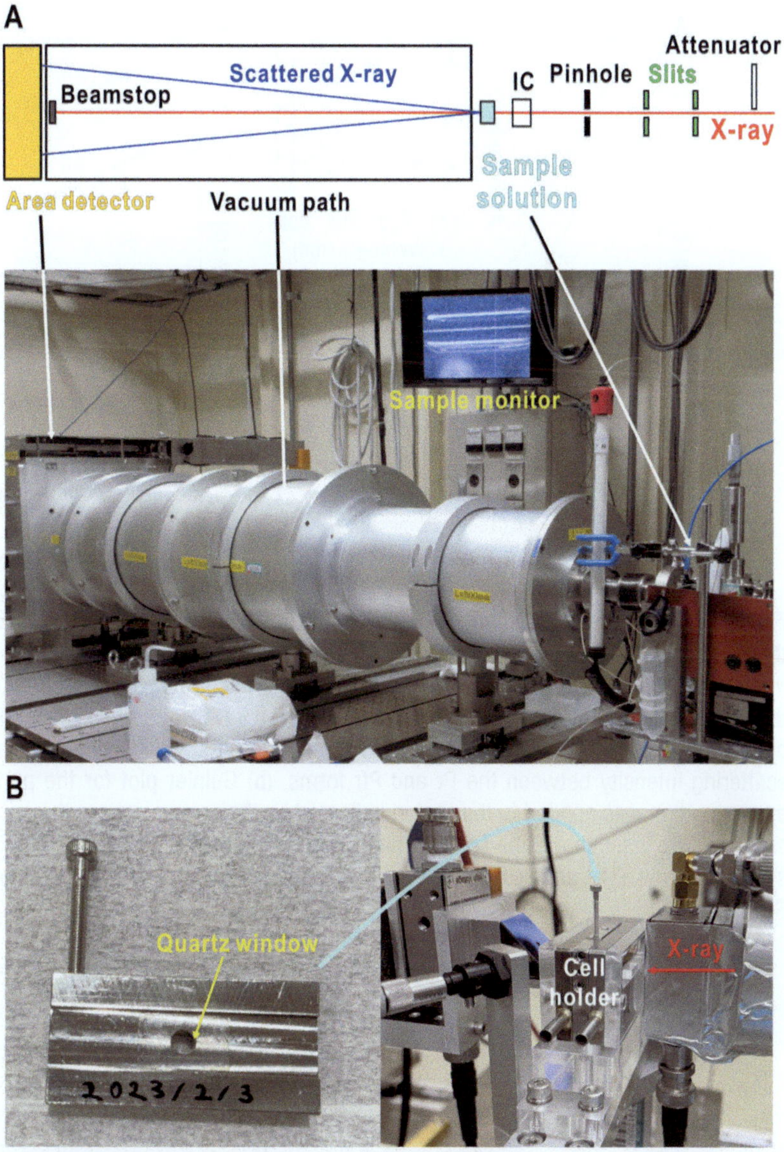

Fig. 6 SAXS apparatus. (**a**) SAXS apparatus in the experimental hutch of BL45XU of SPring-8 [53]. The upper panel schematically illustrates the arrangement of components of the apparatus and optics (*see* **Note 8**). The ionization chamber is abbreviated as IC. (**b**) An example of sample cell used in static measurements. The X-ray path length is 3 mm, and both sides of the hole are covered by two fused-quartz windows of 10 μm thickness. (Published with kind permission of Mao Oide and Masayoshi Nakasako, 2025. All Rights Reserved)

3. Connect the cell holder of the SAXS apparatus (*see* Fig. 6b) to a temperature-controlled circulator, which uses water–alcohol mixture, and adjust the temperature of the sample cell.

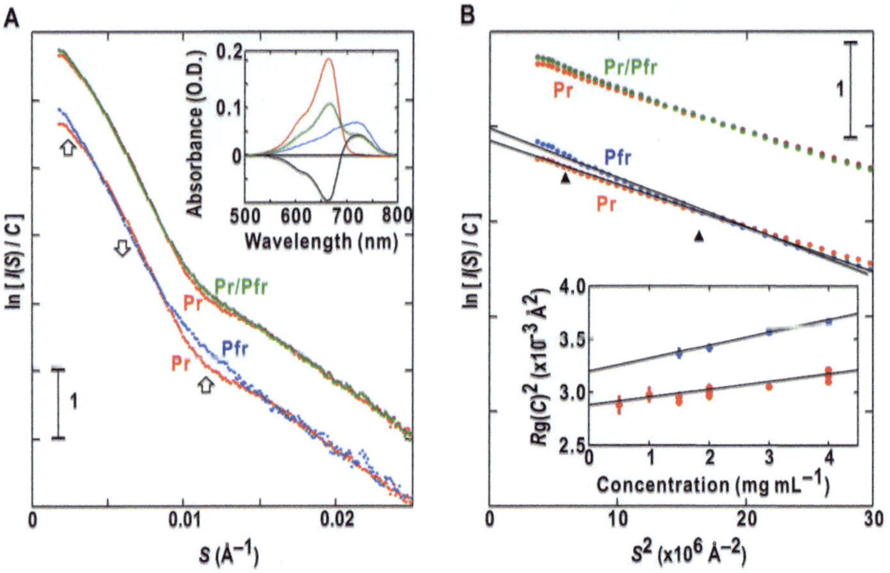

Fig. 7 Static SAXS measurement for LphyA. (**a**) SAXS profiles from LphyA in the Pr form (red symbols) and in the Pr–Pfr steady state (green dots) near the dilution limit. The inset shows absorption spectra of the Pr form (red curve), Pfr-Pr steady state (green), and their difference (black). The populations of the Pr and Pfr forms in the steady state were determined to be 39% and 61% from the absorption spectra, respectively. Using the population ratio, the SAXS profile of the Pfr form (blue dots) was calculated. The arrows indicate changes in scattering intensity between the Pr and Pfr forms. (**b**) Guinier plot for the profiles in panel (**a**). The Guinier approximation was applied to the region indicated by the arrows, where the criteria for the approximation are satisfied. The inset shows the concentration dependencies for the $Rg^2(C)$ of the Pr (red symbols) and Pfr (blue symbols) forms. (The panels are reused after modification from Ref. [36] with permission from John Wiley and Sons)

4. Cover each sample tube storing phytochrome solution with aluminum foil, and set sample tubes in an ice box.
5. Carefully apply sample solution of approximately 30 μL to the sample cell (*see* Fig. 6b) using a micropipette under dim green light.
6. Insert the sample cell into the holder (*see* Fig. 6b). Irradiate the sample cell by far-red (or red) light for measuring the SAXS of the Pr form (or Pr–Pfr photostationary state).
7. Close the door of the experimental hutch to open the main beam shutter of the beamline.
8. Operate a single static measurement using software dedicated for static SAXS measurements. The software will be developed and provided by synchrotron facility (*see* **Note 9**).
9. Close the beamline shutter. Open the experimental hutch and remove the sample cell from the holder.
10. Store the X-ray-irradiated sample solution in a sample tube to examine whether radiation damage to break peptide bonds

occurred during the X-ray exposure by SDS-PAGE, and to measure UV-Vis spectrum for examining radiation-caused bleaching of phytochromes.

11. Wash the sample cell with the sample buffer three times using a micropipette, and completely remove the remaining sample buffer using wicks (*see* **Note 10**).
12. Apply the sample buffer of approximately 30 μL to the sample cell.
13. Measure the scattering pattern in the same manner as in **steps 3** and **4** to record the scattering pattern from the sample solution used as a background in the subsequent analysis (*see* **Note 11**).
14. Calculate the one-dimensional scattering intensity profile by circularly averaging the net scattering pattern. The profile data contains the scattering vector length, *S*, the scattering intensity, *I*(*S*), and the standard deviation of the intensity [square root of *I*(*S*)] (*see* Fig. 1, Eq. 1, and **Note 12**).
15. Analyze the scattering profile using the *ATSAS* program package [45, 46].
16. Display the intensity profile (log[*I*(S)] vs. *S* in Figs. 2a and 7a) and the Guinier plot (log[*I*(S)] vs. S^2 in Figs. 2b and 7b) on the graphical user interface of *ATSAS*.
17. Calculate the *Rg* and *I*(0) values from the Guinier plot (*see* Eqs. 2, 3, and 4 and Figs. 2b and 7b), when the data points near the beamstop area are approximated by a regression line (*see* **Note 13**). If protein molecules are in heavy intermolecular interactions and/or nonspecific aggregation, the data points in the Guinier plot will be out of the Guinier approximation (*see* **Note 14**).
18. Examine the dependences of Rg^2 and $1/I(0)$ (*see* Fig. 3a), after SAXS measurements at different protein concentrations (*see* Fig. 3a and **Note 15**).
19. Calculate the distance distribution function (*see* Eq. 8 and Fig. 3b), for instance, using the *GNOM* program [44] in *ATSAS* (*see* **Note 16**).
20. Restore the low-resolution molecular structure of the protein (*see* Eq. 9 and Fig. 4b) using the ab initio calculation methods [47, 48] (*see* Subheading 3.4).

3.3 Structural Information on Large Phytochrome A from Static Measurement

Here, we describe the results of the static SAXS study on large phytochrome A of pea (*Pisum sativum* cv. Alaska) (LphyA) [36]. LphyA, which lacks the N-terminal 52 amino acid residues (total 1072 residues), was purified from 7-day-old etiolated seedlings. Figure 7a shows SAXS profiles from LphyA in the Pr form under far-red light and in the Pr–Pfr photostationary state under red light. The estimated apparent M_w indicated that LphyA in the

Pr form existed as a dimer in solution. The solution of the photostationary state may be a mixture of Pr/Pr, Pr/Pfr, and Pfr/Pfr dimers. When we assumed the absence of Pr/Pfr dimers, the SAXS profile of the Pfr form can be calculated based on the amount of Pr and Pfr dimers determined from the absorption spectra (the inset of Fig. 7a).

The SAXS intensities of the Pfr form increased from those of the Pr form at $S = 0.002$ and 0.011 Å^{-1} and decreased at $S = 0.006$ Å^{-1}, suggesting red-light-induced conformational changes over the LphyA molecule. The slopes in the Guinier plots (*see* Fig. 7b) and the $Rg(C = 0)$ values of the Pfr (62.6 ± 0.4 Å) and Pr (57.1 ± 0.1 Å) forms indicated that the Pfr form had a larger molecular size than the Pr form, probably due to photoconversion-induced conformational changes over LphyA molecule.

As demonstrated by spectroscopic studies, the photoconverted Pfr form is thermally reverted to the Pr form. It was interesting whether the structural conversion simultaneously proceeded with the spectroscopically observed thermal conversion. In the difference profile at 15 min after the finish of the photostationary state (*see* Fig. 8), the negative peak at $S = 0.005$ Å^{-1}, characterizing the SAXS changes between the Pr and the Pfr states, still remained. This suggested that the conformational changes detectable in the SAXS profiles do not immediately follow the form change that occurs around the chromophore.

3.4 Time-Resolved Measurement and Analysis in the Combinational Use of Size-Exclusion Chromatography and Small-Angle X-Ray Scattering

The combinational use of size-exclusion chromatography and small-angle X-ray scattering (SEC-SAXS) measurement [50, 51] is routinely performed in most of the synchrotron facilities (*see* Figs. 9 and 10). Perform SEC-SAXS measurements and analysis for phytochromes in the following procedure.

1. Modify the experimental hutch into a darkroom as done in static measurements.
2. Connect an SEC column to the SEC system and equilibrate the SEC column using the sample solution for phytochrome.
3. Cover the SEC column with aluminum foil to keep the sample solution in the dark.
4. Inject a concentrated phytochrome solution into the SEC system (*see* **Note 17**).
5. Close the door of the experimental hutch to open the main beam shutter of the beamline.
6. Schedule the SEC elution and SAXS data collection using dedicated software for SEC-SAXS measurement (*see* **Note 18**).
7. Start SEC-SAXS measurement immediately after opening the beamline shutter by the software (*see* **Note 19**).

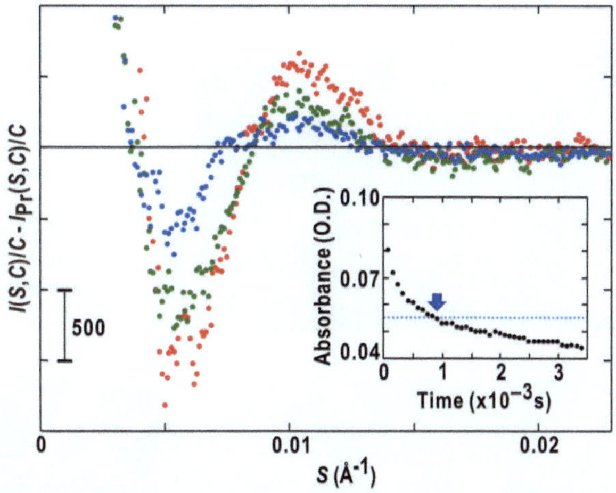

Fig. 8 Time-dependent changes of SAXS profiles of LphyA in the dark reversion from the photostationary state to the Pr form. The difference profiles are calculated by subtracting the profile of the Pr form from those of the Pfr form (red symbols), Pr–Pfr photostationary state (green symbols), and at 15 min after finishing the red-light irradiation (blue symbols). The inset shows the absorption changes at 664 nm during the dark reversion at 298 K. The blue arrow indicates 15 min after the finish of the red-light irradiation. (The panels are reused after modification from Ref. [36] with permission from John Wiley and Sons)

8. Calculate the net scattering profile of phytochrome by circularly averaging the background-subtracted pattern summed in the specified frames (*see* Fig. 10 and **Note 20**). The SAXS analyses applied to the static data are applied to net scattering intensity profiles.

Figure 10 shows an example of SEC-SAXS measurement for glutamate dehydrogenase (GDH) in the unliganded state using the system shown in Fig. 9. The elution from the SEC column appeared as single peak, and the SAXS profile was obtained by accumulating the profiles from specified frames. SEC-SAXS measurements for standard protein solutions are necessary to understand the performance of the SEC-SAXS system.

Figure 11a shows diagrams of red/far-red light irradiation (at a fluence rate of 2000 μmol m^{-2} s^{-1} for 60 s) and X-ray exposure, and the elution profiles in SEC-SAXS measurements for phyB in the Pr form [40]. In contrast to GDH in Fig. 10, the elution profile was composed of three components, designated α, β, and γ. The SAXS profile of the γ component eluted at a concentration of 0.02 mg mL^{-1} (*see* Fig. 11b) was similar to that of LphyA (*see* Fig. 7a). The Guinier plot approximated by a linear regression line yielded the Rg value of 50.5 Å (*see* Fig. 11c), and then the estimated M_w indicated that the component was a dimer. In contrast, the α and β components were polydisperse, as judged from their

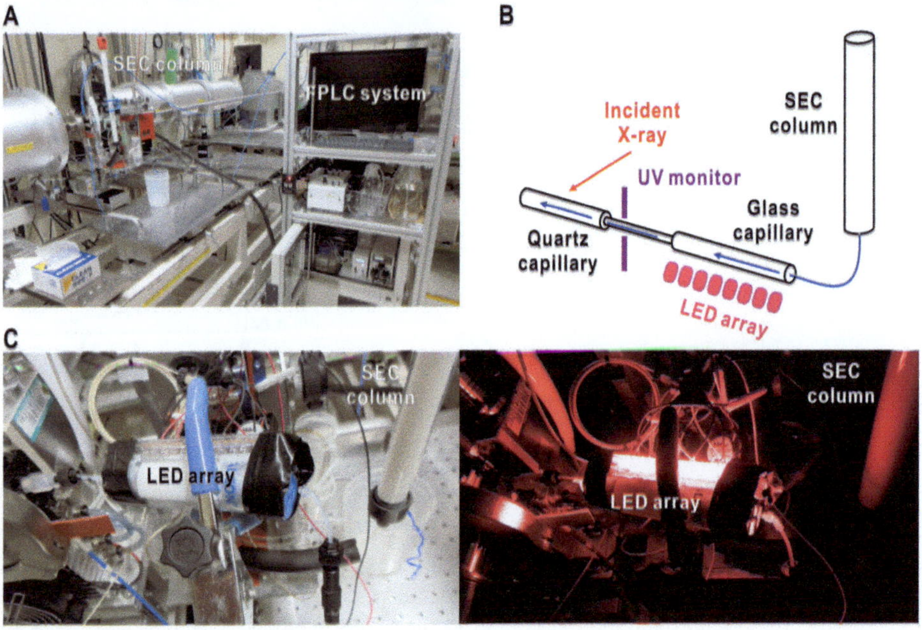

Fig. 9 SEC-SAXS apparatus. (**a**) SEC-SAXS measurement system at BL45XU of SPring-8. The SEC system comprised a pump, an injector, an SEC column, a UV detector, and a degasser. (**b**) Schematic of SEC-SAXS measurement for phyB [40]. The eluent from SEC is supplied to a glass capillary tube for light irradiation using a linear array of light-emitting diodes. The protein concentration in the eluted solution from SEC is measured by the UV detector before X-ray exposure. Finally, the eluted solution flows into a quartz capillary tube for X-ray exposure. (**c**) Magnified views of a custom-made light-irradiation device. The right panel is a view of the light-irradiation device under operation. (Published with kind permission of Mao Oide and Masayoshi Nakasako, 2025. All Rights Reserved)

Guinier plot, where the plotted data in $S^2 < 35 \times 10^{-6}$ Å^{-2} significantly deviated from the regression line for the γ component of the Pr form (*see* Fig. 12c). The inflections of intensities near the forward scattering region suggested nonspecific association of Pr dimers in the α and β components.

To study red-light-induced conformational changes of phyB, the eluent of phyB–Pr solution flowing into the glass capillary was irradiated with red light at a fluence rate of 2000 μmol m^{-2} s^{-1} for 10 s before the X-ray exposure (*see* Figs. 9b, c, and 11d). The irradiation time is sufficient for absorption changes from Pr to Pfr form before reaching the X-ray irradiation area. The elution profile was similar to that of the Pr form and was separated into three components α, β, and γ (*see* Fig. 11d, e). However, the Guinier plot for the scattering profile of component γ was difficult to be approximated by regression line (*see* Fig. 11f), while component γ of the Pr form was monodisperse. This result implies that the photoconversion to the Pfr form in component γ induces nonspecific aggregation of phyB dimers, probably due to the changes in the surface properties of phyB molecules.

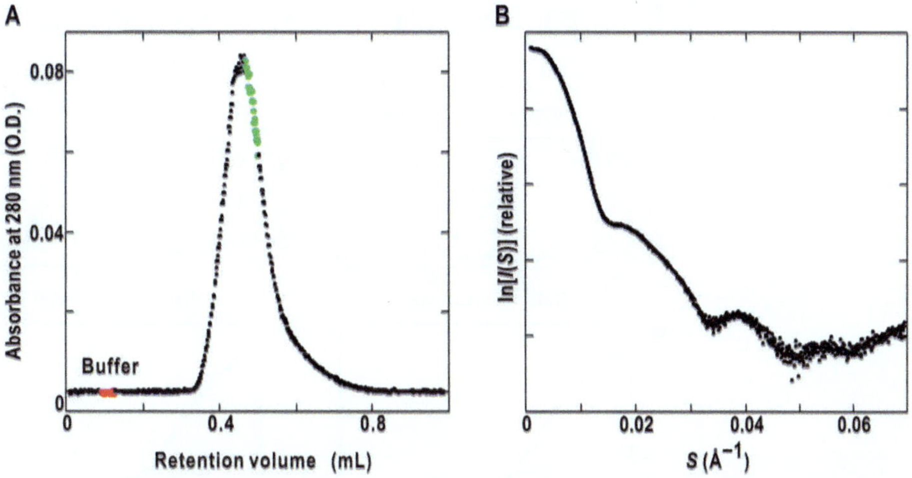

Fig. 10 SEC-SAXS measurement for GDH in the unliganded state. (**a**) The elution profile of GDH from an SEC column (HiLoad 16/600 Superdex 200 PG). The SEC column was used in the SEC-SAXS measurement for phyB as shown in Fig. 11. The scattering patterns from the buffer and GDH solutions were accumulated in the frames indicated by the red and green symbols in the elution profile, respectively. (**b**) The net scattering profile from GDH. (Published with kind permission of Mao Oide and Masayoshi Nakasako, 2025. All Rights Reserved)

3.5 Low-Resolution Molecular Structures of Phytochromes

Here, we describe the calculation procedure to restore the low-resolution structure models of a protein from the SAXS profile using the ab initio calculation methods [47, 48]. As predicted, structures are not always correct; we have to select probable structures from many trial calculations.

1. Prior to executing the ab initio calculations, prepare a file containing the scattering intensity and distance distribution function using the *GNOM* program [44] in the *ATSAS* suite [45, 46].
2. Execute ab initio calculations. It is convenient to execute, in parallel, more than 100 independent calculations starting from different random seeds on a supercomputer (*see* **Note 21**).
3. Compare the obtained low-resolution structure models as assemblies of spheres using some modules in the *ATSAS* suite [45, 46] (*see* **Note 22**).

The low-resolution molecular shapes of LphyA in the Pr and Pfr forms were restored from the SAXS profiles (*see* Fig. 7a) assuming the twofold rotational symmetry in the dimer (*see* Fig. 12 and **Note 23**) [41]. In each of the Pr and Pfr forms, the restored 560 molecular models were separated into ten groups by applying the clustering method after principal component analysis (PCA), and an averaged model was obtained for each group [52] (*see* **Note 22**). For both forms, the butterfly shapes and twisted-tube shapes were major. In the previous study, we focused on the butterfly

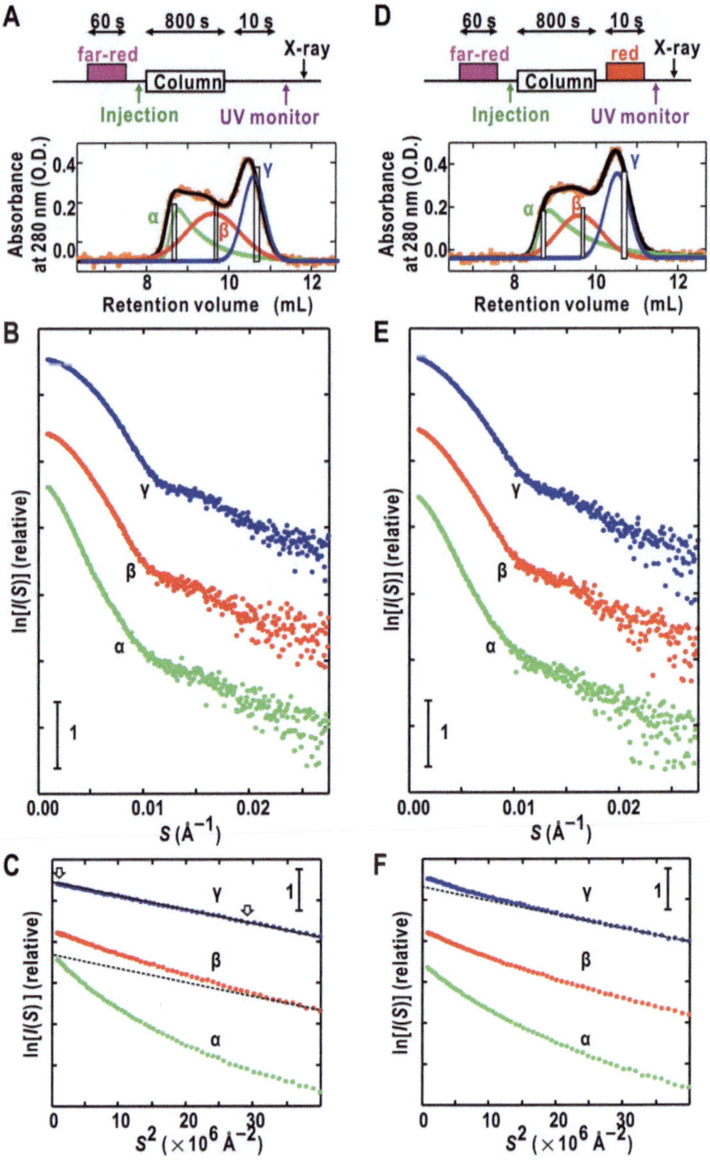

Fig. 11 SEC-SAXS measurements for phyB in the Pr form (**a**–**c**) and in the Pr–Pfr photostationary state (**d**–**f**) [40]. In each of (**a**) and (**d**), the upper panel shows a diagram of light irradiation and X-ray exposure. The lower panel shows the elution profile of phyB monitored at 280 nm (orange circles). Each elution profile was approximated as the sum (black curve) of the α (green curve), β (red), and γ (blue) components. (**b**, **e**) SAXS profiles of the α (green dots), β (red), and γ components (blue) were calculated from the fractions indicated by boxes in the elution profiles in panels (**a**) and (**d**). Profiles are appropriately shifted for clarity. (**c**, **f**) Guinier plot for the profiles of the three components. In panel (**c**), the regression line (dashed line) for the γ component was calculated in the region indicated by arrows. This regression line is also shown for the β component in panel (**c**) and the γ component of the photostationary state in panel (**f**) to demonstrate the inflection of intensity in the small-angle regions caused by nonspecific aggregation of phyB. (The panels are reused after modification from Ref. [40] with permission from John Wiley and Sons)

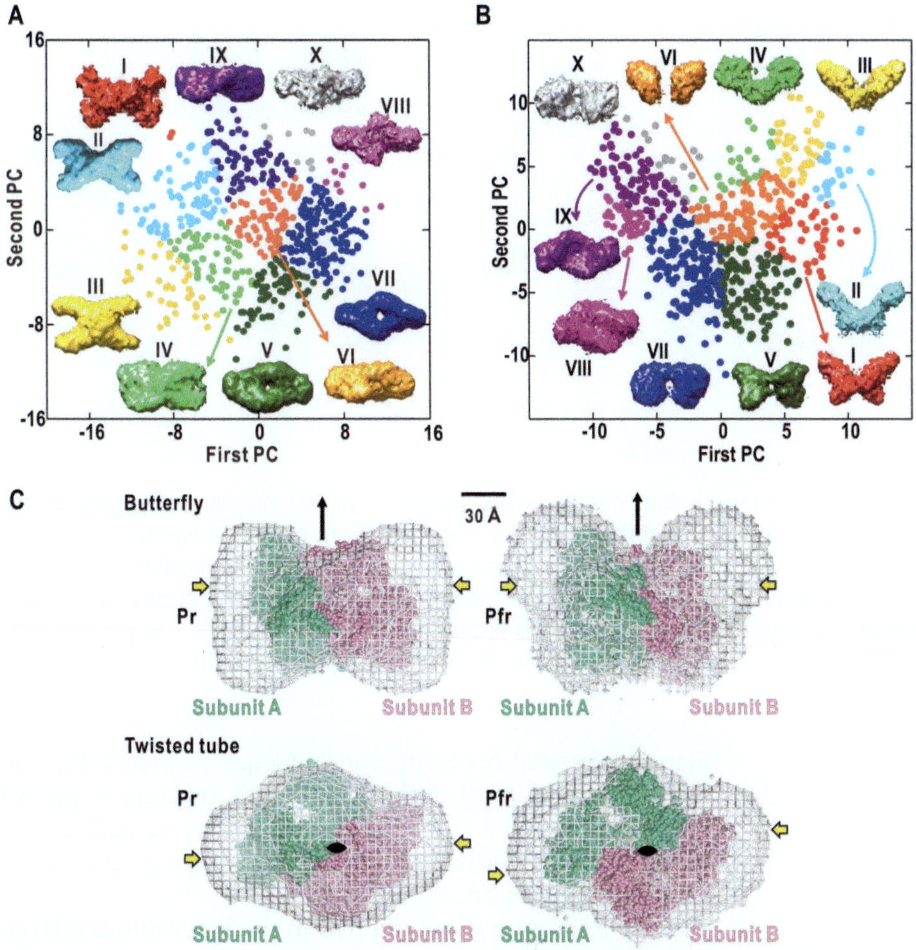

Fig. 12 Molecular models of LphyA in the Pr (**a**) and Pfr (**b**) forms restored from the scattering profiles in Fig. 7a [41]. Each panel shows the distribution of 560 structure models on a plane, which was spanned by the first and second eigenvectors determined by PCA. The 560 models are separated into ten groups (I–X) using K-means clustering. The averaged molecular model of each group was drawn using *UCSF Chimera* [59]. (**c**) The cryo-EM structure model in the Pr form (PDB ID: 8iff) [13] is manually fitted to the representative butterfly and twisted-tube shapes (gray mesh). The yellow arrow indicates the volume lacking cryo-EM structures. Panel c was drawn using *PyMOL* [58]. (Panels (**a**) and (**b**) are reused from Ref. [41] after a modification with permission from Springer Nature)

shapes as the molecular shape of LphyA [41]. However, the recently determined cryo-EM structure in the Pr form [13] may be fitted to the twisted-tube shapes. In Fig. 12c, the cryo-EM model is fitted to the restored shapes. In the shapes of the Pr and Pfr forms, the volumes lacking cryo-EM models may correspond to the domains invisible in cryo-EM analysis.

The molecular size of the Pfr form was larger than that of the Pr form, as expected from the *R*g of the Pfr form being greater than that of the Pr form (*see* Fig. 7b). The differences in the molecular

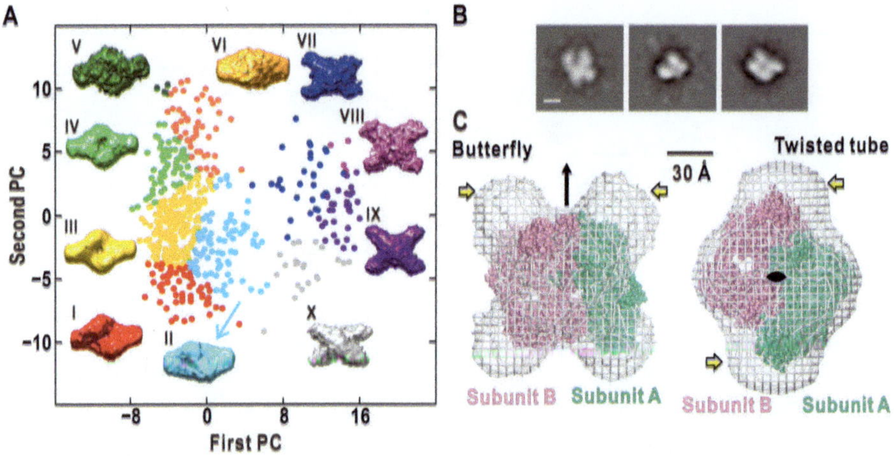

Fig. 13 SAXS models of phyB in the Pr form. (**a**) The distribution of 560 models on the plane spanned by the two principal component vectors obtained by PCA. The models are separated into ten groups (I–X) by the K-means clustering. (**b**) Representative images after the two-dimensional classification for the EM images of negatively stained phyB. The scale bar is 50 Å. (**c**) The cryo-EM structure of phyB in the Pr form [13] is fitted to representative low-resolution models, as well as in the case of LphyA in Fig. 12. The panels (**a**) and (**c**) are drawn as in Fig. 12. (Panels (**a**) and (**b**) are reused after modification from Ref. [40] with permission from John Wiley and Sons)

shapes suggested conformational changes in the Pr–Pfr photoconversion (*see* Fig. 12c). Although the Pfr structure of phyA has not yet been visualized, the SAXS models may be helpful to discuss which domains of the cryo-EM structure change their positions in the photoconversion.

The molecular shape of phyB in the Pr form was restored from the SAXS profile of the γ component in Fig. 11b, assuming the twofold rotational symmetry in the dimer [40] (*see* Fig. 13). As well as in the case of LphyA, the predicted 560 structure models were roughly separated into twisted-tube and cross (or butterfly) shapes (*see* Fig. 13a). This result suggested that the overall structure of phyB in the Pr form was similar to that of LphyA in the Pr form as confirmed by the recent cryo-EM analyses [13, 14]. In the previous study [40], we focused on the cross shape by referring to the EM images of negatively stained phyB. However, the recently determined cryo-EM structure of phyB in the Pr form [14] may be fitted to the twisted-tube shape (*see* Fig. 13b). Just like in the case of LphyA, the volume in the SAXS models indicated by arrows may correspond to the missing parts in the cryo-EM structure of phyB (*see* Fig. 13c).

4 Notes

1. To express phyB for SAXS study [40], we use *Escherichia coli* (*E. coli*) Rosetta2 (DE3) cells, which are transformed by two expression vectors. One is a pET28a vector for expressing full-length phyB with a histidine tag, and the other is a pCDFDuet vector for expressing heme oxygenase from *Synechocystis* sp. PCC 6803 and phytochromobilin synthase from *Adiantum capillus-veneris*. The genes in the latter vector are necessary for phytochromobilin synthesis in *E. coli* cells.
2. Chloramphenicol is dissolved in ethanol.
3. Prepared solution is stored at 243 K.
4. In the purification of phyB, we use a HisTrap column connected to an ÄKTA FPLC system. Then, we set the flow rate to 2.00 mL min^{-1}, and the time for 0%–100% gradient is set to 48 min.
5. In the purification of phyB, we use a HiLoad 16/600 Superdex 200 PG column connected to an ÄKTA FPLC system. The flow rate is set to 0.5 mL min^{-1}.
6. In the purification of phyB, we use a Mono Q column connected to an ÄKTA FPLC system. The flow rate is set to 0.5 mL min^{-1}, and the time for 10%–42% is set to 40 min.
7. When we use Amicon Ultra-15 50 K filter to concentrate phyB, centrifugation at 3000× *g* for 8 min is iterated. The final concentration of phyB is determined using the absorption at 660 nm measured for appropriately diluted phyB solution.
8. In SAXS experiments at a synchrotron radiation facility, X-rays from the synchrotron storage ring are monochromatized at an energy resolution of 10^{-4}. In BL45XU of SPring-8 [53] (*see* Fig. 6), the typical X-ray wavelength used in SAXS experiments is 1.0000 Å. Parasitic scattering from X-ray optics is almost eliminated using two pairs of slits and a pinhole placed upstream of the sample position. The intensity of the incident X-rays is monitored using an ionization chamber placed in front of the sample position and/or a positive-intrinsic-negative (PIN) photodiode placed in front of the detector. When the intensity of the incident X-ray beam is too strong, an aluminum attenuator may be inserted. The cross-sectional size of the incident X-ray beam was smaller than 200 μm^2 at the sample position, and the intensity of the incident X-ray beam was 5×10^{12} photons/s. The vacuum path kept at 10^{-2} Pa is essential to significantly reduce scattering and absorption by air. The length of the vacuum path is set at 2.0 m in SAXS experiments for phytochromes. The direct beam is absorbed by a beamstop to avoid fatal radiation damage to the detector.

9. Typical exposure time is 30 or 60 s. The software controls the open/close of the shutter for X-ray exposure. After finishing the measurement, the scattering pattern will be automatically transferred from the area detector to a storage.
10. Check the irradiation area in the quartz windows of the sample cell (*see* Fig. 6b) using a microscope. If phytochrome aggregation adsorb on the windows, apply 1 M NaOH solution to the cell. After 10 min, remove the NaOH solution and wash the cell using the sample buffer more than ten times.
11. The same sample cell should be used in the measurements for the sample solution and the sample buffer. In addition, as the very small difference in the thickness of the quartz windows has a great influence on the absorption of incident X-rays, the cell must always be set in the holder at the same orientation against the direction of the incident X-ray.
12. When calculating the net scattering pattern by subtracting the background pattern from the sample pattern, the volume excluded by phytochrome molecules must be taken into consideration. This procedure may be done using the measurement software provided by the facility.
13. For phytochrome dimer, the data points in $5 \times 10^{-6} < S^2 < 30 \times 10^{-6}$ Å^{-2} in the Guinier plot should be approximated by a regression line.
14. In our experience, purified LphyA and phyB dimers easily form aggregates (*see* Fig. 11). Scattering from aggregates appears in the very small-angle region ($S < 0.005$ Å^{-1}) and is easily identified in Guinier plots as typically seen in the α and β components in Fig. 11c and all components in Fig. 11f. Aggregation hinders SAXS structural analyses using Eqs. 2–8 and the subsequent calculations for low-resolution molecular structures.

    As the amount of aggregation in solution may depend on the period of time to store, it may be better to measure SAXS immediately after the purification. Therefore, on-site purification at a synchrotron facility is strongly recommended. Prior to SAXS measurements, ultracentrifugation at 145,000× *g* for more than 4 h may be effective to remove aggregates [36]. In addition, to avoid the formation of aggregates, pipetting should be performed mildly. As demonstrated in Fig. 11, SEC-SAXS experiments may be essential for phytochromes from higher plants [40].
15. Rigorously speaking, a monodisperse protein solution must display the linear concentration dependencies of Rg^2 and $1/I(0)$ (*see* Eqs. 5 and 6 and Fig. 3a).
16. From several calculation trials by varying the input maximum dimension, select the function displaying the best score [44].

17. In the SEC-SAXS experiment for phyB, 200 μL of phyB solution at a concentration of 5–7 mg mL^{-1} was injected [40].
18. Parameters necessary to execute SEC-SAXS measurement are the start and end times for a time-resolved measurement, the exposure time for each frame, the flow rate of SEC elution, and so on. To collect background scattering patterns from the sample buffer, start the measurement before the front eluent reaches the X-ray irradiation area. In our SEC-SAXS experiment for phyB [40], the flow rate was set to 0.75 mL/min for Superdex 200 Increase 10/300 column, and the exposure time for each frame was 0.2 s. Exposure time in SEC-SAXS depends on the incident X-ray intensity and the performance of the area detector.
19. The data acquisition system records the following data in each time-resolved frame: the scattering pattern, the intensity of the incident X-ray beam, and the absorbance at 280 nm for determining the concentration of phytochrome.
20. When defining the region for the Guinier approximation, the software may automatically execute on-the-fly estimation of Rg and $I(0)$ for each frame. For eluent including aggregates of phytochromes, the on-the-fly estimation will give nonrealistic values of Rg and $I(0)$.
21. When using the GASBOR suite [48] in the molecular shape calculations, the number of spheres is one of the predominant factors influencing the fitness between the model and observed profiles [26–29, 31, 40, 41, 52].
22. When the predicted low-resolution structure models have various shapes, the models may be classified by multivariate analysis using principal component analysis (PCA) followed by clustering methods [52]. A molecular model expressed using $J \times J \times J$ voxels is assumed as a point in a $J \times J \times J$-dimensional space. PCA determines principal component vectors that most effectively describe the variation of the molecular models in the multidimensional space. Then, the distributions of the molecular models are projected onto the plane spanned by the first two principal component vectors and are classified into several groups using the K-means clustering method. This multivariate analysis has been used in SAXS studies on phototropins [28, 29] and phytochromes [40, 41].
23. When the M_w of a protein estimated by SAXS is twice that calculated from the amino acid sequence, the protein is assumed to form a dimer in solution, like LphyA [41] and phyB [40]. In ab initio calculations for estimating molecular shape, we may assume the twofold rotational symmetry in the dimer [40, 41]. As the position of the twofold axis is unknown in the dimer, ab initio calculations without the symmetry may be helpful to estimate the position of the rotational axis in the molecular shape.

Acknowledgments

The SAXS studies on phytochromes were partly supported by Grants-in-Aid for Scientific Research from the Japan Society for the Promotion of Science (No. jp16H02218, jp22018027, jp20050030, jp1920402, and jp22244054 to M.N., and No. 18J11653 to M.O.) and by grants from the Ministry of Education, Culture, Sports, Science and Technology, Japan (No. jp15076210, jp20050030, jp22018027, jp23120525, jp25120725, jp15H01647, and 17H05891 to M.N.). SAXS experiments were performed at SPring-8 (proposal 2002B-0153, 2003A-0107, 2003B-0142, and RIKEN-20180071).

References

1. Legris M, Ince YÇ, Fankhauser C (2019) Molecular mechanisms underlying phytochrome-controlled morphogenesis in plants. Nat Commun 10:5219. https://doi.org/10.1038/s41467-019-13045-0
2. Drenth J (2007) Principles of protein X-ray crystallography, 3rd edn. Springer, New York. ISBN: 978-1-4757-3092-0
3. Glaeser RM, Nogales E, Chiu W (2021) Single-particle Cryo-EM of biological macromolecules. IOP Publishing, Bristol. ISBN: 978-0-7503-3037-4
4. Henderson R (2015) Overview and future of single particle electron cryomicroscopy. Arch Biochem Biophys 581:19–24. https://doi.org/10.1016/j.abb.2015.02.036
5. Beton JG, Cragnolini T, Kaleel M et al (2020) Integrating model simulation tools and cryo-electron microscopy. WIREs Comput Mol Sci: e1642. https://doi.org/10.1002/wcms.1642
6. Wagner JR, Brunzelle JS, Forest KT et al (2005) A light-sensing knot revealed by the structure of the chromophore-binding domain of phytochrome. Nature 438:325–331. https://doi.org/10.1038/nature04118
7. Yang X, Kuk J, Moffat K (2008) Crystal structure of Pseudomonas aeruginosa bacteriophytochrome: photoconversion and signal transduction. Proc Natl Acad Sci USA 105: 14715–14720. https://doi.org/10.1073/pnas.0806718105
8. Yang X, Ren Z, Kuk J et al (2011) Temperature-scan cryocrystallography reveals reaction intermediates in bacteriophytochrome. Nature 479:428–432. https://doi.org/10.1038/nature10506
9. Yang X, Stojkovic EA, Ozarowski WB et al (2015) Light signaling mechanism of two tandem bacteriophytochromes. Structure 23: 1179–1189. https://doi.org/10.1016/j.str.2015.04.022
10. Burgie ES, Zhang J, Vierstra RD (2016) Crystal structure of Deinococcus phytochrome in the photoactivated state reveals a cascade of structural rearrangements during photoconversion. Structure 24:448–457. https://doi.org/10.1016/j.str.2016.01.001
11. Otero LH, Klinke S, Rinaldi J et al (2016) Structure of the full-Length bacteriophytochrome from the plant pathogen *Xanthomonas campestris* provides clues to its long-range signaling mechanism. J Mol Biol 428:3702–3720. https://doi.org/10.1016/j.jmb.2016.04.012
12. Burgie ES, Bussell AN, Walker JM et al (2014) Crystal structure of the photosensing module from a red/far-red light-absorbing plant phytochrome. Proc Natl Acad Sci USA 111: 10179–10184. https://doi.org/10.1073/pnas.1403096111
13. Wang J, Zhou C, Guan Z et al (2023) Plant phytochrome A in the Pr state assembles as an asymmetric dimer. Cell Res 33:802–805. https://doi.org/10.1038/s41422-023-00847-7
14. Li H, Burgie ES, Vierstra RD et al (2022) Plant phytochrome B is an asymmetric dimer with unique signaling potential. Nature 604:127–133. https://doi.org/10.1038/s41586-022-04529-z
15. Takala H, Björling A, Berntsson O et al (2014) Signal amplification and transduction in phytochrome photosensors. Nature 509:245–248. https://doi.org/10.1038/nature13310
16. Schmidt A, Sauthof L, Szczepek M et al (2018) Structural snapshot of a bacterial phytochrome in its functional intermediate state. Nat Commun 9:4912. https://doi.org/10.1038/s41467-018-07392-7
17. Otero LH, Foscaldi S, Antelo GT et al (2021) Structural basis for the Pr-Pfr long-range signaling mechanism of a full-length bacterial

phytochrome at the atomic level. Sci Adv 7: eabh1097. https://doi.org/10.1126/sciadv.abh1097

18. Malla TN, Hernandez C, Muniyappan S et al (2024) Photoreception and signaling in bacterial phytochrome revealed by single-particle cryo-EM. Sci Adv 10:eadq0653. https://doi.org/10.1126/sciadv.adq0653
19. Burgie ES, Basore K, Rau MJ et al (2024) Signaling by a bacterial phytochrome histidine kinase involves a conformational cascade reorganizing the dimeric photoreceptor. Nat Commun 15:6853. https://doi.org/10.1038/s41467-024-50412-y
20. Glatter O, Kratky O (1982) Small-angle X-ray scattering. Academic Press, London. ISBN: 0122862805
21. Feigin LA, Svergun DI (1987) Structure analysis by small-angle X-ray and neutron scattering. Plenum Press, New York. ISBN: 0-306-42629-3
22. Svergun DI, Koch MHJ, Timmins PA et al (2013) Small angle X-ray and neutron scattering from solutions of biological macromolecules. Oxford University Press. ISBN: 978-0-19-963953-3
23. Nakasako M, Kataoka M, Amemiya Y et al (1991) Crystallographic characterization by X-ray diffraction of the M-intermediate from the photo-cycle of bacteriorhodopsin at room temperature. FEBS Lett 292:73–75. https://doi.org/10.1016/0014-5793(91)80837-s
24. Nakasako M, Iwata T, Matsuoka D et al (2004) Light-induced structural changes of LOV-domain containing polypeptides from Arabidopsis phototropin 1 and 2 studied by small-angle X-ray scattering. Biochemistry 43: 14881–14890. https://doi.org/10.1021/bi0485530
25. Nakasako M, Matsuoka D, Zikihara K et al (2005) Quaternary structure of LOV-domain containing polypeptides of Arabidopsis FKF1 protein. FEBS Lett 579:1067–1071. https://doi.org/10.1016/j.febslet.2004.12.078
26. Takayama Y, Nakasako M, Okajima K et al (2011) Light-induced movement of LOV2 domain in Asp720Asn-mutated LOV2-kinase fragment of Arabidopsis phototropin 2. Biochemistry 50:1174–1183. https://doi.org/10.1021/bi101689b
27. Okajima K, Aihara Y, Takayama Y et al (2014) Light-induced conformational changes of LOV (Light Oxygen Voltage-sensing domain) 1 and LOV2 relative to the kinase domain and regulation of kinase activity in Chlamydomonas phototropin. J Biol Chem 289:413–422. https://doi.org/10.1074/jbc.M113.515403
28. Oide M, Okajima K, Kashojiya S et al (2016) Blue-light-excited LOV2 triggers a rearrangement of the kinase domain to induce phosphorylation activity in Arabidopsis phototropin1. J Biol Chem 291:19975–19984. https://doi.org/10.1074/jbc.M116.735787
29. Oide M, Okajima K, Nakagami H et al (2018) Blue-light excited LOV1 and LOV2 domains cooperatively regulate the kinase activity of full-length phototropin2 from Arabidopsis. J Biol Chem 293:963–972. https://doi.org/10.1074/jbc.RA117.000324
30. Serve O, Kamiya Y, Maeno A et al (2010) Redox-dependent domain rearrangement of protein disulfide isomerase coupled with exposure of its substrate-binding hydrophobic surface. J Mol Biol 396:361–374. https://doi.org/10.1016/j.jmb.2009.11.049
31. Nakasako M, Maeno A, Kurimoto E et al (2010) Redox-dependent domain rearrangement of protein disulfide isomerase from a thermophilic fungus. Biochemistry 49:6953–6962. https://doi.org/10.1021/bi1006089
32. Davies JM, Tsuruta H, May AP et al (2005) Conformational changes of p97 during nucleotide hydrolysis determined by small-angle X-ray scattering. Structure 13:183–195. https://doi.org/10.1016/j.str.2004.11.014
33. Oroguchi T, Hashimoto H, Shimizu T et al (2009) Intrinsic dynamics of restriction endonuclease EcoO109I studied by molecular dynamics simulations and X-ray scattering data analysis. Biophys J 96:2808–2822. https://doi.org/10.1016/j.bpj.2008.12.3914
34. Bock LV, Grubmüller H (2022) Effects of cryo-EM cooling on structural ensembles. Nat Commun 13:1709. https://doi.org/10.1038/s41467-022-29332-2
35. Nakasako M, Wada M, Tokutomi S et al (1990) Quaternary structure of pea phytochrome I dimmer studied with small-angle X-ray scattering and rotary-shadowing electron microscopy. Photochem Photobiol 52:3–12. https://doi.org/10.1111/j.1751-1097.1990.tb01747.x
36. Nakasako M, Iwata T, Inoue K et al (2005) Light-induced global structural changes in phytochrome A regulating photomorphogenesis in plants. FEBS J 272:603–612. https://doi.org/10.1111/j.1742-4658.2004.04508.x
37. Evans K, Grossmann JG, Fordham-Skelton AP et al (2006) Small-angle X-ray scattering reveals the solution structure of a bacteriophytochrome in the catalytically active Pr state. J Mol Biol 364:655–666. https://doi.org/10.1016/j.jmb.2006.09.045

38. Heyes DJ, Hardman SJO, Pedersen MN et al (2019) Light-induced structural changes in a full-length cyanobacterial phytochrome probed by time-resolved X-ray scattering. Commun Biol 2:1. https://doi.org/10.1038/s42003-018-0242-0
39. Papiz MZ, Bellini D, Evans K et al (2019) Light-induced complex formation of bacteriophytochrome RpBphP1 and gene repressor RpPpsR2 probed by SAXS. FEBS J 286: 4261–4277. https://doi.org/10.1111/febs.14973
40. Oide M, Hikima T, Oroguchi T et al (2020) Molecular shape under far-red light and red light-induced association of Arabidopsis phytochrome B. FEBS J 287:1612–1625. https://doi.org/10.1111/febs.15095
41. Oide M, Nakasako M (2021) Red light-induced structure changes in phytochrome A from *Pisum sativum*. Sci Rep 11:2827. https://doi.org/10.1038/s41598-021-82544-2
42. Guinier A, Fournet G (1959) Small-angle scattering of X-rays. John Wiley & Sons, London
43. Fischer H, de Oliveira NM, Napolitano HB et al (2010) Determination of the molecular weight of proteins in solution from a single small-angle X-ray scattering measurement on a relative scale. J Appl Crystallogr 43:101–109. https://doi.org/10.1107/S0021889809043076
44. Svergun DI (1992) Determination of regularization parameter in indirect-transform methods using perceptual criteria. J Appl Crystallogr 25:495–503. https://doi.org/10.1107/S0021889892001663
45. Petoukhov MV, Konarev PV, Kikhney AG et al (2007) ATSAS 2.1 towards automated and web-supported small-angle scattering data analysis. J Appl Crystallogr 40:223–228. https://doi.org/10.1107/S0021889807002853
46. Petoukhov MV, Franke D, Shkumatov AV et al (2012) New developments in the ATSAS program package for small-angle scattering data analysis. J Appl Crystallogr 45:342–350. https://doi.org/10.1107/S0021889812007662
47. Svergun DI (1999) Restoring low resolution structure of biological macro-molecules from solution scattering using simulated annealing. Biophys J 76:2879–2886. https://doi.org/10.1016/S0006-3495(99)77443-6
48. Svergun DI, Petoukhov MV, Koch MHJ (2001) Determination of domain structure of proteins from X-ray solution scattering. Biophys J 80:2946–2953. https://doi.org/10.1016/S0006-3495(01)76260-1
49. Debye PJW (1915) Zerstreuung von Röntgenstrahlen. Ann Phys 351:809–823. https://doi.org/10.1002/andp.19153510606
50. Watanabe Y, Inoko Y (2009) Size-exclusion chromatography combined with small-angle X-ray scattering optics. J Chromatogr A 1216: 7461–7465. https://doi.org/10.1016/j.chroma.2009.02.053
51. Watanabe Y, Inoko Y (2011) Further application of size-exclusion chromatography combined with small-angle X-ray scattering optics for characterization of biological macromolecules. Anal Bioanal Chem 399:1449–1453. https://doi.org/10.1007/s00216-010-4140-7
52. Oide M, Sekiguchi Y, Fukuda A et al (2018) Classification of ab initio models of proteins restored from small-angle X-ray scattering. J Synchrotron Radiat 25:1379–1388. https://doi.org/10.1107/S1600577518010342
53. Fujisawa T, Inoue K, Oka T et al (2000) Small-angle X-ray scattering station at the SPring-8 RIKEN beamline. J Appl Crystallogr 33: 797–800. https://doi.org/10.1107/S002188980000131X
54. Nakasako M, Fujisawa T, Adachi S et al (2001) Large-scale domain movements and hydration structure changes in the active-site cleft of unligated glutamate dehydrogenase from Thermococcus profundus studied by cryogenic X-ray crystal structure analysis and small-angle X-ray scattering. Biochemistry 40:3069–3079. https://doi.org/10.1021/bi002482x
55. Wakabayashi T, Oide M, Nakasako M (2024) CryoEM-sampling of metastable conformations appearing in cofactor-ligand association and catalysis of glutamate dehydrogenase. Sci Rep 14:11165. https://doi.org/10.1038/s41598-024-61793-x
56. Tian G, Xiang S, Noiva R et al (2006) The crystal structure of yeast protein disulfide isomerase suggests cooperativity between its active sites. Cell 124:61–73. https://doi.org/10.1016/j.cell.2005.10.044
57. Nakasako M, Zikihara K, Matsuoka D et al (2008) Structural basis of the LOV1 dimerization of Arabidopsis phototropins 1 and 2. J Mol Biol 381:718–733. https://doi.org/10.1016/j.jmb.2008.06.033
58. DeLano WL. The PyMOL molecular graphics system, version 1.5.0.1. Schrödinger, LLC, New York
59. Pettersen EF, Goddard TD, Huang CC et al (2021) UCSF ChimeraX: structure visualization for researchers, educators, and developers. Protein Sci 30:70–82. https://doi.org/10.1002/pro.3943

Chapter 13

A Simple Modification of a Commercial Plunge-Freezer to Allow Time-Resolved Vitrification for Cryo-EM Analyses of Light-Activated Proteins

Andrea Flores-Ibarra, Monika Gunkel, and Elmar Behrmann

Abstract

Plunge-freezing allows the trapping of potentially complex conformer mixtures for subsequent high-resolution structural analysis by cryogenic electron microscopy. The ultrafast nature of the vitrification process, when combined with a defined activation of the sample, can enable time-resolved structural studies. For light-sensitive samples, this activation can be achieved by controlled illumination of the sample during the plunge-freezing process. Here, we describe a simple approach that enables users to modify commercially available plunge-freezers with a light-emitting diode (LED)-based illumination system, and describe a straightforward protocol for its application to the study of red-light-sensitive phytochromes.

Key words Cryogenic electron microscopy, Structural biology, Vitrification, Time-resolved, Light-activated, Phytochrome

1 Introduction

Vitrification is the process of freezing a liquid sample fast enough as to outpace the formation of crystalline ice [1], thus allowing the preparation of proteins and other macromolecules for high-resolution structure analysis by cryogenic electron microscopy (cryo-EM). Cooling rates have been estimated to be in the order of 10^6 K·s^{-1} or higher [2], which trap and immobilize the proteins under study in an amorphous thin film at similar timescales to those observed for conformational motions in proteins. Recent computational work has implied that vitrification should be able to trap distinct conformations separated by energetic barriers in the order of 10 kJ·mol^{-1} [3]. This ability to preserve potentially complex mixtures of conformers present in a sample can be either a boon or a bane of cryo-EM: While sample heterogeneity complicates data processing, it offers the potential to gain valuable insight into a protein's structure–function relationship. Cryo-EM,

Andreas Winkler (ed.), *Phytochromes: Methods and Protocols*, Methods in Molecular Biology, vol. 2970,
https://doi.org/10.1007/978-1-0716-4791-2_13,

especially when used with 2D or 3D sorting strategies, can untangle conformer mixtures and resolve individual states, though it cannot directly assign biochemical functions to these states. So, ideally, the way vitrification is performed is modified such that it can be used to populate the grid with one or, better, a few specific biochemical states right before plunge-freezing, allowing for time-resolved structural experiments.

Cryo-EM lends itself to time-resolved structural resolution through two approaches: In one, the sample is fast-mixed with an activating component; in the other, the sample is light-activated by defined illumination. For the fast-mixing strategy and depending on the timescale of the reaction studied, the mixing can happen in the test tube before applying the sample to the grid, or directly on the grid, either at the same time or immediately after sample application. In practice, this has been achieved, for instance, using an atomizer to apply an activating ligand onto the sample-loaded grid [4, 5], using mix-and-spray devices [6, 7], or voltage-assisted sprays [8]. However, spraying has the drawback of a harder-to-control sample thickness, due to typical buffer solutions having strong surface tension that tampers with the homogeneous deposition of a thin liquid film on the grid surface.

On the other hand, the light activation strategy is fully compatible with the default vitrification workflow and equipment. Naturally, this requires the protein to be sensitive to light, or made to have light-sensitive functionality by genetic or chemical modification. The latter has been accomplished, for example, by presaturating samples with caged derivatives of adenosine triphosphate (ATP) [9], and guanosine triphosphate (GTP) and calcium [10], and a design for uncaging compounds using an ultraviolet (UV) light-emitting diode (LED) source has been proposed elsewhere [11]. As for proteins with natural light-sensitive groups, relaxation times are often sufficiently long to allow activation in bulk before application on the grid or, if better time resolution is needed, after application on the grid before blotting. This has already proven successful for blue-light-sensitive cryptochromes in our laboratory [12], as well as elsewhere for red-light-sensitive phytochromes [13], and for bacteriorhodopsins, one of the first proteins studied using a broad-spectrum xenon flash lamp [14, 15].

In this protocol, we show that no special equipment is necessary to adapt a standard plunge-freezer to create the conditions necessary for light activation in the visible range of the electromagnetic spectrum. We rely on LED lamps with well-defined emission spectra, allowing us to seamlessly integrate them into light-sensitive experiments. Our system (*see* Fig. 1) is adapted for use with a Vitrobot Mark IV and permits vitrification in under 4 s after stopping the LED. Here, we describe the steps to follow with the equipment available in our laboratory, but most steps can easily be carried out using equipment from different manufacturers. It is

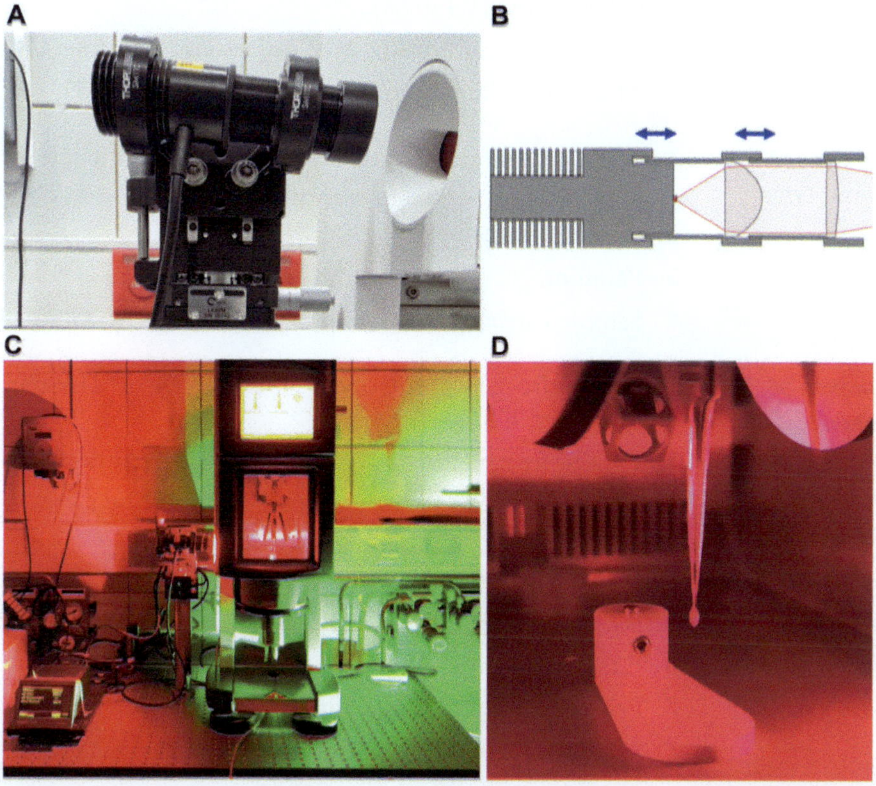

Fig. 1 (**a**) Close-up of the optical assembly and (**b**) schematic drawing of the collimation optic assembly. Adjustable lens tubes (marked by blue arrows) make it easier to get a defined focal point. (**c**) Setup for light activation of a phytochrome sample. Note that the red light on the left side is due to scattered light inside the humidified chamber that is emitted from the left sample port. (**d**) Close-up of the grid during illumination

important to note that although we describe the conditions required to work with a bacterial phytochrome sample, this method can be adapted to a variety of light-sensitive proteins just by using the correct LED wavelengths.

2 Materials

2.1 Modification of a Commercial Plunge-Freezer to Allow for Light Activation of the Sample

All the equipment in this section was obtained from Thorlabs, USA.

1. LED driver (DC2200).
2. Optical table.
3. Mounted LED (660 nm, M660L4).
4. Aspheric condenser lens (ACL2520U).
5. Best form focusing lens (LBF254-150).
6. Optical lens tubes SM1 (1″ threading).
7. Clamps (CP37/M).

8. Optical post (P350/M).
9. Mounting post bracket with quick release clamp (C1515/M).
10. Mechanical X/Y/Z translation stage (LX30/M).

2.2 Preparation of Sample Carrier Grids

1. Glow discharger: Zepto Plasma Cleaner with semiautomated knob control (Diener Electronic, Germany).
2. Small Petri dish.
3. Parafilm.
4. Sample carrier grids (Quantifoil Micro Tools, Germany).

2.3 Preparation of Plunge-Freezer

1. Plunge-freezer: Vitrobot Mark IV (Thermo Fisher Scientific, USA).
2. Plunge-freezer forceps (Thermo Fisher Scientific, USA).
3. Cryo grid boxes (SubAngstrom, USA).
4. Material for handling grid boxes (e.g., tweezers and opening pens).
5. Blotting paper 595 (Plano, diameter 55/20 mm, Germany).
6. Expanded polystyrene (EPS) box to hold cryogen liquids.
7. Primary cryogen: liquid nitrogen.
8. Secondary cryogen: liquid ethane, Grade 5.0.

2.4 Plunge-Freezing with Defined Illumination

1. Phytochrome sample at concentration ~0.7 mg·mL^{-1}.
2. Prepared plunge-freezer and grid(s).

3 Method

In order to prevent premature activation, and possibly photobleaching, it is important to adhere to a strict light regime whenever working with light-activated proteins. This requires sealing of the laboratory rooms against straight light from windows and doors (*see* **Note 1**). Furthermore, it is paramount to know the activation properties of the protein to be studied (*see* **Note 2**) so as to design an ideal illumination system to be used for its activation (*see* **Note 3**).

3.1 Modification of a Commercial Plunge-Freezer for Light Activation

Here, we describe the modification of a Vitrobot Mark IV plunge-freezer. However, this approach is easy to adapt to other plunge-freezers, since the illumination array is outside the actual plunge-freezer. One of the sample application ports is used to get the light from the outside onto the sample carrier grid, the other to apply the sample. This modification needs to be done once.

1. Place the plunge-freezer onto an optical table to ensure that there is a fixed geometry between the optical assembly and the plunge-freezer.
2. Measure the distance between the left pipetting port of the plunge-freezer and the position of the forceps during blotting.
3. Assemble an aspheric condenser lens with a best form focusing lens, using SM1 threaded lens tubes such that the generated focus point coincides with the measured distance (*see* **Note 4**).
4. Connect the focusing optics assembly with a mounted LED of the chosen wavelength (for this sample, an LED with a peak emission of 660 nm).
5. Connect the illumination assembly, using two clamps, with a mechanical X/Y/Z translation stage, for fine adjustment of the focus point (*see* **Note 5**).
6. Fix the X/Y/Z translation stage on the optical table using an optical post.
7. Connect the LED to the LED driver and define the illumination conditions. For phytochrome, we used six 10-s pulses at 1 mA, with a delay between pulses of 10 ms (*see* **Note 6**).

3.2 Preparation of Sample Carrier Grids

Most sample carrier grids have hydrophobic surfaces, preventing the formation of thin liquid films of the typically hydrophilic buffers used for proteins. Glow discharging renders the sample carrier grid surface hydrophilic by using plasma to generate radicals that will react with the grid surface, thereby improving its wetting properties. Grids should be prepared freshly before each vitrification session.

1. Using gloves, take a clean, small petri dish and cover the surface of the petri dish with parafilm (*see* **Note 7**).
2. Place the grids on the parafilm so that the side with the sample carrier film faces upward (*see* **Note 8**).
3. Place the petri dish inside the glow discharger and close the chamber door.
4. Turn the knob to the position "Pump" to start the pump; this will empty the chamber to create a partial vacuum.
5. Turn the knob clockwise to the position "Process"; this will ignite the plasma that cleans and hydrophilizes the grid surface (*see* **Note 9**).
6. Release the vacuum by turning the knob to "Flush," then to "Flood." When the chamber door opens, take out the petri dish with the grids on (*see* **Note 10**).
7. Cover the petri dish with a lid until needed (*see* **Note 11**).

3.3 Preparation of the Plunge-Freezer

To prevent uncontrolled evaporation of buffer after sample application to the grid, the plunge-freezer sample chamber should be equilibrated to a defined temperature and humidity. Moreover, the cryogen needs to be equilibrated to its target temperature. The following process must be carried out before each vitrification session.

1. Assemble the Vitrobot humidifier and plug it into the plunge-freezer; then, fill it with approximately 50 mL of distilled water with a syringe. After filling, pull back the syringe piston slightly to create negative pressure in the humidifier and prevent leakage.
2. Switch on the plunge-freezer and program the following options in the Vitrobot software using the graphical user interface (GUI) on the top: (i) Choose the humidity chamber illumination to OFF; (ii) turn on the humidity option and run it at 100% (*see* **Note 12**); (iii) set chamber temperature to 4 °C; (iv) choose "Skip grid transfer" and "Use foot pedal" (all steps of the vitrification process are then triggered with the pedal); and (v) finally, set blotting force and time. Choose a blotting force of 0 (*see* **Note 13**) and a blotting time in a range between 2–8 s (*see* **Note 14**). We use a blotting time of 3.5 s for over 90% of our samples. When all the settings are done, cover the display to shield the sample from light emitted by the display.
3. Assemble the plunge-freezer EPS box by first placing the grid box holder ring around and afterward the metal beaker into the inner chamber, then the metal frame ("spider") touching the beaker so it can transfer heat to the outer chamber (*see* **Note 15**). Fill the outer chamber with liquid nitrogen and wait for it to cool down. When the assembly is cooled down, slowly blow in ethane gas along the beaker wall to condense the ethane (*see* **Note 16**). Leave the spider in place until the liquid ethane temperature reaches approximately −170 °C; then remove it completely (*see* **Note 17**). Place an open grid box into the grid box holder ring in the outer chamber full of liquid nitrogen; eventually add more nitrogen, adjusting the level to just keep the grid boxes submerged within liquid nitrogen.
4. Assemble the filter paper in the plunge-freezer paddles. Filter paper should be changed every 15 min or after 16 grids (*see* **Note 18**).
5. Pick up a glow-discharged grid with the forceps and connect it with the plunge-freezer rod. Make sure the hydrophilic grid surface is facing the sample application port, that is, the opposite side that is used for illumination (*see* **Note 19**).

3.4 Plunge-Freezing with Defined Illumination

This protocol may be used for a single illumination timepoint as well as for multiple timepoints at different stages of protein activation, since the LED driver can be easily programmed to be on for any time length and with different on/off pulsing options.

1. Switch to safe, working-light illumination before opening the icebox containing the protein (*see* **Note 20**).
2. Dilute the phytochrome sample to 0.5–1.0 mg·mL^{-1} (*see* **Note 21**); then apply 3 μL of sample onto the grid surface without touching the grid with the pipette tip (*see* **Note 22**).
3. Use the LED driver to activate the LED for any set time(s).
4. Press the foot pedal once to blot and plunge the grid, either immediately after the LED turns off to study activation processes or at a calculated delay time to study relaxation processes (*see* **Note 23**). Transfer the grid into the grid box (*see* **Note 24**).
5. Close the grid box with its cover. Store the grid box in liquid nitrogen until needed.

4 Notes

1. There are many sources of light contamination besides the actual activating LED, such as stray illumination coming from windows, from under the door, electronic devices, etc., and care should be taken with all of them. Both aluminum foil and specifically designed blackout foils are ideal for permanently blocking out light from windows; however, they might not agree with local working laws, in which case transparent foils that only block out specific wavelengths, such as those used in lithography labs, can be a viable alternative. Furthermore, to prevent premature activation of the sample, it should be stored inside a sealed opaque icebox until needed.
2. It is advisable to characterize the sample studied prior to plunge-freezing by ultraviolet–visible UV-Vis spectroscopy in order to determine basic activation and relaxation parameters, such as optimal excitation wavelength, energy needed for activation, and time courses of activation and relaxation. It is also beneficial to determine wavelengths that do not interact with the chromophore and thus can be employed for background illumination of the working space. For Subheading 3.1, 3.2, and 3.3 of this method, regular white-light overhead illumination of the room can be safely used.
3. For activation of the sample, we advise employing LEDs with defined narrow emission spectra instead of full- or broad-spectrum lamps. Importantly, any infrared (IR) and UV

content could impact the achievable resolution due to molecular motion created by these wavelength ranges.

4. Due to the large opening angle of LED emitters, collimation optics are needed to focus the complete light output onto the sample grid. If multiple LEDs are to be used, for example, when working at different wavelengths is required, care should be taken to use LEDs with identical opening angles; otherwise, multiple optic collimation assemblies are needed.
5. Depending on the exact geometry, it might be beneficial to fix the optical assembly in a slightly downward-facing position instead of exactly parallel to the surface of the optical table.
6. We regularly check the intensities of the illumination setup to ensure reproducibility by recording spectra plots using a handheld spectrometer (Mavospec Base, Germany). When it is necessary to prevent detector saturation, either a 10% (Thorlabs ND10A) or a 1% (Thorlabs ND20A) transmission density filter is used, fixed to a short SM1 extension tube (Thorlabs SM1L03) glued over the sensor with epoxy cement. The obtained spectra need to be corrected, relying on the documented filter curves in this case.
7. Wearing gloves is essential to prevent any residual skin lipids from contaminating the inner surfaces of the glow discharger or anything placed inside the chamber. Fatty acids can affect the properties of the plasma and change the surface properties of the sample grids.
8. Covering the petri dish with parafilm simplifies grabbing grids with the forceps.
9. Before the first use, and regularly afterward, we test the hydrophilization power of our glow discharger by adding a droplet of water on the hydrophilized surface of the grid. The flatter the droplet becomes, the better the hydrophilization. A macro camera can be used to document this contact angle to ensure reproducibility over time. Exposure to the air plasma should be increased if the grid surface keeps its hydrophobicity (i.e., the droplet keeps a rounder shape) after glow discharging.
10. Make sure there is no strong air current streaming into the glow discharger, as this might blow away the grids. Always place the petri dish at the back of the chamber and, if possible, partially close the air intake valve.
11. Grids lose their hydrophilicity over time after glow discharging. The time the grids can be used after glow discharging depends on various circumstances, most importantly the grid surface material and the humidity in the room. In our laboratory, glow-discharged grids can be safely used within 30 min after glow discharging.

12. Start the humidifier first and set the temperature afterward; this allows for the humidity to reach 100% faster. High humidity (≥90%) is necessary to reduce sample evaporation after blotting, and it makes ice thickness reliant almost exclusively on blotting time.
13. When using the Vitrobot for the first time, or after any modification of any part of the Vitrobot, and at the start of any new experimental design, check the optimal blotting force and blotting time. The optimal blotting force can be found by the *slit method*: Performing the blotting process without a grid (and without the forceps that hold it) allows the light to pass from the top of the humidity chamber, through between the paddles, onto the bottom of the chamber. The blotting force is too low if, during blotting, when the paddles are touching, any sliver of light is visible at the bottom. The blotting force is optimal if, during blotting, the slit between the paddles no longer allows the light through to the bottom. Start with a blotting force where the slit allows the light to be visible at the bottom and increase it until the moment the light is not visible anymore. Afterward, the sample thickness can be optimized by changing the blot time.
14. Even if a blotting time has been optimized for a given sample, any change in the sample conditions, like buffer composition, the presence of new additives, or a change in protein concentration, means the blotting time has to be tested again. A range of blotting times can be tested during any given plunge-freezing session, always keeping reproducibility in mind. If an ice gradient is visible on the grid during data collection, an adjustment by 0.5 s up or down is usually sufficient.
15. Make sure that all parts are properly dried, especially when using the Vitrobot for a second time on the same day; otherwise, frost might make parts stick together.
16. It is paramount to use high-quality ethane only, as this can be a major source of contamination. It is also advisable to place an ethane gas sensor nearby because the gas is combustible even at low concentrations.
17. Ideally, the liquid ethane temperature should be probed by a thermometer. However, if no thermometer is available, it is also possible to cool down the liquid ethane until the inner wall starts getting covered by a layer of white, solidified ethane and then lightly warm it up with room-temperature tweezers if the solid layer becomes too thick.
18. While inside the humidity chamber, the blotting paper will continuously absorb water, slowly reducing its efficiency in wicking away sample during blotting. This can impact the blotting time required to achieve a defined ice thickness.

19. Vitrobot forceps never are perfectly straight and might also get bent over time. To keep blotting parameters reproducible, the grid in the forceps must always be mounted in the same orientation facing the same side port in the Vitrobot. We use the left port for illumination and the right port for application of the sample. Additionally, not only should the forceps always be mounted in the same orientation, but also the grid should always be picked up in the same orientation, with the glow-discharged side facing the sample application port. For us, it was helpful to mark the brass part of the forceps to make sure the grid orientation is always the same.
20. Finding illumination conditions that do not activate phytochromes can be difficult due to their absorption both in the UV/blue and red/far-red range, leaving only green light for "dark" conditions. We used a dimmed three-colored handheld flashlight in the green channel (520 nm peak, emission from 480 to 580 nm) for this. Other sources of light can be mounted in the laboratory as long as they comply with the necessary characteristics.
21. The exact concentration depends on the sample and must be determined empirically. However, based on the molecular weight of typical phytochrome dimers, 0.5–1.0 mg·mL^{-1} is a reasonable starting point. In our experience, fresh dilutions tend to give the best results, therefore we dilute small volumes right before freezing a set of grids. The dilution step can also be used to reduce the concentration of buffer and stabilizing components that might reduce contrast, especially high salt concentrations, detergents, and crowding agents such as glycerol.
22. Touching the grid with the pipette tip can disrupt the delicate carrier film. Approach the surface very carefully and press the pipette plunger slightly such that a small droplet of sample forms at the tip, attach the droplet to the grid surface, and then fully press the plunger to apply the remaining volume. Use a fresh pipette tip every time to avoid the surface tension of the tip to cling on to the sample instead of depositing it on the grid.
23. Note that the time the sample can be left on the grid is limited, and activation or relaxation times should not exceed 15 min.
24. Wearing a filtering facepiece particles (FFP2) mask or a face shield can be helpful during this step to prevent breathing into the cryogen pot, since the humidity in our breath might lead to contamination of the sample. Or hold your breath during grid transfer into the grid boxes.

Acknowledgements

This work was funded by the Ministry of Culture and Science of the State of North Rhine-Westphalia (iHEAD). The authors acknowledge access to the cryo-EM infrastructure of StruBiTEM (Cologne, funded by DFG Grant INST 216/949-1 FUGG), and to the computing infrastructure of the ITCC (Cologne, funded by DFG Grant INST 216/512/1 FUGG).

References

1. Dubochet J, McDowall AW (1981) Vitrification of pure water for electron microscopy. J Microsc 124:3–4. https://doi.org/10.1111/j.1365-2818.1981.tb02483.x
2. Dubochet J, Adrian M, Chang J-J et al (1988) Cryo-electron microscopy of vitrified specimens. Q Rev Biophys 21:129–228. https://doi.org/10.1017/S0033583500004297
3. Bock LV, Grubmuller H (2022) Effects of cryo-EM cooling on structural ensembles. Biophys J 121:148a. https://doi.org/10.1016/j.b-pj.2021.11.1981
4. Berriman J, Unwin N (1994) Analysis of transient structures by cryo-microscopy combined with rapid mixing of spray droplets. Ultramicroscopy 56:241–252. https://doi.org/10.1016/0304-3991(94)90012-4
5. Unwin N, Fujiyoshi Y (2012) Gating movement of acetylcholine receptor caught by plunge-freezing. J Mol Biol 422:617–634. https://doi.org/10.1016/j.jmb.2012.07.010
6. Feng X, Fu Z, Kaledhonkar S et al (2017) A fast and effective microfluidic spraying-plunging method for high-resolution single-particle cryo-EM. Structure 25:663–670.e3. https://doi.org/10.1016/j.str.2017.02.005
7. Fu Z, Indrisiunaite G, Kaledhonkar S et al (2019) The structural basis for release factor activation during translation termination revealed by time-resolved cryogenic electron microscopy. Biophys J 116:574a–575a. https://doi.org/10.1016/j.bpj.2018.11.3090
8. Kontziampasis D, Klebl DP, Iadanza MG et al (2019) A cryo-EM grid preparation device for time-resolved structural studies. IUCrJ 6:1024–1031. https://doi.org/10.1107/S2052252519011345
9. Ménétret J-F, Hofmann W, Schröder RR et al (1991) Time-resolved cryo-electron microscopic study of the dissociation of actomyosin induced by photolysis of photolabile nucleotides. J Mol Biol 219:139–144. https://doi.org/10.1016/0022-2836(91)90554-J
10. Shaikh TR, Barnard D, Meng X, Wagenknecht T (2009) Implementation of a flash-photolysis system for time-resolved cryo-electron microscopy. J Struct Biol 165:184–189. https://doi.org/10.1016/j.jsb.2008.11.007
11. Yoder N, Jalali-Yazdi F, Noreng S et al (2020) Light-coupled cryo-plunger for time-resolved cryo-EM. J Struct Biol 212:107624. https://doi.org/10.1016/j.jsb.2020.107624
12. Vu HH, Behrmann H, Hanić M et al (2023) A marine cryptochrome with an inverse photo-oligomerization mechanism. Nat Commun 14:6918. https://doi.org/10.1038/s41467-023-42708-2
13. Bódizs S, Mészáros P, Grunewald L et al (2024) Cryo-EM structures of a bathy phytochrome histidine kinase reveal a unique light-dependent activation mechanism. Structure 32:1952–1962.e3. https://doi.org/10.1016/j.str.2024.08.008
14. Subramaniam S, Gerstein M, Oesterhelt D, Henderson R (1993) Electron diffraction analysis of structural changes in the photocycle of bacteriorhodopsin. EMBO J 12:1–8. https://doi.org/10.1002/j.1460-2075.1993.tb05625.x
15. Nagel G, Kelety B, Möckel B et al (1998) Voltage dependence of proton pumping by bacteriorhodopsin is regulated by the voltage-sensitive ratio of M1 to M2. Biophys J 74:403–412. https://doi.org/10.1016/S0006-3495(98)77797-5

Chapter 14

Structures of Myxobacterial Phytochrome Revealed by Cryo-EM

Tek Narsingh Malla, Emina A. Stojković, and Marius Schmidt

Abstract

Phytochromes are red-light photoreceptors first identified in plants, with homologs found in bacteria and fungi that regulate a variety of critical physiological processes. They undergo a reversible photocycle between two distinct states: a red-light-absorbing phytochrome (Pr) form and a far-red light-absorbing phytochrome (Pfr) form. This Pr-to-Pfr photoconversion controls the activity of a C-terminal enzymatic domain, typically a histidine kinase (HK). The molecular mechanisms underlying light-induced regulation of HK activity in bacteria remain poorly understood, as structures of unmodified bacterial phytochromes with HK activity were unknown until recently. Here, we report details of using cryogenic electron microscopy (cryo-EM) to determine the structure of an intact (full-length) bacteriophytochrome.

Key words Cryo-EM, CryoSPARC, Protein structure, Photoreceptor enzyme, Bacteriophytochrome, Histidine kinase, Myxobacteria

1 Introduction

Sensing and responding to changing environmental conditions are crucial for the survival of living organisms. In prokaryotes, environmental signals are typically detected through a two-component signaling system, consisting of a histidine kinase (HK) and a response regulator (RR) [1]. These highly modular two-component systems have been adapted into diverse cellular signaling circuits. In photosynthetic and nonphotosynthetic bacteria, phytochrome photoreceptors are integrated into such signaling mechanisms, responding to variations in light spectrum, intensity, and direction [2]. Bacteriophytochrome photoreceptors (BphPs) are homodimeric photoreceptor enzymes that consist of four domains as depicted in Fig. 1a. By absorption of red light, they are driven into a photocycle (*see* Fig. 1b). The BphPs can be found in two stable states with distinct structures, the red-absorbing phytochrome (Pr) state and the far-red-absorbing phytochrome

Andreas Winkler (ed.), *Phytochromes: Methods and Protocols*, Methods in Molecular Biology, vol. 2970,
https://doi.org/10.1007/978-1-0716-4791-2_14,

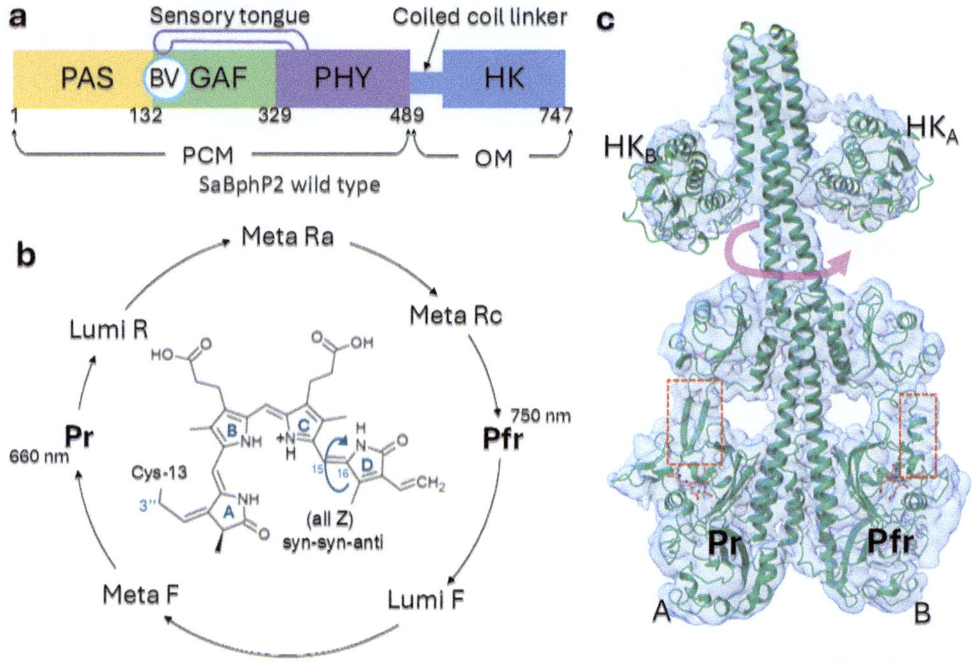

Fig. 1 The *S. aurantiaca* BphP2. (**a**) Scheme of the domain architecture of the wild-type SaBphP2. Sequence numbers are marked. The BV chromophore position is indicated. PAS, GAF, and PHY domains form the PCM which is connected to the output effector module (OM) which is a histidine kinase (HK). The PHY domain has a long extension marked as the sensory tongue that is required for complete Pr/Pfr photoconversion. (**b**) The reversible photocycle with intermediates. When the Pr state is illuminated with red light (640–700 nm), it converts to the Pfr state. When illuminated with far-red light (750 nm), the Pfr state can be driven to Pr. The structure of BV is shown with the four pyrrole rings in the (all Z) syn-syn-anti configuration. The $C_{15} = C_{16}$ double bond between rings C and D that undergoes Z-to-E isomerization is marked by an arrow. (**c**) Structure of the full-length SaBphP2 Pr/Pfr heterodimer as determined by cryo-EM [3]. The conformations of the sensory tongues are marked with the red dashed boxes. The structure in the Pfr state is distinguished by the telltale α-helix conformation. When the second subunit, here subunit A, transitions to Pfr, the HK lobes rotate (red arrow)

(Pfr) state, respectively (*see* Fig. 1b). Causative for the structural transition is the Z-to-E photoisomerization of a central chromophore which is typically a biliverdin (BV) in BphPs (*see* Fig. 1b). The Pr-to-Pfr transition results in a substantial rotation of the HK lobes (*see* Fig. 1c) that might ultimately be involved in RR binding and dissociation from the BphP.

In a recent single-particle cryogenic electron microscopy (cryo-EM) experiment, structures of a full-length, unmodified BphP from the myxobacterium *Stigmatella aurantiaca* (SaBphp2) were determined [3]. The full-length SaBphP2 consists of the photosensory core module (PCM) and includes linker and HK domains (*see* Fig. 1a). Although the structure of the PCM could be determined to near-atomic resolution (3.2 Å), the density for the all-important HK enzymatic domain has been weak, except in maps derived from particles found in a heterodimeric state. In the heterodimer, one subunit is in the Pr state and the other in the photoactivated Pfr state (*see* Fig. 1c). The density of two HK moieties was good enough that an amino acid backbone model for the HK lobes could be inserted. Details for obtaining the structure of the Pr/Pfr heterodimer from single-particle cryo-EM investigations are reported below.

The protocol is written for nonexpert users looking to start cryo-EM data processing. The actual cryo-EM data collection procedure requires robust training and is usually handled by the staff at cryo-EM facility.

2 Materials

1. Purified wild-type SaBphP2: 10 mg/mL in stabilization buffer (20 mM Tris–HCl pH 8.0, 50 mM NaCl).
2. UltrAuFoil holey grids (300 mesh R1.2/1.3, QUANTIFOIL) to prepare sample for cryo-EM investigations.
3. FEI Vitrobot automatic plunge-freezer to vitrify the sample on the grids.
4. AutoGrid cartridges (Thermo Fisher Scientific) to store the frozen grids.
5. Thermo Fisher Scientific Titan Krios electron microscope equipped with a Gatan BioQuantum K3 energy filter direct electron detector camera to image the vitrified sample.
6. High-performance computers with graphics processing units (GPUs) to analyze the data.
7. Software: Leginon to collect the data, Globus Connect to transfer data, cryoSPARC software to process the data, ChimeraX, Coot, ISOLDE, and Phenix to determine structure and produce publication-style images.

3 Methods

3.1 SaBphP2 Preparation

1. Express full-length N-terminally His_6-tagged SaBphP2 in *Escherichia coli*, and purify the protein using Co^{2+} TALON affinity chromatography. Refer to Woitowich et al. and Sanchez et al. [4, 5] for a detailed protocol that is used for shorter constructs. Use an identical protocol to express and purify the full-length protein. Avoid freeze–thaw cycles during purification as the temperature changes impact protein stability.
2. Concentrate the SaBphP2 to 10 mg/mL and store in stabilization buffer (20 mM Tris–HCl, 50 mM NaCl, pH 8). Freeze in

20 μL aliquots and ship the frozen protein on dry ice to the New York Structural Biology Center (NYSBC).

Cryo-EM facilities like NYSBC accept both purified protein and frozen grids from the users. Most laboratories like ours which do not have access to plunge-freezers send the purified protein to the cryo-EM facility, and the facility will prepare the grids. The laboratories with access to plunge-freezers may prepare their own grids the following way and send frozen grids. The following steps can be used to replicate the grids used for data collection of Pr/Pfr heterodimer [3].

3.2 Grid Preparation

1. Glow discharge UltrAuFoil holey grids (300 mesh R1.2/1.3, QUANTIFOIL) for 10 s in oxygen and argon gases.
2. Set up the FEI Vitrobot automatic plunge-freezer at a chamber humidity of 95%, temperature of 4 C, blotting time of 1 s, and incubation time of 10 s.
3. Pipette 3 μL of diluted SaBphP2 solution (0.3 mg/mL) on the grid for plunge-freezing. Adjust the blotting and incubation time to obtain grids with optimum ice thickness.
4. Store the frozen grids in liquid nitrogen. The frozen grids can be shipped to cryo-EM facility.

3.3 Cryo-EM Imaging

These steps are usually performed by the trained staff members at the cryo-EM facilities. Any users trained to perform these actions must consult with facility management to replicate the conditions and parameters presented here.

1. Clip the frozen grids into AutoGrid cartridges (Thermo Fisher Scientific). Transfer the AutoGrid cartridge into the autoloader of the Thermo Fisher Scientific Titan Krios electron microscope.
2. Screen the grids to determine the grids with good particle distribution and optimum ice thickness.
3. Use facility-specific software and procedures to set up data collection. The NYSBC uses the Leginon system [6]. A good norm is to collect ~10,000 movies. Depending on the quality and concentration/distribution of the protein particles in the grids, the number of recorded movies can be increased or decreased as appropriate.
4. Collect movies with suitable parameters. The imaging parameters are determined by the camera used for data collection. The Gatan K3 camera used in Malla et al. [3] produces high quality data at magnification of 105,000× at a pixel (px) size of 0.844 Å/px with exposure time of 2000 ms, 50 frames per movie, 40 ms per frame, and a total dose of ~60 $e^-/Å^2$.

5. Transfer the raw movies to the workstation where the data will be processed. The data transfer depends on the network speed and the size of the data. Globus Connect (www.globus.org) is an efficient and effective way of transferring large scientific data across institutions. The transfer of data from NYSBC to the University of Wisconsin–Milwaukee (UWM) High-Performance Computing Cluster via Globus Connect, which totaled ~6 terabytes for ~17,000, required several days.

3.4 Data Processing

There are several programs to process single-particle cryo-EM data such as cryoSPARC [7], RELION [8], Scipion3 [9], and many more. Due to the relative ease of installation and maintenance, we use cryoSPARC. Aside from some steps that may use different algorithms, the general processing pipeline is similar across all platforms. Table 1 lists important parameters required for full-length SaBphP2 single-particle data processing. Figure 2 displays its progress graphically.

1. Open the cryoSPARC user interface. Create a new project. Enter an appropriate project title and select a location where the processed and intermediate files will be stored. Create a workspace within the project. As many workspaces can be created within a project as needed. This is useful to organize

Table 1
Parameters and statistics for cryo-EM analysis

CryoSPARC-processing parameters	
Raw images	Pixel size: 0.844 Å, voltage: 300 kV, spherical aberration: 2.7 mm, total exposure dose: 56.39 $e^{-}/Å^{2}$
Manual curate exposure	CTF fit resolution cutoff: 7 Å Relative ice thickness cutoff: 1.1 Total full-frame motion distance cutoff: 50 pixels
Blob picker	Particle Ø from 70 to 220 Å
Template picker	Particle Ø 195 Å
Extract from micrographs	Extraction box size: 360 pixels
Initial number of particle images	1,520,224
2D classification	Initial iteration: 100 classes Later iterations: 50–100 classes, batch size per class: 200, initial classification uncertainty factor: 5, number of online-EM iteration: 40
Ab-initio reconstruction	Number of ab-initio classes: 5
Final number of particle images	215,374
Gold-standard Fourier shell correlation (GSFSC) resolution	3.75 Å

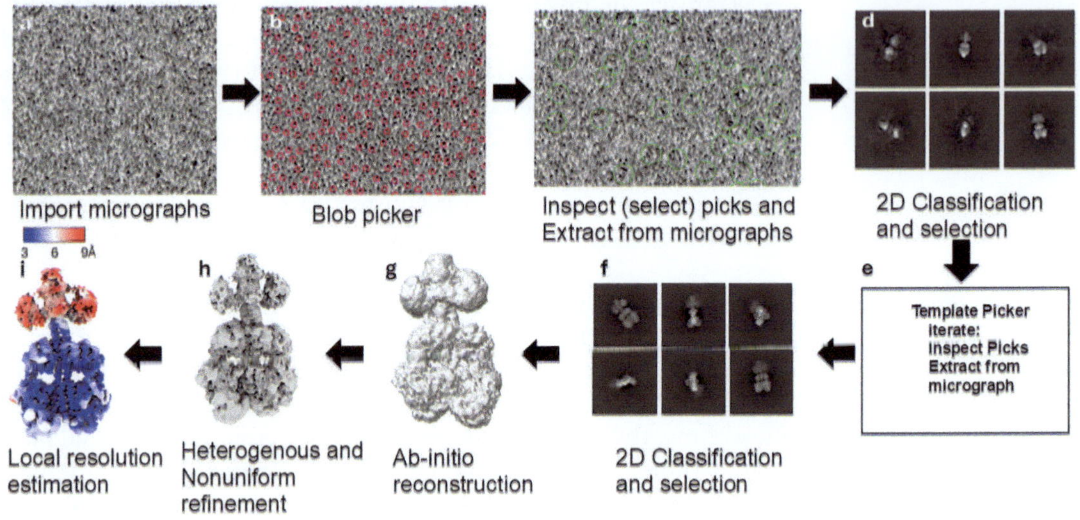

Fig. 2 Bootstrapping cryoSPARC data processing sequence highlighted in panels (**a**) to (**i**). Progress is indicated by arrows. Individual steps are indicated below panels (**a–d**) and (**f–i**). The color bar in panel (**i**) encodes the local resolution of the corresponding cryo-EM map. Iterations that involve the template picker (panel **e**) are explained in the text

the data processing performed with similar but separate parameters.

2. Set up "import movies" job. There will be options to select the movies data path and gain reference path which must be correctly filled. Toggle the "Flip the gain reference file in Y?" to true. Other mandatory parameters in this step include raw pixel size (Å), accelerating voltage (kV), spherical aberration (mm), and total exposure dose ($e^-/Å^2$). This information is provided by the cryo-EM facility (*see* Table 1). The other available parameters can be set to default.

3. Align and average the movie frames into a single-frame micrograph with the "patch motion correction" job. Connect the output of "import movies" job as the input of this motion correction job.

4. Create a "patch CTF estimation job" and connect the micrograph output of the "patch motion correction job" into its input. This job estimates several contrast transfer function (CTF) parameters of the micrographs, including defocus, CTF fit resolution, ice thickness, etc. (*see* Table 1).

5. Set up "manually curate exposure" job to filter out the bad quality micrographs. The quality metrics are derived from the previous CTF estimation job. Change the threshold for each of the quality metrics until a desired subset of data is achieved.

6. Set up "Blob picker" utility in cryoSPARC and connect the output from previous job consisting of accepted micrographs

to its input. Set the appropriate "minimum particle diameter" and "maximum particle diameter" values (*see* Table 1). It is better to underestimate the minimum diameter and overestimate the maximum diameter of the target protein. Blob picker automatically picks particles (*see* Fig. 2b) by looking for a Gaussian signal based on these parameters.

7. Set up "Inspect particle picks" job to filter the particles picked by the previous particle picker job. Use the slider in the power histogram to select the target particles. Aggregates, ice crystals, and carbon edges have high power score while empty picks have low power score. Use the normalized cross-correlation (NCC) score slider to filter out the particles based on the NCC score. The true particles generally have high NCC scores.
8. Set up the "Extract from micrographs" job (*see* Fig. 2c) and connect the particle output of the previous job to its input. The mandatory parameter for this job is the extraction size. The box size must be at least 1.5–2× the diameter of the particle (*see* Table 1).
9. Set up a "2D classification" job and connect the "particles extracted" output from the previous "Extract from micrographs" job to its input. For the first round, set the number of 2D classes to 100, the batch size per class to 200, and the number of online expectation-maximization (online-EM) iterations to 40, and keep other parameters as default (*see* Table 1). The default parameters which are half the value of each of these parameters are generally good enough. However, these updated values have shown improved results for small or low signal-to-noise ratio (SNR) particles. This job generates two outputs: all particles and 2D class averages (*see* Fig. 2d).
10. The good 2D class averages and the corresponding particles are selected with the "select 2D classes" job. Create the "select 2D classes" job and connect the outputs of the previous 2D classification job to its inputs. Here, the classes can be filtered based on the relative resolution, effective classes assigned (ECA) value, and the number of particles in each class. Select the good-looking classes that represent multiple orientations of the particles.
11. Repeat **step 9** using the output particles of **step 10** as input to further filter out the good classes. This process of selecting good 2D classes/particles and running 2D classification again can be iterated as many times as necessary until clean 2D class averages are obtained (*see* Fig. 2e).
12. The resulting good 2D class averages are used as "ab-initio" templates that are used to repeat picking particles from the entire set of curated micrographs (from **step 5**). For this, set up a "template picker" job (*see* Fig. 2). Connect the selected

micrograph output of **step 5** and selected 2D classes of **step 11** as the inputs for the "template picker" job. Unlike the "blob picker," the template picker only requires the average particle diameter as the mandatory parameter (*see* Table 1).

13. Repeat **step** 7 through **step 11** to obtain good 2D class averages (*see* Fig. 2f) and the corresponding particles. It is expected that at the end of this stage a set of high-quality particles is picked.
14. Set up "ab-initio reconstruction" (*see* Fig. 2g) to generate 3D volume(s) directly by connecting the output from the previous step (the set of high-quality particles) to the input of this step. It is beneficial to change the number of generated 3D classes to more than one even if there is only one distinct structure present. If the parameter is set to 1, the program will otherwise combine all particles, even the subquality particles in one class which may produce a low-quality 3D volume. By allowing several 3D volumes (*see* Table 1), the program has some freedom to exclude the low-quality particles from a particular 3D volume.
15. Set up "heterogenous refinement" job (*see* Fig. 2h) to further classify particles and refine the initial models generated by the "ab-initio reconstruction." The output of the previous "ab-initio reconstruction" job: Multiple 3D volumes and their corresponding particles are connected as input to the "heterogenous refinement" job. All parameters can be kept as default. Inspect the output volumes and choose the best one for further refinement.
16. Set up a "homogenous refinement" or "non-uniform refinement" job. The objective of both of these jobs is same except that the "non-uniform refinement" is suitable for particles with more disorder, flexibility, and reduced SNR. Use the best-looking output volume and the corresponding particles (*see* Table 1) from **step 16** as the input to this job. Set the symmetry that matches the symmetry present in the particle. For a particle with C2 symmetry, enforcing the value is roughly equivalent to doubling the number of particles. For cryo-EM data of the SaBphP2 Pr/Pfr heterodimer (*see* Fig. 1c), for obvious reasons, application of C2 symmetry is discouraged (*see* **Note 1**).
17. The refinement job in **step 17** produces a refined 3D map. This utility also produces a sharpened map which is much better for model building rather than the regular unsharpened map. The local resolution can be determined by the "local resolution estimation" feature in cryoSPARC and displayed in ChimeraX [10] as shown in Fig. 2i.

In most cases, these steps are sufficient to generate a 3D volume from which a high-quality model can be built. There are other utilities in cryoSPARC which will be useful for more advanced processing. These include "3D variability analysis" to explore heterogeneity in the dataset, or "local refinement" to focus on a subregion of the map, etc. More documentation is available at www.guide.cryosparc.com.

3.5 Cryo-EM Structure Determination

1. Download the refined 3D map from cryoSPARC and load the map in ChimeraX.
2. Load an atomic model in ChimeraX. The initial model can be a previously published X-ray structure or an AlphaFold-predicted model.
3. Manually move the atomic model toward the map. Use visual cues to best align the model to the map. Accuracy is not important at this stage.
4. Use the "fitmap" command in ChimeraX to fit the atomic model into the map. This step can be refined further by selecting a subset of atoms in the model. The program will use only the selected atoms for fitting.
5. Identify and isolate the sections of the model which are well resolved in the cryo-EM maps. Use the selective fitting option to position the individual sections into the corresponding regions of the cryo-EM map. Save the partly built atomic model.
6. Load the partly built atomic model and the cryo-EM map in Coot. Use the "add residue" utility in Coot to manually insert the remaining residues.
7. Once the complete model is obtained, refine the model further by alternating rounds of real-space refinement in Phenix v1.21 [11] and model building in Coot [12] and ISOLDE [13]. The final model (*see* Fig. 1c) is obtained by a real-space refinement against the unsharpened map in Phenix (*see* **Note 2**).

4 Notes

1. The structure of the Pr/Pfr heterodimer as determined by single-particle cryo-EM is the first with contiguous density in the HK lobe region, which could be used to determine at least a low-resolution structure (9 Å) of the HK lobes. The Pr/Pfr heterodimer is insofar remarkable as its existence has been predicted previously for plant phytochromes [14] and also observed previously in a crystal structure of an unrelated (modified) BphP with diguanylyl cyclase activity [15, 16]. By photoactivating all subunits, a Pfr/Pfr homodimer is obtained that

distinguishes itself mainly by a large rotation of the HK lobes (*see* Fig. 1c, red arrow). This rotation may be a decisive part of the signaling mechanism of the BphP [17]. Using cryo-EM, molecular structures are solved from solution. Large-scale structural relaxations that are initiated after photoactivation of the phytochrome are supported as shown [3]. The ability to access a large conformational space can, however, also result in structural heterogeneity that locally reduces the achievable resolution such as that observed for the HK domain which is much lower (9 Å) than that of SaBphP2 core (3.2 Å).

2. The SaBphP2 cryo-EM structures reported by Malla et al. [3] lay the foundation for time-resolved cryo-EM (TR-cryo-EM) applications [18–21]. The formation of the Pr/Pfr intermediate as well as the transition to the Pfr/Pfr homodimer could be investigated in "real-time." We expect to observe the β-sheet to α-helix transition of the sensory tongue (*see* Fig. 1a for its location) as well as the large reorientations of the PAS (Per-Arndt-Sim), GAF (cGMP phosphodiesterase/adenylyl cyclase/FhlA), and PHY (phytochrome-specific GAF-related) domains and the rotation of the HK lobes. For TR-cryo-EM, the grid-preparation and the plunge-freezing process described in Subheading 3.1 must be modified. With the Spotiton technique [19, 22], the application of protein solution to the grid, blotting, and plunge-freezing are all unified into one automatic process. When the grid is illuminated by an intense light source such as a (pulsed) laser during plunging, a photoreaction can be initiated. The reaction stops once the grid is frozen. The time resolution is given by the shortest time possible between laser illumination and freezing which should be on the order of 20 ms with the Spotiton technique. Processes that occur on longer times in the photocycle (*see* Fig. 1b) can then be investigated.

Acknowledgment

The research was funded by BioXFEL (NSF-STC-1231306), by BioXFEL Center sub-award 6227 and NSF MCB Collaborative RUI research grant 2423601. We thank Megan McCracken at UWM HPC service for help with installation of cryoSPARC and Globus Connect on the UWM HPC cluster.

References

1. Stock AM, Robinson VL, Goudreau PN (2000) Two-component signal transduction. Annu Rev Biochem 69:183–215. https://doi.org/10.1146/annurev.biochem.69.1.183
2. Furuya M (1993) Phytochromes: their molecular species, gene families, and functions. Annu Rev Plant Physiol 44:617–645

3. Malla TN, Hernandez C, Muniyappan S, Menendez D, Bizhga D, Mendez JH et al (2024) Photoreception and signaling in bacterial phytochrome revealed by single-particle cryo-EM. Sci Adv 10(32):eadq0653. https://doi.org/10.1126/sciadv.adq0653
4. Woitowich NC, Halavaty AS, Waltz P, Kupitz C, Valera J, Tracy G et al (2018) Structural basis for light control of cell development revealed by crystal structures of a myxobacterial phytochrome. IUCrJ 5(Pt 5):619–634. https://doi.org/10.1107/S2052252518010631
5. Sanchez JC, Carrillo M, Pandey S, Noda M, Aldama L, Feliz D et al (2019) High-resolution crystal structures of a myxobacterial phytochrome at cryo and room temperatures. Struct Dyn-Us 6(5):054701. https://doi.org/10.1063/1.5120527
6. Suloway C, Pulokas J, Fellmann D, Cheng A, Guerra F, Quispe J et al (2005) Automated molecular microscopy: the new Leginon system. J Struct Biol 151(1):41–60. https://doi.org/10.1016/j.jsb.2005.03.010
7. Punjani A, Rubinstein JL, Fleet DJ, Brubaker MA (2017) cryoSPARC: algorithms for rapid unsupervised cryo-EM structure determination. Nat Methods 14(3):290–296. https://doi.org/10.1038/nmeth.4169
8. Scheres SHW (2012) RELION: implementation of a Bayesian approach to cryo-EM structure determination. J Struct Biol 180(3):519–530. https://doi.org/10.1016/j.jsb.2012.09.006
9. Conesa P, Fonseca YC, Jimenez de la Morena J, Sharov G, de la Rosa-Trevin JM, Cuervo A et al (2023) Scipion3: a workflow engine for cryo-electron microscopy image processing and structural biology. Biol Imaging 3:e13. https://doi.org/10.1017/S2633903X23000132
10. Pettersen EF, Goddard TD, Huang CC, Meng EC, Couch GS, Croll TI et al (2021) UCSF ChimeraX: structure visualization for researchers, educators, and developers. Protein Sci 30(1):70–82. https://doi.org/10.1002/pro.3943
11. Liebschner D, Afonine PV, Baker ML, Bunkoczi G, Chen VB, Croll TI et al (2019) Macromolecular structure determination using X-rays, neutrons and electrons: recent developments in Phenix. Acta Crystallogr D Struct Biol 75(Pt 10):861–877. https://doi.org/10.1107/S2059798319011471
12. Casanal A, Lohkamp B, Emsley P (2020) Current developments in Coot for macromolecular model building of electron cryo-microscopy and crystallographic data. Protein Sci 29(4):1069–1078. https://doi.org/10.1002/pro.3791
13. Croll TI (2018) ISOLDE: a physically realistic environment for model building into low-resolution electron-density maps. Acta Crystallogr D Struct Biol 74(Pt 6):519–530. https://doi.org/10.1107/S2059798318002425
14. Klose C, Venezia F, Hussong A, Kircher S, Schafer E, Fleck C (2015) Systematic analysis of how phytochrome B dimerization determines its specificity. Nat Plants 1:15090. https://doi.org/10.1038/nplants.2015.90
15. Gourinchas G, Heintz U, Winkler A (2018) Asymmetric activation mechanism of a homodimeric red light-regulated photoreceptor. Elife 7:e34815. https://doi.org/10.7554/eLife.34815
16. Gourinchas G, Etzl S, Gobl C, Vide U, Madl T, Winkler A (2017) Long-range allosteric signaling in red light-regulated diguanylyl cyclases. Sci Adv 3(3):e1602498. https://doi.org/10.1126/sciadv.1602498
17. Malla TN, Muniyappan S, Menendez D, Ogukwe F, Dale AN, Clayton JD et al (2025) Exploiting fourth-generation synchrotron radiation for enzyme and photoreceptor characterization. IUCrJ 12:36. https://doi.org/10.1107/S2052252524010868
18. Torino S, Dhurandhar M, Stroobants A, Claessens R, Efremov RG (2023) Time-resolved cryo-EM using a combination of droplet microfluidics with on-demand jetting. Nat Methods 20(9):1400–1408. https://doi.org/10.1038/s41592-023-01967-z
19. Dandey VP, Budell WC, Wei H, Bobe D, Maruthi K, Kopylov M et al (2020) Time-resolved cryo-EM using Spotiton. Nat Methods 17(9):897–900. https://doi.org/10.1038/s41592-020-0925-6
20. Frank J (2017) Time-resolved cryo-electron microscopy: recent progress. J Struct Biol 200(3):303–306. https://doi.org/10.1016/j.jsb.2017.06.005
21. Lorenz UJ (2024) Microsecond time-resolved cryo-electron microscopy. Curr Opin Struct Biol 87:102840. https://doi.org/10.1016/j.sbi.2024.102840
22. Budell WC, Allegri L, Dandey V, Potter CS, Carragher B (2021) Cryo-electron microscopic grid preparation for time-resolved studies using a novel robotic system, spotiton. J Vis Exp 168:e62271. https://doi.org/10.3791/62271

Chapter 15

Hydrogen–Deuterium Exchange Coupled to Mass Spectrometry-Based Analysis of Phytochrome Photoreceptors

Maximilian Fuchs and Andreas Winkler

Abstract

Red light sensing phytochromes typically have at least two functionally distinct conformations, a red-absorbing Pr state and a far-red-absorbing Pfr state. Understanding the structural differences between these functional states is key to appreciating aspects of light signal integration at the molecular level. Through potentially different interactions with downstream effectors, they can also lay the foundation for gaining better insights at an organismal level.

The multidomain architecture of many full-length phytochromes makes them challenging targets for detailed structural analysis using classical X-ray crystallography or nuclear magnetic resonance approaches. While recent advances in cryo-electron microscopy offer a promising alternative, the inherently dynamic nature of photoreceptors can also be challenging for single-particle analysis. In this chapter, we describe an alternative approach for the structural analysis of phytochromes, hydrogen–deuterium exchange coupled to mass spectrometry (HDX-MS). This in-solution technique permits investigation of the conformational dynamics of proteins via the readout of their secondary structure stability. The ease of using different light regimes to enrich either Pr or Pfr states in solution is illustrated using the model bacteriophytochrome system from *Deinococcus radiodurans*. However, HDX-MS also offers the possibility of extracting the corresponding structural information from ensembles with partial Pr or Pfr states.

Key words HDX-MS, Conformational dynamics, Light activation, Signal integration, Bacteriophytochrome, BphP

1 Introduction

Phytochromes typically function as sensors of red or far-red light and integrate this information of ambient light conditions into downstream signaling cascades to regulate important cellular functions in cyanobacteria, bacteria, fungi, and plants [1]. In order to perform their function, they need to exist in at least two functionally different states. For bacterial, fungal, and plant phytochromes, these two conformations are typically referred to as the Pr and Pfr states, absorbing in red and far-red, respectively. Characteristics of

Andreas Winkler (ed.), *Phytochromes: Methods and Protocols*, Methods in Molecular Biology, vol. 2970, https://doi.org/10.1007/978-1-0716-4791-2_15,

these two metastable ground-state conformations are structural differences in the surrounding of the light-sensing bilin cofactors [1]; however, the conformational dynamics of the bilin environments and associated structural and functional elements also play an important role in the signal integration mechanism [2–5].

Here, we describe a method that allows the characterization of in-solution conformational dynamics of full-length phytochromes in particular and photoreceptors in general that can easily be extended to include complexes with functionally relevant ligands and/or target effector proteins. Since the strength of the hydrogen bonds between the peptide amide hydrogens and carbonyl oxygens in alpha helices and beta sheets correlates with secondary structure stability, the rate of hydrogen–deuterium exchange can be employed as a proxy for the conformational dynamics of proteins [6, 7]. Due to the inherent mass difference between hydrogen and deuterium, mass spectrometry offers a straightforward readout of the deuterium exchange kinetics. Especially upon including protease digestion in the workup routine, the conformational dynamics of specific structural elements can be assigned using hydrogen–deuterium exchange coupled to mass spectrometry (HDX-MS). More importantly, however, the comparison of light-responsive systems in their dark-adapted states with the corresponding light-activated or photostationary states (PSSs) allows the identification of structural elements involved in light signal integration and communication to downstream functional regions. The versatility of performing these experiments also in the presence of specific ligands or substrates, as well as variants of the protein of interest, provides access to structural information of different functionally relevant states that is otherwise difficult to obtain. To illustrate the workflow, we show representative results for a paradigm bacterial model system, bacteriophytochrome photoreceptor (BphP) from *Deinococcus radiodurans* (*Dr*BphP) [8, 9], in its dark-adapted and PSS conformation.

2 Materials

Prepare all solutions using ultrapure water and chemicals of liquid chromatography–mass spectrometry (LC-MS) grade. The solutions must be prepared and pH/pD adjusted at the specified temperatures.

2.1 Solutions and Buffers

1. Reference buffer (1×): 10 mM 4-(2-hydroxyethyl)-1-piperazineethanesulfonic acid (HEPES), pH 7.0, 50 mM NaCl, 2 mM $MgCl_2$, 20 °C (*see* **Note 1**).

   Prepare the desired volume by dissolving the components in H_2O. Adjust the pH using concentrated sodium hydroxide solution. The pH should be set at 20 °C. Fill up to the final volume with H_2O after the pH is adjusted.

2. Deuteration buffer (1×): 10 mM HEPES, pD 7.0, 50 mM NaCl, 2 mM $MgCl_2$, 20 °C (*see* **Note 2**). The preparation is similar to the reference buffer, but the components are dissolved in D_2O. Adjusting to pD 7.0 can be done with a standard glass electrode. The pD can be approximated using the following eq. [10]:

$$pD = pH\ \text{reading} + 0.41 \quad (1)$$

Adjust the buffer pD with concentrated sodium hydroxide dissolved in D_2O (*see* **Note 3**).

3. Quench buffer (1×): 200 mM ammonium formate (NH_4HCO_2), pH 2.6, 1.5 M urea (*see* **Note 4**), 0 °C. For 250 mL, dissolve 3.18 g ammonium formate and 22.52 g urea in 200 mL H_2O and cool to 0 °C in an ice water bath. Use formic acid to adjust the pH. After adjusting the pH, fill up to the final volume. The solution can be divided into aliquots and stored at −20 °C until use.
4. Mobile phase A: 0.6% (v/v) formic acid in H_2O.
5. Mobile phase B: 0.6% (v/v) formic acid in acetonitrile.
6. Mobile phase C: 0.1% (v/v) formic acid, 1% methanol in H_2O.
7. Pepsin column storage solution: 0.1% (v/v) formic acid in H_2O.
8. Pepsin column washing solution: 1.5 M guanidine hydrochloride, 4% (v/v) acetonitrile, 0.8% (v/v) formic acid.

2.2 Equipment

The following equipment is used for the experiments in our laboratory, but alternative suppliers are available for all items.

1. Shimadzu Nexera HPLC (modified for high-performance liquid chromatography [HPLC] separations to be run at 0.5 °C—*see* Subheading 2.2.1).
2. Laboratory refrigerator unit with glass door and flexible shelves; suitable for enclosing the HPLC system and solvents.
3. Bruker impact II quadrupole time-of-flight (QTOF) mass spectrometer.
4. Pepsin column: Waters® Enzymate™ BEH (2.1 × 30 mm).
5. Analytical column: Shim-pack Arata Peptide C18 2.2, μm (2.0 × 50 mm).
6. Trap column: Shim-pack GISS-HP(G) C18 (2.1 × 10 mm).
7. Thorlabs mounted LEDs (e.g., M660L4—660 nm, M810L5—810 nm, M780L3—780 nm).

8. Thorlabs LEDD1B T-Cube LED (light-emitting diode) Driver.
9. Thorlabs SM2F32-A—Adjustable Collimation Adapter, 350–700 nm.
10. Thorlabs SM2P50-B—Adjustable Collimation Adapter, 650–1050 nm.
11. Thermal Incubator for Eppendorf tubes (VWR Thermal Shake Lite or equivalent).
12. Open vessel for liquid nitrogen use—ideally a 3 L open nitrogen dewar.
13. 100 μL Gastight Hamilton syringe for injections.
14. Ultraviolet–visible (UV-Vis) spectrophotometer.

2.2.1 HDX-MS Setup

Commercial HDX-MS setups can be acquired from Waters, Thermo Fisher Scientific, and other vendors. In combination with robotic systems, these offer automated sample preparation and injection. However, HDX-MS experiments can also be performed effectively with home-built systems and manual injections, as described below and schematically visualized in Fig. 1.

In our HDX-MS setup, we integrated Shimadzu Nexera HPLC pumps (one LC-40B XR binary gradient pump and one LC-20AD single solvent delivery pump), the DGU-405 solvent degassing

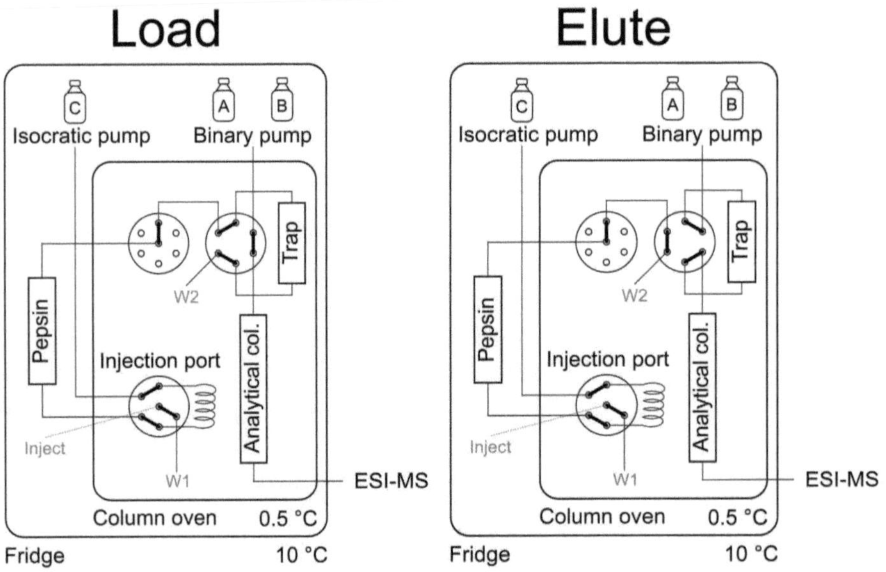

Fig. 1 Schematic representation of the home-built HDX-MS setup. The differences in the flow path between the load and elute phases of the separation method are highlighted. During the load phase, the isocratic pump transports the sample from the injection loop via the pepsin column to the trap column. Switching to the elution phase connects the binary gradient pump with the trap column and the analytical column. A gradient is used to elute the peptides from the column and transport them toward the mass spectrometer

unit, and the CTO-40C column oven into a laboratory refrigerator unit. The CTO-40C is equipped with a Rheodyne manual injection port for injecting the individual HDX timepoints. All of these parts communicate with a CBM-20A system controller on the outside of the refrigerator, which also allows for the integration of the HPLC into the Bruker HyStar software and communication between the two systems using the Agilent Instrument Control Framework (ICF) plug-in.

The refrigerator unit is set to 10 °C, and the CTO-40C column oven is recalibrated to allow for temperature control at 0.5 °C. This is important to reduce the back-exchange from deuterons to protons during the HPLC workup routine of the HDX reactions. All chromatographic steps (sample injection, desalting, and separation of peptides) occur at this low temperature. The only exception is the digestion of the protein samples on the pepsin column. It is positioned outside of the column oven to allow for more efficient pepsin cleavage at 10 °C (*see* Fig. 1).

The final step of eluting the peptides from the analytical column into the mass spectrometer is enabled by a direct connection of polyether ether ketone (PEEK) tubing (red color, ID 0.005″) from the column exit port to the electrospray ionization (ESI) source of the mass spectrometer.

2.3 Software

Bruker HyStar and Agilent ICF plug-in for integration of the Shimadzu HPLC system.

Bruker otofControl.

Bruker DataAnalysis or CompassXport for mzXML export.

Hexicon 2 for HDX-MS data analysis [11].

3 Methods

In order to efficiently perform the key HDX-MS experiment, some preparatory steps are required. Therefore, the Methods section is divided into specific subsections that allow the reader to address the individual points separately. Each section contains a short justification for why these tasks are important to be carried out and what to consider in the application for other systems to be analyzed. Some aspects are also connected between subchapters; however, for better readability, we have separated them into individual chapters. For example, MS/MS identification, HDX reactions, and HPLC gradient details all refer to specific aspect of the respective other chapters.

3.1 Characterization of Pr and Pfr States Using UV-Vis Spectroscopy

A spectroscopic characterization should be performed prior to the HDX experiments to verify that the protein samples are in a defined dark or light state and are also stable for the duration of the deuterium exchange reaction. In addition, the illumination time necessary to reach the PSS must be determined (*see* **Note 5**). The

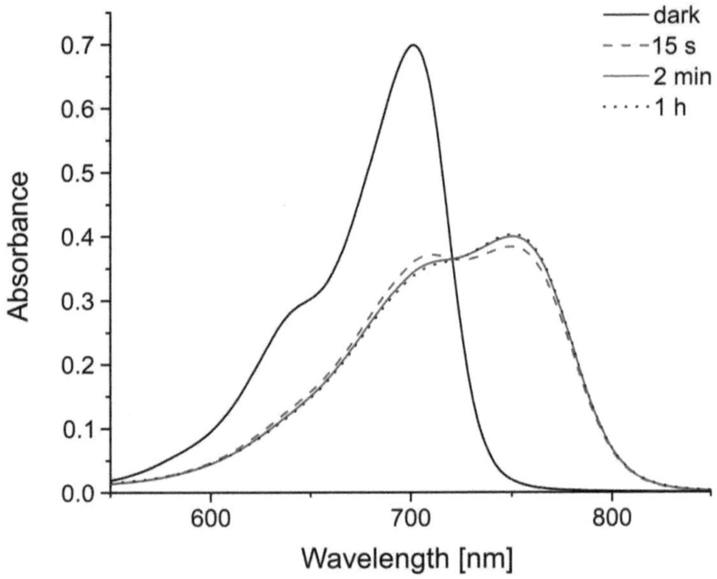

Fig. 2 UV-Vis absorption spectrum of *Dr*BphP from *Deinococcus radiodurans.* The shift from dark (Pr) to light state (Pfr, 2 min vs. 1 h illumination time) is shown for conditions resembling the experimental setup for the HDX-MS reactions. For better visibility of subtle differences, we show only a close-up of the Q-band, which is red-shifted upon illumination with a 660-nm LED for the indicated times

characterization does not need to be performed in deuteration buffer, but an equivalent aqueous composition should be used. All measurements need to be performed at the same temperature as the HDX reactions, in our case, at 20 °C (*see* **Note 6**). Make sure to work under safelight conditions that are appropriate for your system of interest; in our case, indirect dim green light can be used to allow for efficient sample handling without actinic light partially activating, and thus negatively affecting the samples.

1. Dilute the protein stock solution to 5 μM in reference buffer and transfer it to a quartz cuvette. Record an initial spectrum and check the 750 nm absorbance (Pfr indicator). If substantial amounts of protein are in the Pfr state, illuminate with far-red light (780 or 810-nm LED) to fully populate the Pr state.
2. Using a 660-nm LED (*see* **Note 7**), determine the time needed to fully populate the photostationary state. Initially, illuminate the sample for 15 s (irradiance: 2 mW cm^{-2}) and record a spectrum.
3. Repeat **step 2** until no further spectral changes can be observed (*see* **Note 8**).
4. Once the PSS is obtained, the protein's stability under constant illumination must be tested. Therefore, the maximally required

illumination time (in our case, typically, a 1-h incubation is the last timepoint in the deuterium exchange experiment) is simulated. Record a spectrum after 1 h (*see* **Note 9**) of constant illumination and check for spectral changes (*see* **Note 10** and Fig. 2).

3.2 Peptide Coverage Optimization

A key aspect of the method is to identify enough peptide fragments to cover the entire protein sequence. To address the changes in conformational dynamics in a particular region, it is necessary to generate enough peptides during the pepsin digestion step. Since the protein's deuteration does not influence the digestion, it is possible to perform test injections with protein in reference buffer, to identify the ideal amount of protein to be used and/or the optimal contact time between the pepsin column and the sample. Another possibility to influence the coverage is the addition of urea to the quench buffer (*see* **Note 4**). It will assist protein denaturation and can help to increase coverage for individual structural elements or domains.

1. Prepare 30 μL of protein solution with a concentration of 10 μM in reference buffer.
2. For each test injection, mix 5 μL of protein solution with 45 μL ice-cold quench buffer.
3. Inject 10, 30, or 50 μL of the mixture into your HPLC system (this will yield 10, 30, and 50 pmol of protein per injection). Details of HPLC parameters are provided in Subheading 3.7.
4. Follow the chromatograms to observe and compare signal strength. If low amounts of protein injections yield comparable results, this is preferable for the full experiment.
5. Analyze the peptide coverage using the automated data processing software Hexicon 2. Compare the *mzxml* file of the desired injection amount (e.g., 30 pmol of protein) against itself by importing it twice and arbitrarily assigning a fictional timepoint and the replicate group 1 (*see* Subheading 3.9) in one case. This will provide a *yaml* file that allows the visualization of peptide map plots showing the coverage of peptides across the whole protein (more details are provided in Subheading 3.9).
6. Depending on the outcome, either increase the contact time by lowering the flow rate of the HPLC system during the initial loading phase, or reduce the contact time by increasing the flow rate (*see* **Note 11**).
7. If the changes in contact time do not allow full coverage of all protein domains, optimize the quench buffer by including chaotropic agents, for example, urea.

3.3 MS/MS Identification

Once you have identified suitable quench and pepsin digestion conditions to obtain good coverage of the whole protein, at least a substantial fraction of the analyzed peptides should be confirmed by MS/MS sequencing. Due to the low specificity of pepsin, many overlapping peptides are generated, and a mass-only-based peptide assignment, as performed by Hexicon 2 and most other HDX analysis packages, runs the risk of wrong assignments and, hence, also interpreting false positives. Redundancy in similar peptides covering the same region and MS/MS identification helps to counteract this potential issue.

1. Quench and digest an aliquot of your protein using the optimized parameters identified in Subheading 3.2. The amount used for the MS/MS identification should also be comparable to the standard HDX reactions in order to obtain the same pepsin fragmentation pattern.
2. Inject your sample into the HPLC system and perform a gradient similar to what is described in Subheading 3.6. To improve MS/MS identification, however, the gradient should be extended from 4–5 to 20 min. This provides more time for MS/MS fragmentation spectra to be acquired and eventually increases the coverage of MS/MS-identified peptides.
3. Use standard MS/MS settings for your mass spectrometer; the AutoMS option in Bruker OtofControl is well suited for identification of precursor ions and intensity-based fragmentation.
4. Open the file in DataAnalysis and export the fragment spectra as *mgf* file for uploading as an MS/MS ion search in Mascot [12].
5. Copy/paste the *peptide summary result* of the top hit into a text file and save for subsequent inclusion in Hexicon 2 runs (*see* Subheading 3.8). Adjust the file structure according to the software's readme file, if needed (*see* **Note 12**).

3.4 HDX Sample Preparation

The amount of protein needed for the experiment depends on the analyzed protein, but roughly 30 pmol per injection should give a strong signal. This is determined by the test injections described in Subheading 3.2. The following instructions correspond to 30 pmol protein per injection as used in the *Dr*BphP experiments.

1. Prepare at least five 2 μL aliquots of 175 μM protein in Eppendorf tubes for each functional state to be analyzed (i.e., *Dr*BphP measurements under dark and light conditions require at least ten aliquots)—per functional state we typically prepare: one aliquot for each replicate (triplicates at least), one for the undeuterated reference, and one backup sample (*see* **Note 13**). Use your standard storage buffer to dilute the samples. If the initial spectroscopic characterization has

revealed mixed Pr/Pfr states (*see* Subheading 3.1), use appropriate LEDs (780 or 810 nm) to fully populate a defined state and carefully handle the sample under indirect dim green light conditions.

2. Ensure the protein solution is placed at the bottom of the tubes.
3. Freeze the tubes in liquid nitrogen and store them at −80 °C until you perform the reactions.

3.5 HDX Reactions

Automated systems for timing deuteration reactions exist. However, we report an experimental approach without such automated robotic systems. Therefore, the individual samples must be injected manually. Once enough reaction timepoints have been collected, the injections can be done in parallel to the incubation of reactions. Keep in mind that the safelight conditions have to be maintained for the dark-adapted samples throughout the reaction. Dark-adapted and photostationary state reactions differ during the initial equilibration step. For the latter, illuminate the sample during the initial thermal equilibration, while safelight conditions need to be maintained during the equilibration of dark-adapted samples. Adjust the length of this step according to the time needed to reach the PSS (based on the results from the characterization described under Subheading 3.1). In our case, we use 2 min as defined in Subheading 3.1 and Fig. 2.

1. Thaw a 2 μL aliquot and equilibrate in a thermoblock set to 20 °C for 2 min, under safelight settings for dark-state measurements or with parallel red light illumination for light-state experiments. The distance of the LED to the sample is set to 15 cm to allow for efficient handling of the sample.
2. Add 38 μL deuteration buffer and ensure homogeneous mixing.
3. After ~8 s, withdraw 4 μL of the reaction and mix 1 + 1 with ice-cold quench buffer (*see* **Note 14**) once the timer reaches 10 s by pipetting up and down 3–4 times (*see* **Note 15**). Immediately freeze the sample in liquid nitrogen to minimize back-exchange. This will be your first timepoint.
4. Similarly, take samples for 45, 180, 900, and 3600 s timepoints to complete one time series. Close the lid of the Eppendorf tube between taking samples after the 45 s timepoint to minimize condensation effects.
5. Repeat this workflow, starting at **step 1**, twice to obtain experimental triplicates (*see* **Note 16**).
6. Before injecting each sample into the HPLC, thaw the aliquot by pipetting 50 μL ice-cold quench buffer onto the frozen

pellets. Pipet up and down until the pellet is thawed completely (*see* **Note 17**).

7. Use an ice-cold Hamilton syringe to withdraw 50 μL sample and inject it into the HPLC port.
8. Follow the chromatograms of individual injections to observe obvious outliers. At this stage, additional replicates can be performed if needed using the excess protein aliquots available. Make sure to have three usable measurements for each time-point to allow for subsequent statistical analysis.

3.6 Reference Sample Preparation

Prepare an undeuterated reference sample by performing an equivalent dilution to Subheading 3.5 for a protein aliquot with reference buffer. No specific incubation time is needed; therefore, directly dilute your sample with quench buffer and inject into the HPLC system.

3.7 HPLC Gradient Details

The following description outlines the most important settings for the separation method used in our setup. The times and flow rates for the steps mentioned are influenced by the volume of the connecting PEEK or metal tubing. Additional steps like washing the flow path and the trap column are necessary to reduce carryover and might need to be adjusted for each individual setup.

1. Set the flow rate of the isocratic pump carrying mobile phase C to 0.3 mL/min to flush the injection loop containing the sample and to bring the latter in contact with the pepsin column for online digestion. By changing the flow rate during this initial step, you can influence the contact time between the sample and the column (*see* **Note 11**). The cleavage efficiency of pepsin is not only affected by the contact time but also by pressure. Therefore, the relationship between efficiency and contact time is not always linear.
2. After 1.5 min, increase the flow rate to 0.5 mL/min. This completes the loading of the sample onto the trap column (before switching to the higher flow rate, the whole sample should have passed the pepsin column) and eventually also allows washing off any undesired compounds.
3. Terminate the loading and washing phase of the trap column by switching to the corresponding valve setting (*see* Fig. 1) after 3 min. This initiates the elution phase by putting the trap column in line with the binary pump and the analytical column. Set the binary pump to deliver mobile phases A and B; during the elution phase, a gradient from 10% to 45% mobile phase B is run over a period of 4.5 min. The analytical column is used to improve separation of the peptides before entering the ESI-MS (*see* **Note 18**).

4. Wash the column during each individual injection using a short gradient from 45% to 95% of mobile phase B as part of the HPLC program.
5. Finally, equilibrate the trap and analytical columns to 10% B to prepare them for the next sample.
6. Test for carryover between individual injections by running a blank injection (quench buffer only) after the first sample. If you observe less than 10% carryover (*see* **Note 19**), you can proceed with your different timepoint injections. It is good laboratory practice to randomize individual injections rather than following the order of timepoints taken, to prevent systematic effects influencing your outcome.
7. Once you have finished all injections, make sure to store your pepsin column in the appropriate storage solution and your chromatographic equipment according to your standard operating procedures.

3.8 Parameters of the Mass Spectrometer

The settings on the mass spectrometer depend a lot on your specific system and the HPLC flow rates employed. Recommendations on how to set critical MS parameters for Bruker electrospray hybrid quadrupole time-of-flight devices like the impact II (as used in our specific case) or maXis are provided below.

1. For data acquisition control, use positive ion mode and calibrate the device for a mass range from 500 to 3000 m/z with a repetition rate of two spectra per second. To reduce the amount of data to be analyzed subsequently, use a threshold of 45 counts per 1000 summations (*see* **Note 20**).
2. Considering the HPLC flow rate of 0.3 mL/min, adjust the nebulizer and dry gas settings to 2.6 bar and 11 L/min, respectively. The dry gas temperature should be set to 200 °C. Adjust as needed if signal strength drops significantly and/or liquid droplets accumulate in the source.
3. All other settings can remain at their standard values, optimized for the current state of your funnel and collision cell, the detector lifetime, etc.

3.9 HDX Data Analysis

For HDX-MS data analysis, a range of software packages have been developed over the years [13]. Below we describe the typical workflow for the software package Hexicon 2 [11], which provides a straightforward workflow and a flexible and easy-to-use interface for data analysis and visualization.

1. The data from each HDX timepoint and the reference need to be exported as line spectra in *mzXML* format. Depending on your data analysis software, this can either be done directly or by using applications like CompassXport (Bruker).

2. Open Hexicon 2 and import the data files of one functional state (dark or light) individually or all at once. In the single file import scenario, you are prompted to define the respective timepoint of each measurement and assign a replicate group upon importing. The standard convention for the replicate groups is: 0—reference, 1—first timepoint, 2—second timepoint, and so on. In case of importing all *mzxml* files at once, the timepoints and replicate groups need to be defined in the respective boxes of the dataset window after successfully loading the files.
3. Load the text-based *mascot* file for the respective Mascot search report (*see* Subheading 3.3) in the corresponding line. Follow the guidelines in the Hexicon 2 readme file to have the correct file architecture.
4. Load the text-based protein sequence file in the next line. Make sure you have an amino acid only sequence file corresponding exactly to the protein used in the experiment. Any purification tags or cloning artifacts should be included to allow the software to look for the respective peptide sequences. Covalent protein modifications with cofactors (such as biliverdin in the case of phytochromes) or phosphorylation can also be defined, if needed (*see* **Note 21**).
5. You can optionally provide a reference protein sequence file that contains only the corresponding database entry for your protein of interest. Thereby all residues are labeled with their numbers as used in reference publications.
6. Save the data import contents for quick loading of the corresponding data for subsequent Hexicon 2 runs with potentially altered and optimized run settings (*see* **step 8** below).
7. The next tab, *Filters*, allows you to specify proteins of a mixture that should be ignored; for example, in the case of analyzing protein complexes. This is not needed in the specific case of a single protein analysis.
8. On the next tab, *Parameters*, the above-mentioned settings of each run need to be specified. Adjust the retention time window according to the time frame where peptides are analyzed in the mass spectrometer. Based on our dead volumes and connector lengths, this is between 140 and 480 s, typically. Also, specify a resolution that reflects the current analyses—we use 40,000. The calibration error should be specified in a realistic window, even though Hexicon 2 performs an internal recalibration—10 ppm should be on the safe side. The noise quantile setting depends on your stringency of filtering noise using the threshold option in the data acquisition step. Typically, 0.05 should be good, but this might need optimization in both directions.

9. Last but not least, you need to specify an export path where your data are stored, and you can optimize the advanced settings, if needed. Typically, lower values result in more peptides being identified, but potentially also more noise and wrong assignments being exported.
10. Start the Hexicon 2 run on the next tab and let the workflow proceed until the end. This generates a *yaml* file that can then be opened in the HXViewer executable of the Hexicon 2 package.
11. Repeat this procedure from **step 2** onward for each functional state of interest and save as a separate *yaml* file—from our analysis, we eventually get *Dr*BphP_dark.yaml and *Dr*BphP_light.yaml.

3.10 HDX Data Interpretation and Reporting

The most interesting aspect of performing HDX measurements is their interpretation and the way you use it to enrich your functional analyses of your protein of interest. A comprehensive description of how to interpret HDX data, however, goes beyond this method chapter, and the reader is referred to review articles, for example [7]. Here, we provide a few example figures that Hexicon 2 can generate based on the two *yaml* files created in the first part of the analysis workflow (*see* Subheading 3.8).

1. Open both *yaml* files in the HXViewer application of Hexicon 2 and give them meaningful names, as these will be shown in the analyses.
2. Process both data files to reduce the number of sequence conflicts using Hexicon 2's inbuilt feature based on pepsin cleavage probabilities—"*Processing—Resolve sequence conflicts.*"
3. Double-click on a peptide and assign a color for this peptide series. We use black for the dark-state peptides. This opens a window with the deuterium incorporation over time for this respective peptide. Activate the software-estimated distribution of deuterated species by clicking on the histogram button at the bottom of the plot. Filter outlier peptides (potential wrong assignments, overlapping intensities from other signals, and noisy signals) by going through all peptides and checking for smooth deuteration distributions as well as physically meaningful deuterium incorporation kinetics. Right-click outliers to inactivate them and remove them from reporting. Repeat this for the second functional state—we use red for *Dr*BphP_light peptides.
4. To visualize the intrinsic conformational dynamics of the protein, go to "*Visualization—Peptide map*" and select, for example, the dark dataset as reference. No target is required in this case.

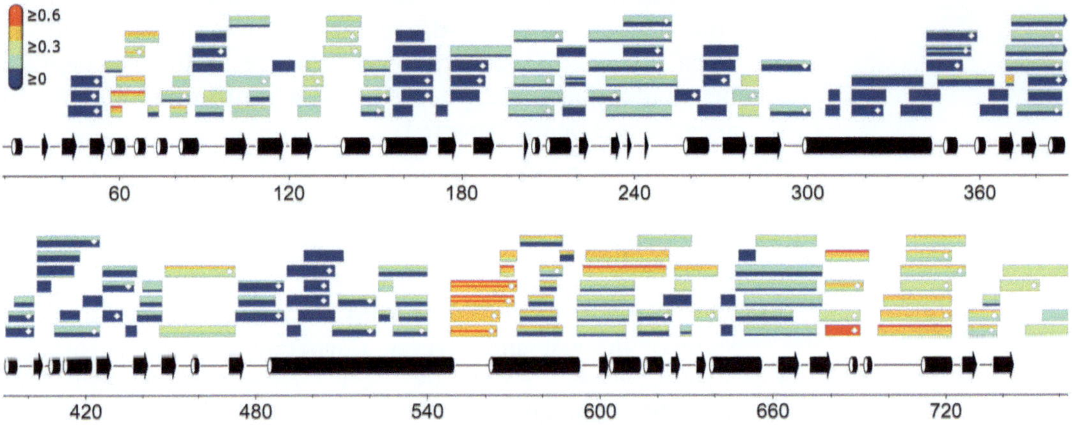

Fig. 3 Deuteration map of *Dr*BphP in its dark state. The graph shows the relative deuterium incorporation for each peptide at each labeling timepoint. Individual timepoints of each peptide are colored from bottom up. Colors correspond to the amount of deuterium incorporated normalized to peptide length. The scale bar settings have been adjusted to account for a back-exchange of ~30%. Regions in blue incorporate fewer deuterons compared to regions in orange and red. White diamond symbols indicate MS/MS-identified peptides. The arrow and cylinder symbols below the peptides correspond to the secondary structure according to the cryo-electron microscopy (cryo-EM) structure of *Dr*BphP, and the numbers reflect the positions of the primary sequence

5. In the newly opened window, set the dynamic range to 0.75 to account for a maximum incorporation of 70–75% deuterium under the assumption of 25–30% back-exchange during the workup routine.
6. Open a *dssp* description file (according to the Hexicon 2 instructions) by clicking on the black arrow at the bottom right. This allows us to visualize secondary structure elements in addition to the primary sequence in these so-called deuteration plots. An example for the dark state of *Dr*BphP is shown in Fig. 3. This reveals regions in blue with low relative deuterium incorporation—that is, stable core regions of structured domains—versus orange or red colors for regions with substantial incorporation of deuterium due to increased conformational dynamics or being unstructured.
7. To visualize only regions affected by the perturbation between two functional states, that is, *Dr*BphP_light relative to *Dr*BphP_dark, again select "*Visualization—Peptide map,*" but use the *Dr*BphP_light *yaml* file as target dataset in this case.
8. In the newly opened window, set the dynamic range according to the amplitude of changes seen between the two states. For *Dr*BphP, 2.8 is a good setting when using seven colors (to be set at the right bottom of the plot) for visualizing effects of illumination in interesting regions and obtaining simple numbers for individual bin descriptions in the color-bar legend, but

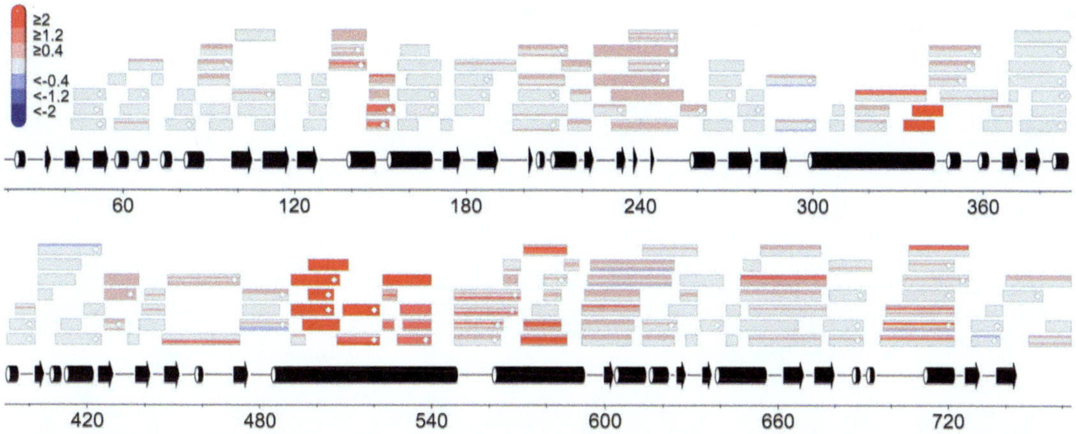

Fig. 4 Difference map emphasizing changes in deuterium exchange kinetics between dark and light states. The two functional states of *Dr*BphP are compared by subtracting the deuterium incorporation in the light state from that of the corresponding dark state. Blue regions correspond to a stabilization in the light state, while red regions reflect increased conformational dynamics. The absolute differences are colored according to the bar legend. Each box corresponds to one peptide with individual colors from bottom to top corresponding to the increasing timepoints. White diamond symbols indicate MS/MS-identified peptides. The arrow and cylinder symbols below the peptides correspond to the secondary structure according to the cryo-EM structure of *Dr*BphP, and the numbers reflect the positions of the primary sequence

this can be adjusted according to your needs. Figure 4 shows such a comparison plot where red regions indicate elements with increased deuterium uptake in the light-activated state, and blue regions correspond to a stabilization of the respective element due to red light irradiation.

9. Right-clicking on regions of interest and selecting "*Show time series*" permits visualization of the respective peptides' deuterium incorporation plots of dark and light datasets side by side. Use black for dark-state data and red for the respective light-state peptides, and scroll through the list of peptides. Interesting peptides can then be analyzed in more detail and exported for publications. An example of four interesting regions is shown in Fig. 5.

10. All relevant plots can be exported as line-art PDF and can then be imported into your favorite illustration application to prepare publication-quality figs.

11. Last, but not least, prepare a summary table that reports on the key parameters and the statistical analysis of your HDX experiments as described in Masson et al. [7]. An example table for the analyses described in this chapter is provided below (*see* Table 1). A Python-based script to facilitate the extraction and comparison process between two datasets (based on two Hexicon 2 *yaml* files) can be made available upon request.

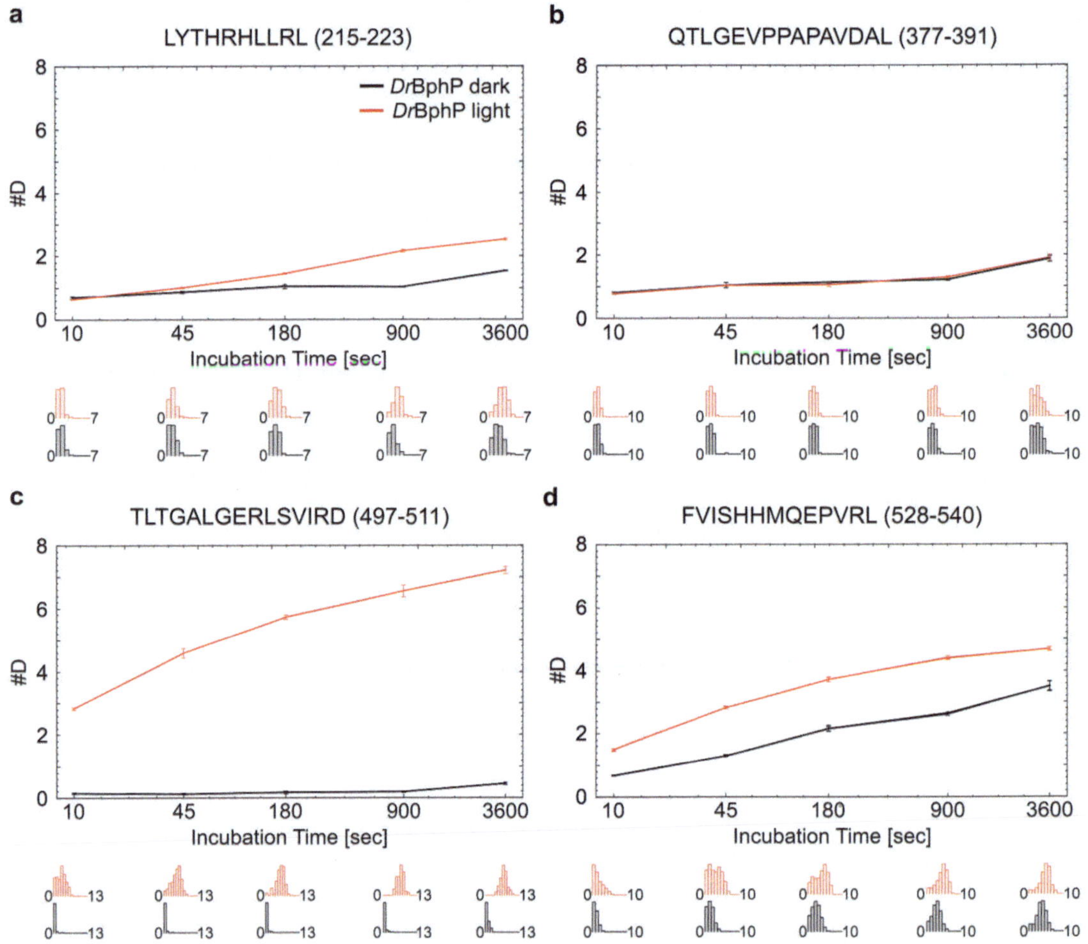

Fig. 5 Selected peptides of the *Dr*BphP analysis showcasing different scenarios when comparing two functional states. In all panels, the deuterium incorporation is plotted against time for dark- (black lines) and light-state (red lines) measurements. Below these graphs, the Hexicon 2-estimated deuterium level of each peptide is plotted for each timepoint and functional state. Above each graph, the corresponding peptide sequence and position in the protein is shown. (**a**) Although both states start with similar deuteration levels, the light state incorporates more deuterons over time, which shows that illumination leads to increased conformational dynamics in an otherwise well-ordered region. This peptide shows up in red for the later timepoints in Fig. 4. (**b**) A well-structured region with low conformational dynamics and hence slow deuterium incorporation in both dark and light states—representative for grey peptides in Fig. 4. (**c**) A peptide with drastic increase in conformational dynamics for the light state, corresponding to a structural change from very rigid to almost unstructured. Dark red peptide in Fig. 4. (**d**) Increased deuterium incorporation in the light state over the course of the time series. The light state also shows a bimodal deuterium incorporation distribution, which likely corresponds to a mixed population with an altered conformation of a part of the peptide that features rapid exchange kinetics. Also a red peptide in Fig. 4, but different functional interpretation compared to panels **a** and **c**

Table 1
Detailed overview of the acquired HDX datasets and their comparison

	*Dr*BphP	
Light conditions	**Dark**	**Light (660 nm)**
HDX reaction details	10 mM HEPES, 50 mM NaCl, 2 mM $MgCl_2$, pD = 7.0, 20 °C	
HDX time course (s)	10, 45, 180, 900, 3600	
HDX control samples	Unlabeled control	
Back-exchange	Not measured	
# of peptides	234	226
Sequence coverage	96%	96%
Average peptide length/ average redundancy	14/4.2	14/4.0
Replicates	3	3
Repeatability (average SD for each timepoint)	0.05, 0.06, 0.09, 0.05, 0.13	0.06, 0.07, 0.10, 0.08, 0.09
Significant difference in HDX	ΔHDX >0.3 D	

4 Notes

1. Use the optimal buffer system for your system of interest. HDX-MS is also compatible with complex buffer mixtures. Just ensure you keep it as simple as possible, so as not to make the experiments unnecessarily expensive due to the need for more complex deuterated chemicals to prepare the deuteration buffer. If you have to include substantial amounts of additives, for example, glycerol, make sure to include the corresponding deuterated compound with all exchangeable hydrogens replaced by deuterium; in the latter case, glycer(ol-d3) would be needed.
2. Prepare more buffer than you need for one experiment. Excess buffer can be frozen and stored for later use, increasing reproducibility and comparability between experiments even over longer intervals.
3. In order to prevent overshooting the pD value, you can estimate the amounts of acid/base needed by preparing an equivalent amount of aqueous buffer initially and quantifying the amounts used to obtain the desired values.
4. The urea included in the quench buffer is optional and depends on the protein of interest. Initial testing for peptide coverage

should be performed without including urea. Only if inefficient digestion of individual parts of the protein is observed, and variation of contact times with the pepsin column (*see* **Note 11**) also does not improve the situation, should increasing concentrations of urea be tested. Apart from urea, other chaotropic substances like guanidine HCl can also be used. Double-check the manual of your specific pepsin column to check which concentrations of chaotropic agents are tolerated. Reducing agents can also be added to the quench buffer to handle disulfide bridges, if present.

5. The illumination setup should resemble the setup you use during the HDX experiments. This refers to the distance between the sample and the LED source, light intensity, and protein concentration.
6. The temperature of the experiment is an additional variable that can be used to enhance the understanding of your system. Lower temperatures result in slower deuterium exchange and can, therefore, be interesting for areas that show particularly fast deuterium exchange. In contrast, higher temperatures might be interesting for areas that show medium-slow exchange with only moderate uptake of deuterium and consequently only subtle observable effects of your perturbation of interest at standard temperatures.
7. Use the LED, which gives the highest contribution of Pfr in the photostationary state.
8. Depending on the protein of interest, the required red light activation times to obtain a steady-state of forward (Pr to Pfr) and backward (Pfr to Pr) rates (the photostationary state—PSS) can vary substantially between seconds to minutes.
9. The longest timepoint of the experiment is determined by the dynamics of the system and the features one is interested in. For photoreceptors with relatively high intrinsic conformational dynamics, 1 h is typically enough to see the elements affected by illumination.
10. For red light illumination, we have not observed any negative effects of 1-h illumination at an irradiance of 2 mW cm^{-2} so far. Depending on the wavelength of the employed LED and the irradiance required to effectively populate the light-activated state, this should, however, be confirmed for each experimental system. In case of observable bleaching or other unexpected spectral changes, the irradiance should be lowered until a constant spectrum is observed during the duration of the experiment. In that case, the time needed to reach the photostationary state might also need readjustment. If your system of interest has a very slow thermal relaxation to the dark state, one could also omit the constant illumination during the

deuterium exchange reaction; in this scenario, the stability control experiment could be omitted, but might still be valuable to check for aggregation or other effects of long incubations.

11. The contact time between protein and pepsin column defines the cleavage pattern and influences the observed sequence coverage. Too-long contact times can result in many short peptides, which are sometimes less informative than longer ones, because the first two amide hydrogens are quickly back-exchanged to hydrogen during the workup of the HDX reactions. Therefore, aim for an average peptide length of 10–15 amino acids. Too-short contact times, on the other hand, may reduce the coverage by incomplete digestion of the protein; this can be seen by very long peptides eluting toward the end of the HPLC gradient.
12. You can also combine multiple MS/MS identification runs into one joint *mascot* file for Hexicon 2. Just make sure to remove redundant entries and to have unique identifiers for each peptide. To improve the coverage, you can put already identified peptides into an exclusion list and/or specifically extract m/z values from Hexicon 2 to be included in an inclusion list for your MS/MS measurements.
13. The number of aliquots you prepare will depend on the number of measurements you want to perform. Generally, you should do triplicate measurements for both Pr (dark) and photostationary state (light). However, we recommend preparing additional aliquots in case of a user error. This ensures that you have the same stock for all measurements.
14. Prepare 0.5 mL Eppendorf tubes with 4 μL of quench buffer beforehand and put them on ice. Label the tubes according to timepoint.
15. To make the process more efficient, it is recommended to perform the reactions for the triplicates in parallel, with slightly different starting times. The handling, especially of short timepoints, should initially be practiced using water and dummy aliquots to optimize the procedure. To reduce the amount of handling in the initial steps, the timer is started at the equilibration stage. In case of 2 min equilibration, just let the timer continue to run and mix sample with deuteration buffer at the 2:00 reading. Quenching of the first timepoint is then performed at a reading of 2:10.
16. A foam rack in the liquid nitrogen dewar can help you track your reactions/injections. If no suitable design is at hand, you can easily make one yourself using any Styrofoam lid.
17. The thawing process should be performed as consistently as possible. Consistently wipe down the condensate from the

tube to have a clear view of the samples. Do not constantly heat up the sample in your fingers during the process, as this will negatively affect the back-exchange and result in excessive loss of your deuterium signal.

18. The length of the analytical column is a compromise between efficient separation of peptides and short runtimes. In our hands, a 5 cm column allows for sufficiently well-separated peptides at the average resolution of 40,000–50,000 of our impact II mass spectrometer in a relatively short gradient of 4–5 min.
19. Substantial carryover is an issue because the sample peptides continue to lose their incorporated deuterium (back-exchange). Thereby carryover of signal from one injection to the next falsifies your extracted isotope distribution. Reasons for carryover can be manifold and sometimes difficult to identify. In case you observe carryover, watch out for precipitation of your sample upon quenching (trapping on in-line filters and slow continuous release of protein), efficient washing of the flow path, injection port, and columns between injections. If you observe carryover, also more intense cleaning of the pepsin column might be indicated. Use the pepsin column washing solution offline or manually inject 100 μL of it between individual runs.
20. Depending on your choice of sample amounts loaded and background signals in your mass spectrometer system, we recommend to vary the threshold parameter to eventually end up with line spectra *mzxml* files that have ~30–100 Mb per run.
21. Hexicon 2 offers a versatile modification option, where the user can define any chemical modification, provided the chemistry of the linkage and the added compound is known. Follow the guidelines in Hexicon 2 software to modify the modifications file (modifications.yaml) as needed.

Acknowledgments

We thank Heikki Takala (University of Jyväskylä) for providing the plasmid of *Dr*BphP. We also thank Bernd Gesslbauer (University of Graz) for Mascot analysis of the MS/MS runs of *Dr*BphP. We appreciate helpful suggestions for the manuscript text from Cornelia Böhm. M.F. acknowledges support as an associate student from the Austrian Science Fund (*FWF*) grant https://doi.org/10.55776/DOC130. This research was funded in whole or in part by the Austrian Science Fund (FWF) grants [DOIs: https://doi.org/10.55776/P32022 and https://doi.org/10.55776/P34387] to A.W. For open access purposes, the authors have applied a CC BY public copyright license to any author-accepted manuscript version arising from this submission.

References

1. Hughes J, Winkler A (2024) New insight into phytochromes: connecting structure to function. Annu Rev Plant Biol 75(1):153–183. https://doi.org/10.1146/annurev-arplant-070623-110636
2. Song C, Mroginski MA, Lang C et al (2018) 3D structures of plant phytochrome a as Pr and Pfr from solid-state NMR: implications for molecular function. Front Plant Sci 9:498. https://doi.org/10.3389/fpls.2018.00498
3. Gourinchas G, Heintz U, Winkler A (2018) Asymmetric activation mechanism of a homodimeric red light regulated photoreceptor. Elife 7:e34815. https://doi.org/10.7554/eLife.34815
4. Tran QH, Eder OM, Winkler A (2024) Dynamics-driven allosteric stimulation of diguanylate cyclase activity in a red light-regulated phytochrome. J Biol Chem 300(5):107217. https://doi.org/10.1016/j.jbc.2024.107217
5. von Horsten S, Straß S, Hellwig N et al (2016) Mapping light-driven conformational changes within the photosensory module of plant phytochrome B. Sci Rep 6:34366. https://doi.org/10.1038/srep34366
6. Hodge EA, Benhaim MA, Lee KK (2020) Bridging protein structure, dynamics, and function using hydrogen/deuterium-exchange mass spectrometry. Protein Sci 29(4):843–855. https://doi.org/10.1002/pro.3790
7. Masson GR, Burke JE, Ahn NG et al (2019) Recommendations for performing, interpreting and reporting hydrogen deuterium exchange mass spectrometry (HDX-MS) experiments. Nat Methods 16(7):595–602. https://doi.org/10.1038/s41592-019-0459-y
8. Multamäki E, Nanekar R, Morozov D et al (2021) Comparative analysis of two paradigm bacteriophytochromes reveals opposite functionalities in two-component signaling. Nat Commun 12(1):4394. https://doi.org/10.1038/s41467-021-24676-7
9. Wahlgren WY, Claesson E, Tuure I et al (2022) Structural mechanism of signal transduction in a phytochrome histidine kinase. Nat Commun 13(1):7673. https://doi.org/10.1038/s41467-022-34893-3
10. Glasoe PK, Long FA (1960) Use of glass electrodes to measure acidities in deuterium oxide. J Phys Chem 64:188–189
11. Lindner R, Lou X, Reinstein J et al (2014) Hexicon 2: automated processing of hydrogen-deuterium exchange mass spectrometry data with improved deuteration distribution estimation. J Am Soc Mass Spectrom 25(6):1018–1028. https://doi.org/10.1007/s13361-014-0850-y
12. Perkins DN, Pappin DJC, Creasy DM et al (1999) Probability-based protein identification by searching sequence databases using mass spectrometry data. Electrophoresis 20(18):3551–3567. https://doi.org/10.1002/(SICI)1522-2683(19991201)20:18<3551::AID-ELPS3551>3.0.CO;2-2
13. Stofella M, Grimaldi A, Smit JH et al (2024) Computational tools for hydrogen-deuterium exchange mass spectrometry data analysis. Chem Rev 124(21):12242–12263. https://doi.org/10.1021/acs.chemrev.4c00438

Chapter 16

Encapsulation of Phytochrome Proteins in Trehalose Matrices: Practical Considerations for the Use in Solid-State NMR Spectroscopy

Valentin Rohr, Wolfgang Gärtner, Jörg Matysik, and Chen Song

Abstract

Conventional in situ cryotrapping of phytochrome (PHY) intermediates has proven experimentally challenging for solid-state nuclear magnetic resonance (NMR) studies. To overcome this hindrance, we introduced the preparation of dehydrated disaccharide–PHY glasses, which allow transient photocycle intermediates to be isolated for NMR measurement at room temperature (RT). Here, we provide details of the incorporation of PHYs into a glassy matrix formed by trehalose. The protocol has been in routine use in our solid-state NMR studies for diverse PHYs having different molecular weights, bilin chromophore structures, and lifetimes of photoproducts. More recently, this protocol has been optimized for the isolation of individual transient reaction intermediates of a cyanobacteriochrome (CBCR) photosensor. We also provide information on Nuclear Magnetic Resonance(NMR) analysis to gain structural and dynamic information of the trehalose-embedded PHY proteins.

Key words Trehalose, Phytochromes, Protein-embedding matrix, Biopreservation, Dynamic confinement, Bilin chromophores, Photoproducts, Intermediates

1 Introduction

Bilin-based photosensors represented by red/far-red reversible PHYs are widespread in nature [1]. The recently discovered subfamily cyanobacteriochromes (CBCRs) extended remarkably the spectral absorbance range (from 330 to 750 nm), indicating that bilin photochemistry is much more versatile than expected [2]. CBCRs carry, as a common identification motif besides a wide variety of domain arrangements, the bilin-binding cGMP-specific phosphodiesterases, cyanobacterial adenylate cyclases, and formate hydrogen lyase transcription activator FhlA (GAF) domain. Photoactive GAF domains are small and soluble proteins with promising properties for optogenetic applications, although details on their photochemistry are not fully elucidated [3].

Andreas Winkler (ed.), *Phytochromes: Methods and Protocols*, Methods in Molecular Biology, vol. 2970,
https://doi.org/10.1007/978-1-0716-4791-2_16,

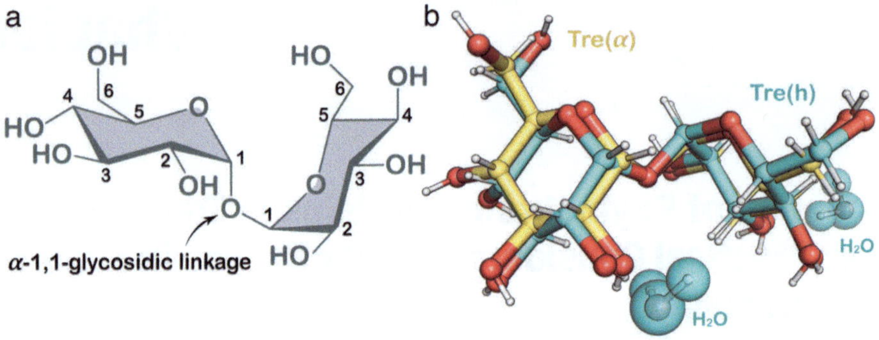

Fig. 1 Trehalose. (**a**) Skeletal formula of trehalose ($C_{12}H_{22}O_{11}$). The molecule consists of two glucose units linked with an α-1,1-glycosidic bond. (**b**) Crystal structures of two polymorphic forms of trehalose which may play a central role in the mechanisms of bioprotection. The anhydrous α-form (designated here as Tre(α), cambridge crystallographic data centre (CCDC) deposition ID 668079, orange yellow) and the dihydrated form ($C_{12}H_{22}O_{11}{\cdot}2H_2O$, designated here as Tre(h), Cambridge Crystallographic Data Centre (CCDC) deposition ID 668078, cyan) [13]. The respective molecules are shown in ball-and-stick form. Crystal waters from Tre(h) are shown as cyan spheres

Solid-state magic-angle spinning (MAS) NMR was applied for the first time to the PHY photosensor in 2006 [4]. Since then, our MAS NMR studies on PHYs used frozen solutions (FSs) of concentrated proteins (e.g., >0.7 mM for two-dimensional [2D] measurements) routinely [5–7]. MAS NMR on FSs (in the presence of glycerol) provides in principle a simple and relatively natural access to analyzing protein folding and dynamics [8]. However, characterization of slr1393g3 (the third GAF domain of red/green CBCR slr1393 from *Synechocystis* 6803) red-absorbing Pr dark state by UV-Vis and fluorescence spectroscopy revealed an unusual temperature-dependent spectral shift [9]. This result indicates that not all PHYs are amenable to preparation of FSs. Moreover, the low-temperature approach to trap photocycle intermediates within the MAS probe with in situ illumination is less feasible because the highly concentrated samples frozen at cryogenic temperatures prevent penetration of actinic light beyond a few hundred μm into the sample, thus creating a mixture of photostates [10]. It is also worth noting that the NMR linewidth of frozen protein solutions broadens with lowering of temperature, hampering attempts to solve the high-resolution structure by MAS NMR [11]. We have recently developed a simple sample preparation method to circumvent these problems via the incorporation of PHY/CBCR proteins into amorphous matrices formed by trehalose [12].

Trehalose is a nonreducing disaccharide made up of the two α-glucose units that are linked through a glycosidic bond at the anomeric positions (*see* Fig. 1). The sugar is known for its ability to form a homogeneous glassy matrix in which the hosted protein is evenly distributed. This enables the preservation of protein

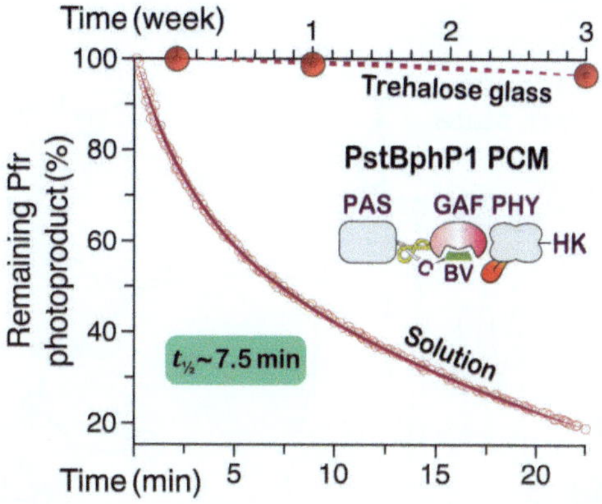

Fig. 2 Stabilization of short-lived photoproduct conferred by incorporation into TGs. Thermal reversion of *Pst*BphP1 photosensory core module (PCM, consisting of PAS (*P*er-*A*RNT-*S*im)-GAF-PHY (*PHY*-specific) tridomain) in solution (open circles with the fitting line in dark red) and in TG (solid circles) [12]. The calculated half-life time ($t_{1/2}$) of the Pfr photoproduct in solution is given

integrity under adverse conditions such as high salt content, extensive dehydration, and freezing to cryogenic temperatures [14, 15]. In the case of PHYs/CBCRs, trehalose glasses (TGs) are effective in inhibiting inherent thermal reversion of photoproducts [12, 16]. As an example shown in Fig. 2, the incorporation into dehydrated TGs promotes a tremendous stabilization of the short-lived far-red-absorbing Pfr photoproduct of the *Pst*BphP1 bacteriophytochrome from *Pseudomonas syringae* which, having a half-life time ($t_{1/2}$) of only about 7.5 min in solution (at 20 °C), can now be stored at room and even higher temperatures for weeks without undergoing any significant reversion. The advantages of trehalose matrices also allow isolation of individual transient reaction intermediates and long-term preservation under noncryogenic conditions (and even with the exposure to ambient light). This sample preparation approach is particularly beneficial for the MAS NMR technique, by which our attempts to trap the entire photocycle within the precooled MAS probe have failed so far [17]. Moreover, a much larger amount of sample in solid amorphous matrices relative to that in FS state can be loaded into the MAS rotor which boosts the sensitivity of NMR measurements. Taking the CBCR slr1393g3 containing [^{15}N]-δ-aminolevulinic acid (ALA)-labeled chromophore in the Pr dark state as an example, the trehalose preparation yields a 20-fold enhancement of the ^{15}N NMR resonance intensity relative to the FS which means a 400-fold reduction of experimental time (*see* Fig. 3a).

Here, we present a detailed protocol for expression, purification, and labeling of PHY/CBCR proteins and the subsequent preparation of dehydrated TG samples. This protocol is adapted from the approach originally described for the stabilization of

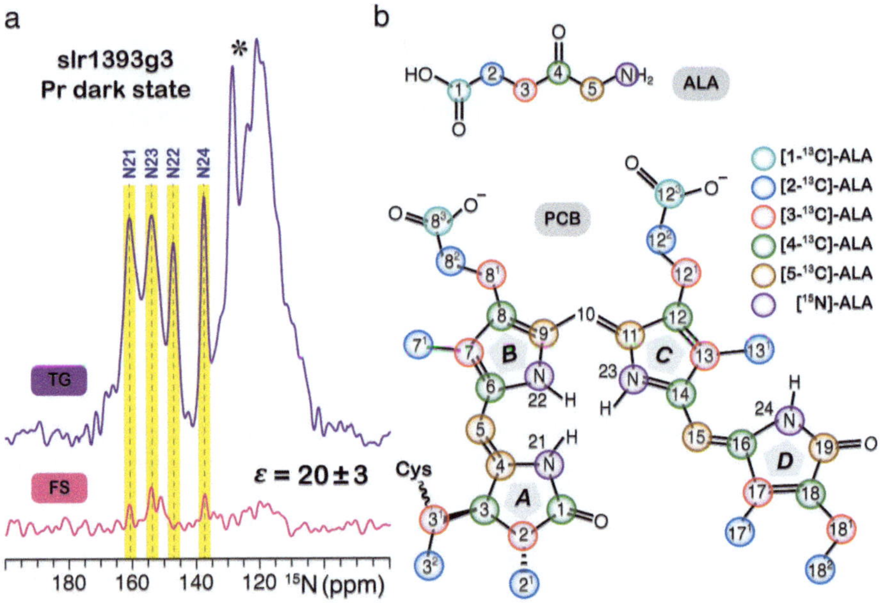

Fig. 3 Sensitivity enhancement of TG-embedded samples in MAS NMR. (**a**) ^{15}N CP spectra of slr1393g3 shown in the Pr dark state resulting from the TG with a T/P molar ratio of 50:1 (dark violet) and from FS (reddish magenta) (*unpublished data*). For comparison, the ^{15}N data for this protein were recorded in both preparations with identical acquisition and processing parameters at −23 °C. The factor ε represents the increase in the signal-to-noise (S/N) ratio of the sample as TG relative to that of the FS. Resonances marked with an asterisk indicate amide nitrogens of the protein backbone originating from ^{15}N in natural abundance. (**b**) Labeling and numbering schemes for the bilin precursor ALA (*top*) and for the PCB chromophore (*bottom*). Utilization of [^{15}N]-ALA during biosynthesis of the bilin chromophore yields ^{15}N incorporation at tetrapyrrole nitrogens (N21–N24) [12]

short-lived PHY/CBCR photoproducts [12] and further optimized for generation and isolation of intermediates of the red/-green CBCR slr1393g3 [17]. By following the protocol outlined here, the photocycle states of trehalose-embedded PHY/CBCR samples in both forward and backward reactions can be isolated in sufficient quantity to allow advanced MAS NMR investigation, irrespective of molecular weight, domain architecture, bilin structure, and photoproduct lifetimes. As references, typical MAS spectra of slr1393g3 TG samples with the ^{13}C- or ^{15}N-labeled phycocyanobilin (PCB) chromophore are shown in this protocol.

2 Materials

All solutions should be prepared using deionized water purified to obtain a resistivity of 15.4 MΩ/cm (equivalent to a conductivity of 0.065 μS/cm) at 20 °C. Precool the buffers, filters, and vials to 4 °C prior to protein handling. The pH adjustment should be made using (highly) concentrated NaOH or HCl solutions.

2.1 Bacterial Growth

1. Glass culture flasks: One 250-mL and two 3-L glass culture flasks sterilized by autoclaving.
2. Lysogenic broth (LB) medium: Dissolve 10 g tryptone, 5 g yeast extract, and 5 g NaCl per L water; prepare at least 50 mL; and sterilize by autoclaving. Adjust pH to 7.0.
3. M9 minimal medium (M9): Dissolve 33.9 g Na_2HPO_4, 15.0 g KH_2PO_4, and 2.5 g NaCl per L of water; prepare at least 540 mL; and sterilize by autoclaving. Adjust pH to 7.0.
4. Optional: Antibiotics (*see* **Note 1**).
5. Glucose: 6 g D-Glucose anhydrous.
6. NH_4Cl: 1.5 g NH_4Cl powder.
7. $MgSO_4$ solution: With 1 M $MgSO_4$, prepare at least 3 mL and sterilize by autoclaving.
8. $CaSO_4$ solution: With 1 M $CaSO_4$, prepare at least 150 μL and sterilize by autoclaving.
9. Ice: Ice flakes, prepare about 4 kg.
10. Sterile filters.
11. Shaking incubator: Shaking incubator set to 37 °C.

2.2 Expression of PHY/CBCR Proteins and Isotopic Labeling of the Chromophore

1. Isopropyl β-D-1-thiogalactopyranoside (IPTG) solution: With 0.8 M IPTG, prepare at least 0.5 mL and sterilize by sterile filtration.
2. $FeSO_4$ solution: 12.5 mM $FeSO_4$, prepare at least 1.5 mL and sterilize by autoclaving.
3. ALA: Bilin precursor ALA with different isotopic labeling patterns (e.g., [5-^{13}C]-ALA or [^{15}N]-ALA; *see* Fig. 3b), prepare at least 20 mg.
4. Sterile filters.
5. Shaking incubator: Shaking incubator set to 17 °C.
6. Cell density meter.
7. Centrifuge: Centrifuge with rotor and centrifuge bottles to collect bacteria from a 3 L culture (e.g., Beckman Coulter Avanti J-20 XP with JA-10 rotor).
8. Optional: −20 °C freezer.

2.3 Bacterial Lysis

1. Lysis buffer: 10 mM KH_2PO_4, 10 mM K_2HPO_4, and 500 mM NaCl, prepare at least 40 mL. Adjust pH to 8.0.
2. Ice: Ice flakes, prepare about 0.5 kg.
3. Scale: Scale with an accuracy of ±1 mg.
4. Shaking incubator: Shaking incubator set to 4 °C.
5. Lysis tool: Ultrasonic homogenizer (e.g., Biosafer650-92) or alternative lysis tool.

6. Centrifuge: Centrifuge with rotor and centrifuge bottles to collect cell fragments (e.g., Thermo Scientific Multifuge X1R with Fiberlite F15-6x 100y rotor).

2.4 Column Purification

1. Filtered lysis buffer: Lysis buffer (*see* Subheading 2.2, **item 1**), prepare about 1 L by filtering through a 0.22-μm-pore-size membrane filter.
2. Filtered elution buffer: 10 mM KH_2PO_4, 10 mM K_2HPO_4, 500 mM NaCl, and 500 mM imidazole, prepare about 500 mL. Adjust pH to 8.0 and filter through a 0.22-μm-pore-size membrane filter.
3. Dialysis buffer: 20 mM Tris–HCl and 150 mM NaCl, prepare at least 4 L. Adjust pH to 7.0.
4. Filtered water: Water, prepare about 200 mL by filtering through a 0.22-μm-pore-size membrane filter.
5. Filtered water–ethanol solution: Water with 20 vol% ethanol mixture, prepare about 200 mL by filtering through a 0.22-μ m-pore-size membrane filter.
6. Fast protein liquid chromatography (FPLC): FPLC device and fraction collector (e.g., Bio-Rad NGC with multiwavelength UV-Vis detector and two pumps).
7. Sample tubes for fraction collector.
8. Chromatography column: 5 mL HisTrap column (and size exclusion column if necessary).
9. Superloop: 100 mL sample loading superloop.
10. Dialysis membrane: 1 kDa cutoff dialysis membrane.
11. Ultra centrifugal filter: Ultra centrifugal filter with 10 kDa MWCO.
12. Centrifuge: Centrifuge with rotor and centrifuge bottles to accommodate centrifugal filters (e.g., Heraeus Labofuge 400R with tilt rotor).
13. UV-Vis spectrophotometer: UV-Vis spectrophotometer equipped with temperature regulation (e.g., Shimadzu UV-1900i with temperature regulation system TCC-240A) and cuvettes.

2.5 TG Preparation

1. Trehalose stock solution: 1.2 M trehalose, prepare 10 mL (*see* **Note 2**).
2. Drying gas: N_2 gas or dried compressed air (with pressure dew point <−40 °C).
3. TG preparation buffer: With 20 mM Tris–HCl and 20 mM NaCl, prepare at least 40 mL. Adjust pH to 7.2.
4. TG drying vial: Petri dish or glass plate.

5. Light sources: High-power LEDs (e.g., Roithner Lasertechnik H2A1-H650 LEDs, 650 ± 17 nm, 40 mW, 350 mA, and 30° viewing angle for generating slr1393g3 green-absorbing Pg photoproduct).
6. Illumination chamber: An in-house illumination chamber (*see* **Note 3**).

2.6 NMR Sample Packing

1. MAS NMR rotor: We typically use Bruker 3.2 mm regular wall zirconia MAS rotor (Inside diameter (I.D). of 2.2 mm) and 4 mm zirconia MAS rotor (Inside diameter(I.D). of 3.0 mm) with a disposable Kel-F insert (*see* **Note 4**) and Kel-F drive cap (*see* **Note 5**). Alternative rotors are available from Varian/ Agilent.
2. Micro-laboratory spoon and spatula.
3. Agate mortar and pestle set (e.g., Cole-Parmer)
4. Rotor packing tool: We typically use Bruker packing toolkits for 3.2- and 4 mm rotors including sample packer, filling funnel, cap set tool, and cap remover. Alternative tools are available from Varian/Agilent.
5. Packing tools for insert: We typically use Bruker packing toolkits for Kel-F insert of 4 mm rotor including blunt-ended screwdriver, fork screwdriver, and extraction screw. Alternative tools are available from Varian/Agilent.

2.7 NMR Analysis

1. NMR spectrometer: We typically use a Bruker Avance NEO 600 MHz NMR spectrometer (^{1}H: 600.13 MHz, narrow bore) equipped with a 3.2 mm double-resonance MAS probe and a Bruker Avance III 400 MHz NMR spectrometer (^{1}H: 400.13 MHz, wide bore) equipped with a 4 mm double-resonance MAS probe. Comparable NMR magnets and MAS probes are available from Varian/Agilent.

3 Methods

3.1 Bacterial Growth

1. Inoculate 50 mL LB medium in a 250 mL culture flask with selected colonies of the required *Escherichia coli* strain (*see* **Note 6**). Incubate at 37 °C with shaking at 180 rpm for about 12 h.
2. For two 3 L culture flasks, mix 270 mL M9 minimal medium, 1230 mL water, 6 g sterile filtered glucose, 1.5 g sterile filtered NH_4Cl, 3 mL 1 M $MgSO_4$ solution, and 150 μL 1 M $CaCl_2$ solution. Inoculate each flask with 10 mL of the culture from the previous step. Incubate at 37 °C with shaking at 180 rpm until OD_{650} is about 1.
3. Cool cultures in ice bath for 30 min.

3.2 Expression of PHY/CBCR Proteins and Isotopic Labeling of the Chromophore

1. Add 1.5 mL 12.5 mM $FeSO_4$, 0.5 mL 0.8 M IPTG, and 20 mg isotopically labeled ALA (e.g., [5-^{13}C]-ALA or [^{15}N]-ALA; *see* Fig. 3b) to start gene expression. Keep light-sensitive protein in darkness (*see* **Notes** 7 and **8**).
2. Shake at 17 °C with 120 rpm during gene expression for about 16 h.
3. Transfer the cultures to centrifuge bottles and collect bacteria by centrifugation at 5900 × *g* (equivalent to 57,800 m/s^2) for 15 min at 4 °C.
4. Discard the supernatant (freeze the bacterial pellet at −20 °C for storage, *see* **Note 9**).

3.3 Bacterial Lysis

1. Resuspend bacterial pellets in precooled lysis buffer at 4 °C (5 mL/g of wet pellet mass).
2. Sonicate the resuspended bacterial pellets while cooling in an ice bath (7 min/g of wet pellet mass, 1 s pulse per 2 s pause, 390 W pulse electric input power on ultrasound homogenizer). Set the temperature alarm to 2 °C and place the temperature gauge in the ice bath.
3. Transfer the lysate to centrifuge bottles and collect cell fragments by centrifugation at 24,700 × *g* (equivalent to 242,000 m/s^2) for 1 h at 4 °C (*see* **Note 10**).
4. Discard the pellet. Proceed with column purification of the supernatant (*see* Subheading 3.4).

3.4 Column Purification by HisTrap

1. At the FPLC UV-Vis absorption detector unit, set the detection wavelength to the expected absorption maximum (λ_{max}) of a PHY/CBCR photostate (e.g., 652 nm for the slr1393g3 Pr dark state). Set the second detection wavelength to 280 nm (if available).
2. Perform pump wash for pump B with the inlet tube in water.
3. Equilibrate the 5 mL HisTrap column with five column volumes of water at a flow rate of 5 mL/min. Set pressure alarm to 0.5 MPa (*see* **Note 11**).
4. Perform pump wash for pump A with an inlet tube in lysis buffer and for pump B with an inlet tube in elution buffer.
5. Equilibrate the 5 mL HisTrap column with five column volumes (25 mL) of lysis buffer at a flow rate of 5 mL/min. Use that flow rate for all following steps.
6. Spin the supernatant of the bacterial lysis at 24,700 × *g* (equivalent to 242,000 m/s^2) for 10 min at 4 °C to remove debris. Keep 20 μL aliquot for SDS-PAGE.

7. Connect a superloop to the FPLC device and inject the sample into the superloop.
8. Load the sample from the superloop onto the equilibrated HisTrap column: Pump A must be in lysis buffer. Use a flow rate of 5 mL/min to load the sample.
9. After loading the sample, wash the column with five column volumes (25 mL) of lysis buffer. Begin a two-step elution with 10% and then 14% elution buffer.
10. Elute the PHY/CBCR proteins with a step elution at 18% elution buffer. Collect fractions.
11. Cut a 1-kDa dialysis membrane into an appropriate length for dialysis. Rinse the membrane with water and dialysis buffer.
12. Transfer the sample into the prepared dialysis membrane and seal it.
13. Dialyze the sample against 2 L of dialysis buffer for 12 h at 4 °C with stirring. Replace dialysis buffer after 6 h.
14. Run SDS-PAGE to verify protein purity. If the purity is not considered satisfactory, repeat the procedure from **step 6** with a size exclusion column instead of a HisTrap column.
15. Check photoreversibility of PHY/CBCR proteins (e.g., slr1393g3) by following the irradiation protocol of λ_{max}(dark) → λ_{max}(photoproduct) → λ_{max}(dark) using an array of appropriate LEDs (e.g., 530 and 650 nm for Pr dark state and Pg photoproduct of slr1393g3, respectively) for 2 min at RT to ensure full photoconversion.
16. Concentrate the protein using a 10 kDa MWCO centrifugal filter spun at 8000 × g (equivalent to 78,600 m/s^2) for 10 min at 4 °C (e.g., up to 60 mg/mL). Mix the filter content after each refill. Discard the filtrate.

3.5 General TG Preparation

1. Replace buffer with the TG preparation buffer (*see* Subheading 2.5, **item 3**) using a 10 kDa MWCO centrifugal filter spun at 8000 × g (equivalent to 78,600 m/s^2) for 10 min at 4 °C. Discard the filtrate.
2. Mix the appropriate volume of the stock solution of 1.2 M trehalose, prediluted in the TG preparation buffer (*see* Subheading 2.5, **item 3**) with the target protein solution at a given T/P molar ratio (between trehalose and the hosted protein). Ensure the mixture is in the form of a homogeneous solution (*see* **Note 12**).
3. Distribute the trehalose–protein mixture solution without bubbles on a Petri dish or glass plate.

4. Connect the N_2 gas supply and light source to the illumination chamber. Place the Petri dish into the chamber (*see* **Notes 3** and **13**).
5. Dry the mixture on a Petri dish overnight under continuous N_2 gas flow and under illumination with light of the appropriate wavelength (*see* **Notes 14–18**).
6. Examine the absorption properties and the stability of the obtained TG samples using UV-Vis spectroscopy prior to and after the NMR acquisition. Ensure the surface of the TG faces the incident light beam (*see* **Note 19**).

3.6 TG-Trapping Photocycle Intermediates of slr1393g3 (See Fig. 4)

1. Prepare the TG sample harboring the Pr dark state at 100% occupancy following the steps described under Subheading 3.5.
2. Illuminate the dried Pr TG with 650 nm light for 5 h at RT, resulting in the formation of the primary intermediate state of the forward reaction, Lumi-R.
3. Cool the Lumi-R TG down to −20 °C with saturated illumination of 650 nm light for 25 min for generating the late intermediate state of the forward reaction, Meta-R (*see* **Notes 20** and **21**).
4. Prepare the TG sample harboring the Pg photoproduct at 100% occupancy following the steps described under Subheading 3.5 (*see* **Note 22**).

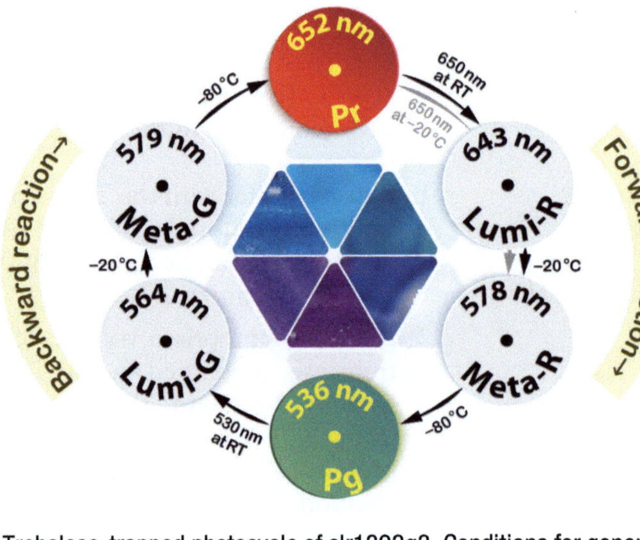

Fig. 4 Trehalose-trapped photocycle of slr1393g3. Conditions for generating and trapping photocycle intermediates in both forward and backward reactions of slr1393g3 are indicated. Photographs of all six TG samples under ambient light and temperature conditions to present their absorption "colors" are assembled at the center of the cycle

5. Illuminate the dried Pg TG with 530 nm light for 3 h at RT to form the primary intermediate state of the back reaction, Lumi-G.
6. Cool the Lumi-R TG down to −20 °C for generating the late intermediate state of the back-reaction, Meta-G (*see* **Note 23**).

3.7 NMR Analysis

1. Gently crush the TG sample dried on the Petri dish (Fig. 5) using a micro-laboratory spoon and spatula.
2. Grind the collected sample to a fine powder with a mortar and pestle.
3. Load the TG sample into a MAS rotor (*see* **Note 24**).
4. Regular probehead performance check (*see* **Note 25**).
5. Insert the MAS rotor into the probehead.
6. Start the sample spinning and let the pneumatic unit stabilize the MAS at a desired rotation frequency (*see* **Note 26**).
7. Match and tune the probehead for the inserted sample (*see* **Note 27**).
8. Apply a well-optimized ^{13}C cross-polarization (CP) setup (*see* **Note 28**).
9. Record a ^{13}C CP spectrum with the TG sample (e.g., slr1393g3 with [5-^{13}C]-ALA-labeled PCB; *see* Fig. 3b) as shown in Fig. 6a (*see* **Note 29**).
10. To achieve one-to-one assignment of the ^{13}C resonances of PCB carbon atoms in the CP spectrum (*see* Fig. 6a), we acquired 2D ^{13}C–^{13}C dipolar-assisted rotational recoupling

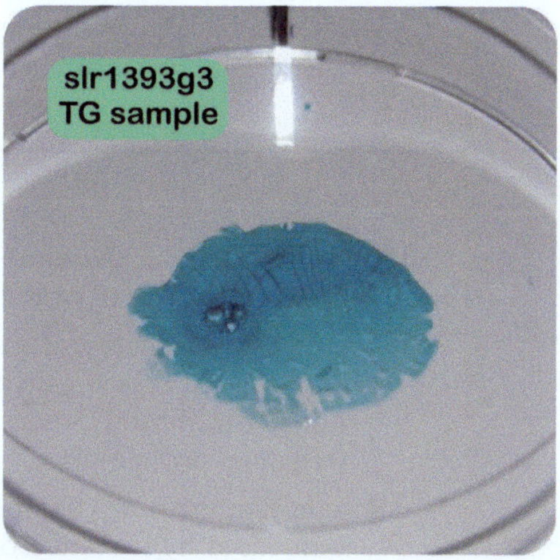

Fig. 5 Trehalose-embedded slr1393g3 glass sample in the Pr dark state. For NMR use, the slr1393g3 TG sample was formed on a Petri dish under continuous N_2 gas flow and illuminated with 530 nm green light from an array of LEDs

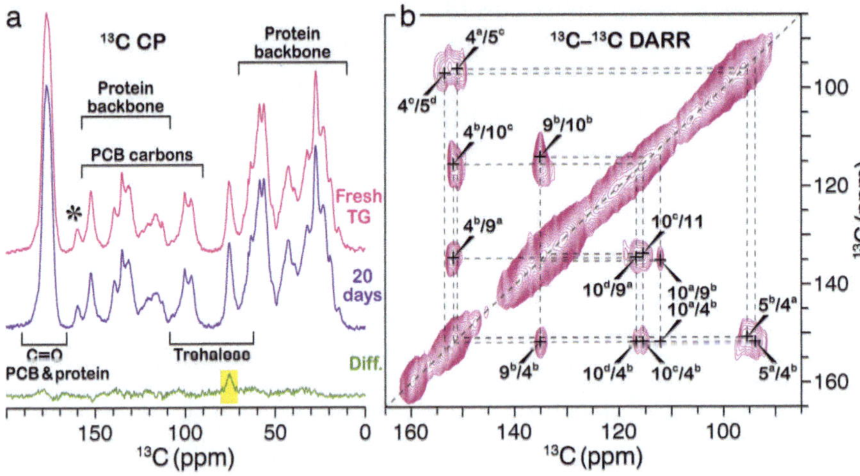

Fig. 6 Isolation and long-term preservation of individual transient photocycle intermediates in TGs for RT MAS NMR measurements. (**a**) ^{13}C CP spectra of freshly prepared slr1393g3 labeled with [5-^{13}C]-ALA in the Lumi-R intermediate state trapped in the trehalose glassy matrix (reddish magenta) and after 20 days stored at RT (dark violet) using a rotation frequency of 20 kHz. The spectral difference after the 20-day storage at RT is given at the bottom (lime-green curve). The spinning sideband is labeled with an asterisk. Narrowing of the trehalose resonance (increased peak intensity) at 72.5 ppm (shaded in yellow) would arise from a slow conversion of the amorphous trehalose into the crystalline polymorph over the period of storage [12]. (**b**) Two-dimensional (2D) DARR spectrum of the freshly prepared Lumi-R sample for complete ^{13}C assignment of the [5-^{13}C]-ALA-labeled PCB chromophore (for numbering, *see* Fig. 3b). For most of the PCB carbons showing multiple chemical shifts, their ^{13}C resonances are superscripted with a, b, c, and so on from the high- to the low-field side

(DARR) data (*see* Fig. 6b). The experiment setup was based on one-dimensional (1D) ^{13}C CP parameter optimization (*see* **Note 28**) and the calculation of a weak radio-frequency (RF) pulse applied during the mixing time (τ_{mix}) for recoupling the dipolar interactions (*see* **Note 30**).

4 Notes

1. Antibiotics like chloramphenicol and kanamycin A are recommended to prevent or reduce the growth of undesired bacteria.
2. The 1.2 M trehalose stock solution should not be stored below 10 °C, as that will cause precipitation, which may then slow down the drying process and modify the T/P ratio of the trehalose–protein glass.
3. The illumination chamber was built with aluminum foil, an unfocused light source of LEDs, and a drying gas hose. The light source and the gas hose can be placed next to each other in the chamber. A diffuser made of milk glass can be used to yield a more homogeneous illumination.

4. The active sample volumes are 30 μL for both 3.2 mm rotor and the 4 mm rotor with Kel-F insert.
5. The Bruker Vespel SP-1 cap does not fit well into the 4 mm rotor with a Kel-F insert due to the longer neck relative to that of the Kel-F cap.
6. For slr1393g3 holoprotein production, BL21(DE3) chemically competent *Escherichia coli* cells were cotransformed with the chosen plasmids pET-30 encoding slr1393-GAF3 (amino acids 439–596) from cyanobacterium *Synechocystis* 6803 and pACYC carrying the PCB-generating enzymes hol-pcyA [12].
7. To obtain the corresponding protein sample containing PCB at natural abundance, we used the LB medium for the main culture instead of the suggested medium as described in Subheading 3.1, **step 1**.
8. A dim safe light (e.g., 450–460 nm blue light for slr1393g3) can be used when necessary to secure the visibility. Alternatively, infrared visualization with a charge-coupled device (CCD) camera and head-mounted goggles (e.g., Trivisio) can be used to work in total darkness.
9. For freezing concentrated protein samples at −20 °C, glycerol or other cryoprotectants can be added to a concentration of 2–5 vol%.
10. The pellet of the cell fragments should go to nearly white after lysis. If the greenish and bluish colors are present in the pellet, repeat lysis for this pellet to increase yield.
11. Do not reset the pressure alarm when the column is connected. If the pressure exceeds 0.5 MPa, reduce the flow to prevent column damage.
12. For molar ratios ranging from 1000:1 to 200:1, the high viscosity of these glassy matrices is not sufficient to prohibit the thermal reversion of short-lived photoproducts (such as from *Pst*BphP1) [12]. For MAS NMR studies, TG samples having T/P molar ratios below 100:1 were usually used, in which the hosted proteins will not undergo significant thermal reversion and denaturation at RT for weeks (Fig. 2).
13. The compressed air with pressure dew point <−40 °C can be an alternative gas for N_2. However, N_2 is drier in nature and contains less impurities than compressed air. This not only helps shorten the drying time but also regulate the quality of the formation of TGs.
14. For optimal illumination conditions, factors like light source power and its distribution, illumination distance, bandwidth, and viewing angle may be taken into account and conditions optimized. It is recommended to use proteins at natural abundance for this purpose (*see* **Note 7**).

15. Close overlap of the absorption bands of dark state and photoproduct (and the high extinction coefficient for the photoproduct) would limit photoproduct occupancy (e.g., 70–85% for cyanobacterial Cph1 PCM [18]). A cautious choice of the illumination wavelength based on the absorbance properties of both states and their difference would be beneficial to maximize photoproduct occupancy in the photoequilibrium mixture.
16. The drying gas flow and the shape of the nozzle should be optimized. Ideally, the gas flow should be as high as possible to allow fast drying without causing an inhomogeneous thickness distribution of the dehydrated trehalose–protein films.
17. Leaving the TG sample under a dry N_2 gas atmosphere for long periods of time (days) may produce additional drying.
18. The TG sample can be redissolved in lysis buffer or TG preparation buffer as described in Subheadings 2.3 and 2.5, respectively.
19. If there is concern about sample heterogeneity of the dried TG, it is recommended to run the same sample in triplicate to determine an average spectrum by choosing different sampling regions on the TG film.
20. Given to the inverse temperature effect of the TG preparation [12], the transition to the late Meta-R intermediate can be initialized upon cooling down the primary Lumi-R TG, in contrast to cryotrapping of PHY intermediates in solutions [10, 19].
21. An alternative way to generate the Meta-R intermediate is to illuminate the Pr TG with 650 nm light at −20 °C. This allows the Pr transition to proceed beyond the primary Lumi-R intermediate [17].
22. The final transition from Meta-R to form the Pg photoproduct proceeds well upon further cooling of the TG sample to −80 °C.
23. Similar to the final transition to form Pg, the Meta-G intermediate in the backward reaction can be fully converted back to Pr upon cooling of the TG sample to −80 °C.
24. For the full loading of a 3.2 mm rotor (or a Kel-F insert for a 4 mm rotor), about 132 mg TG sample of slr1393g3 (25.8 kDa), having a molar ratio of 50:1 T/P, was used.
25. The check comprises the magic-angle setting (using KBr), the shims (using H_2O), and the S/N performance (using glycine·HCl).

26. The maximum stable rotation frequencies of Bruker 3.2 and 4 mm zirconia rotors in our laboratory are 22 and 13.5 kHz, respectively.
27. Multiple fake resonances may show up during the X tuning. It is thus good to keep track of which nucleus was tuned last.
28. First load a standard parameter set for adamantane and record a 1D ^{13}C CP spectrum. Optimize the length of the ^{1}H pulse to get a proper 90° excitation pulse. Calculate the power of the RF pulse. Sweep the ^{1}H spin lock power for the optimal ^{1}H to ^{13}C Hartmann–Hahn matching condition. Calculate the decoupling power using the RF power determined above.
29. All MAS NMR spectra of TG samples can be acquired at RT.
30. Calculate the power level of the RF irradiation during the τ_{mix} period which corresponds to n times the rotation frequency.

Acknowledgments

We thank Dr. L. Köhler (Universität Leipzig) for valuable discussions regarding this and other work on photosensors from our laboratory.

References

1. Hughes J, Winkler A (2024) New insight into phytochromes: connecting structure to function. Annu Rev Plant Biol 75(1):153–183. https://doi.org/10.1146/annurev-arplant-070623-110636
2. Rockwell NC, Lagarias JC (2024) Cyanobacteriochromes: a rainbow of photoreceptors. Ann Rev Microbiol 78(1):61–81. https://doi.org/10.1146/annurev-micro-041522-094613
3. Jang J, Tang K, Youn J et al (2023) Engineering of bidirectional, cyanobacteriochrome-based light-inducible dimers (BICYCL)s. Nat Methods 20(3):432–441. https://doi.org/10.1038/s41592-023-01764-8
4. Rohmer T, Strauss H, Hughes J et al (2006) ^{15}N MAS NMR studies of Cph1 phytochrome: chromophore dynamics and intramolecular signal transduction. J Phys Chem B 110(41): 20580–20585. https://doi.org/10.1021/jp062454+
5. Song C, Rohmer T, Tiersch M et al (2013) Solid-state NMR spectroscopy to probe photoactivation in canonical phytochromes. Photochem Photobiol 89(2):259–273. https://doi.org/10.1111/php.12029
6. Xu Q-Z, Bielytskyi P, Otis J et al (2019) MAS NMR on a red/far-red photochromic cyanobacteriochrome All2699 from *Nostoc*. Int J Mol Sci 20(15):3656. https://doi.org/10.3390/ijms20153656
7. Altmayer S, Köhler L, Bielytskyi P et al (2022) Light- and pH-dependent structural changes in cyanobacteriochrome AnPixJg2. Photochem Photobiol Sci 21(4):447–469. https://doi.org/10.1007/s43630-022-00204-4
8. Hu K-N, Havlin RH, Yau W-M et al (2009) Quantitative determination of site-specific conformational distributions in an unfolded protein by solid-state nuclear magnetic resonance. J Mol Biol 392(4):1055–1073. https://doi.org/10.1016/j.jmb.2009.07.073
9. Buhrke D, Battocchio G, Wilkening S et al (2020) Red, orange, green: light- and temperature-dependent color tuning in a cyanobacteriochrome. Biochemistry 59(4): 509–519. https://doi.org/10.1021/acs.biochem.9b00931
10. Rohmer T, Lang C, Bongards C et al (2010) Phytochrome as molecular machine: revealing chromophore action during the Pfr → Pr photoconversion by magic-angle spinning NMR spectroscopy. J Am Chem Soc 132(12):

4431–4437. https://doi.org/10.1021/ja9108616

11. Siemer AB, Huang K-Y, McDermott AE (2012) Protein linewidth and solvent dynamics in frozen solution NMR. PLoS One 7(10): e47242. https://doi.org/10.1371/journal.pone.0047242
12. Köhler L, Gärtner W, Matysik J et al (2022) Long-term preservation of short-lived photoproducts of phytochromes at room temperature. ChemPhotoChem 6(3):e202100220. https://doi.org/10.1002/cptc.202100220
13. Nagase H, Ogawa N, Endo T et al (2008) Crystal structure of an anhydrous form of trehalose: structure of water channels of trehalose polymorphism. J Phys Chem B 112(30): 9105–9111. https://doi.org/10.1021/jp800936z
14. Möbius K, Savitsky A, Malferrari M et al (2020) Soft dynamic confinement of membrane proteins by dehydrated trehalose matrices: high-field EPR and fast-laser studies. Appl Magn Reson 51(9):773–850. https://doi.org/10.1007/s00723-020-01240-y
15. Cordone L, Cottone G, Cupane A et al (2015) Proteins in saccharides matrices and the trehalose peculiarity: biochemical and biophysical properties. Curr Org Chem 19(17): 1684–1706. https://doi.org/10.2174/1385272819666150429232426
16. Köhler L, Trunk F, Rohr V et al (2025) Rotameric heterogeneity of conserved tryptophan is responsible for reduced photochemical quantum yield in cyanobacteriochrome slr1393g3. ChemPhysChem 26(2):e202400453. https://doi.org/10.1002/cphc.202400453
17. Köhler L, Gärtner W, Salvan G et al (2023) Photocycle of a cyanobacteriochrome: a charge defect on ring *C* impairs conjugation in chromophore. Chem Sci 14(23):6295–6308. https://doi.org/10.1039/d3sc00636k
18. Hughes J (2010) Phytochrome three-dimensional structures and functions. Biochem Soc Trans 38(2):710–716. https://doi.org/10.1042/BST0380710
19. Schwinté P, Gärtner W, Sharda S et al (2009) The photoreactions of recombinant phytochrome CphA from the cyanobacterium *Calothrix* PCC7601: a low-temperature UV–vis and FTIR study. Photochem Photobiol 85(1): 239–249. https://doi.org/10.1111/j.1751-1097.2008.00426.x

Part IV

Addressing Phytochrome Functions and Applications

Chapter 17

Analyzing In Vitro Autokinase Activity of Microbial Phytochromes Under Well-Defined Light Conditions

Christina Huber, Merle Strack, Rolf Diller, and Nicole Frankenberg-Dinkel

Abstract

Microbial phytochromes primarily function as sensor histidine kinases in two-component regulatory systems, with light-triggered conformational changes driving varying autokinase outputs. Analyzing these responses under different light conditions is key to understanding phytochrome functions in host organisms.

Here, we introduce a novel method for providing quantitative correlation between kinase activity and calculated Pr/Pfr fractions as determined by spectroscopic measurements. The approach includes data analysis for calculating Pr/Pfr content using UV/Vis absorption spectroscopy. It also introduces a Kinase Illumination Module (KIM) that creates a controlled light environment during autokinase assays, replicating conditions from UV/Vis spectroscopy to ensure consistent photostationary states. The method provides a versatile platform for investigating the dynamics of phytochrome signaling pathways, which is applicable to a wide range of experimental conditions.

Key words Defined Pr/Pfr-ratios, Steady-state conditions, Autokinase assay, Light-dependent kinase, Light-regulated kinase, Photoreception, Signal transduction, Light sensing

1 Introduction

Microbial phytochromes often possess a histidine kinase domain as a regulatory output domain and thus are part of a bacterial two-component signaling system [1, 2]. To understand their biological function, investigation of the light-dependent autokinase activity is crucial. In vitro kinase assays with radioactive-labeled ATP under different light conditions have furthered our understanding of the relationship between phytochrome photostates and autokinase activity. For example, studies revealed that a key property of

Christina Huber and Merle Strack contributed equally to this work.

Andreas Winkler (ed.), *Phytochromes: Methods and Protocols*, Methods in Molecular Biology, vol. 2970,
https://doi.org/10.1007/978-1-0716-4791-2_17,

bacterial phytochromes is that the Pr form exhibits the highest autokinase output in both prototypical [3–7] and *bathy* phytochromes [8–11].

Previous studies on the autokinase activity of bacterial phytochromes have focused mainly on pure Pr or Pfr forms of the proteins [3–7, 10, 12, 13]. Determining the light conditions (i.e., wavelength, intensity and duration of illumination, or darkness [14, 15]) to achieve "pure" Pfr and Pr can be considered standard practice. However, the literature frequently provides only limited information on how the state of interest (Pr, Pfr, or mixed state) was realized experimentally. Given the highly dynamic nature of light-sensitive phytochromes, this deficiency limits the significance of the autokinase assay results for the correlation with the respective photostate.

To satisfy the requirements for a correlation, the same photostate must be established in the spectrophotometer sample and the autokinase assay sample. Therefore, the first step is to characterize the photostate using a suitable method (e.g., by analysis of Q-band absorption spectra [16–22]), while in the second step, the same illumination conditions must be ensured for the kinase assay.

Here, we present a set of both experimental and analytical methods that realizes this concept. First, we provide a simple protocol for determining the Pfr and Pr content of a sample under individually selected but well-defined light conditions using absorption spectroscopy data. Second, we describe the steps of an autokinase assay employing a specially designed, home-built three-dimensional (3D)-printed setup for protein kinase assays, called Kinase Illumination Module (KIM). The KIM enables the illumination of a phytochrome sample in a reaction tube from underneath with a replaceable light source that can be transferred from the UV/Vis spectrophotometer. Additionally, other wavelengths of interest can easily be adapted to the experiment, offering the study of Pr/Pfr mixtures and expanding the types of phytochromes that can be studied, for example, red/green cyanobacteriochromes (CBCRs). For the latter, the corresponding forms are the *15Z* (cis) configuration in the dark-adapted state and the *15E* (trans) photoproduct (*see* also [23]). This optimized and standardized protocol bridges the gap between two well-established methods—UV/Vis-spectroscopy and autokinase assays—and, by providing sets of correlated data, furthers our understanding of the functions of phytochromes [8].

2 Materials

Prepare and dilute all solutions freshly using ultrapure autoclaved water and chemicals of American Chemical Society (ACS) grade or higher. It is important that the solutions are prepared, pH-adjusted, and stored at the specified temperature.

2.1 Autokinase Assay

1. Kinase buffer (1×): 50 mM Tris–HCl, pH 8.0; 100 mM NaCl; 5 mM $MgCl_2$; 10% (v/v) Glycerol (*see* **Note 1**). Make up to 1 L with cold water and store the buffer at 4 °C.
2. ATP mix (5×): Stock solutions 10 μCi/μL [γ-^{32}P]-ATP ([γ-^{32}P]-ATP (SRP-301), 500 μCi (as provided from the supplier Hartmann Analytic), and 100 mM ATP (non-radioactive). For the ATP mix (5×), use 1:10 dilution of the stock solutions in water (i.e., 20 μL water, 2.5 μL non-radioactive ATP solution (100 mM solution ATP ≥ 99%, (Carl Roth)) and 2.5 μL [γ-^{32}P]-ATP to a microtube and mix it (*see* **Note 2**). Store the solution on ice and use it without delay.
3. Sodium dodecyl sulfate–polyacrylamide gel electrophoresis (SDS-PAGE): 10% acrylamide (w/v) usually sufficient for most bacterial phytochromes.
4. SDS sample buffer (4×): 0.1 M Tris–HCl pH 6.8, 8% (w/v) SDS, 40% (v/v) glycerol, 10% (v/v) β-mercaptoethanol, and 1% (w/v) bromophenol blue.
5. SDS running buffer (10×): 0.25 M Tris–HCl pH 8.8, 1.92 M glycine, 1% (w/v) SDS) (*see* **Note 3**).
6. Staining solution: 10% (v/v) acetic acid, 30% (v/v) ethanol, and 0.25% (w/v) Coomassie brilliant blue.
7. Destaining solution: 10% (v/v) acetic acid and 30% (v/v) ethanol.
8. Stop mix (1×): 1.5 μL 0.5 M ethylenediaminetetraacetic acid (EDTA), 2.5 μL 4× sample buffer. (Prepare a single-stop mix for each time point you want to observe the kinase activity in a microtube.)
9. Phosphorimager.
10. SDS-PAGE chamber.
11. Centrifugal filter units: Microspin™ G-50 Columns (Cytiva).
12. Prestained protein size standard.
13. Kinase Illumination Module (KIM): *see* below.
14. Photodiodes: *see* below.

2.2 UV/Vis Absorption Spectroscopy

1. Diode array spectrometer (*see* **Note 4**).
2. Quartz cuvette.
3. Photodiodes: *see* below.

2.3 Photodiodes

Blue LED: 426 ± 70 nm, 0.65 mW, and 0.4 mW/cm^2 (*Everlight* DLE-038-046).

Far-red laser diode: 791 ± 2 nm, 200 mW, and 61 mW/cm^2 (*Panasonic* LNCT28PS01WW).

Red laser diode: 667 ± 2 nm, 100 mW, and 27 mW/cm^2 (*Panasonic* LNCT28PS01WW) (*see* **Note 5**).

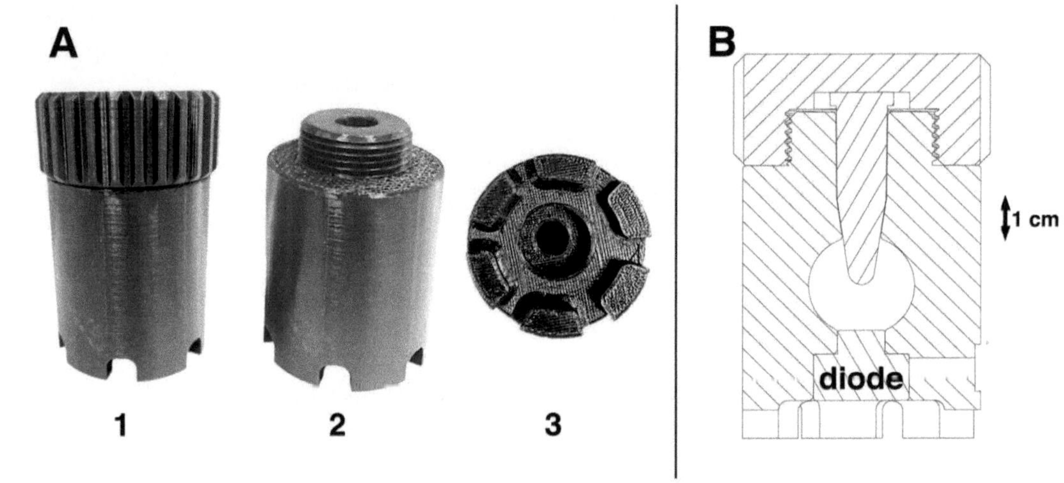

Fig. 1 Photograph (**a**) and technical drawing (**b**) of the new illumination setup used for the radioactive kinase assays. It consists of a main housing (**a**2) with a screw cap (**a**1), three-dimensional (3D)-printed from black polylactide, that protects the sample from unwanted ambient light. A photodiode (with specifically chosen optical emission properties) can be inserted easily from the bottom (**a**3, **b**). The phytochrome sample itself is located in a reaction tube (**b**) [8]. The 3D model (file formats: F3D, OBJ, and STL) for printing of the KIM is published on GitHub (*see* **Note 6**)

2.4 Illumination Setup (Kinase Illumination Module, KIM)

The cylindrical illumination setup (*see* Fig. 1) is 3D-printed (Original Prusa i3 MK3S 3D printer), using black polylactide, and consists of a main housing and a screw cap (*see* **Note 6**). In the center, the pod is hollowed out in such a fashion that the light from the diode can illuminate the tip of the reaction tube (and sample within) completely (*see* **Note 7**). The opaque housing wall protects the reaction tube from any unwanted light. The reaction tube is inserted via the top into an accurately fitting chamber, fixing its position. The samples are therefore always at the same distance from and in the same orientation to the light source. The light source is customizable: Photodiodes of chosen emission intensity and wavelength are electrified according to the manufacturer's instructions and then installed in a casing that fits the appointed slot at the bottom. The photodiode is fixed in place with a set screw. The base of the pod has several openings so that the diode cables can be easily fed in and excess heat is dissipated to the exterior instead of being trapped in the mount.

3 Methods

3.1 Establishing and Characterizing the Desired Photostationary State Using UV/Vis-Spectroscopy

The characterization of a given phytochrome photostate (Pr/Pfr mixture) requires an appropriate spectroscopic representation of the corresponding parent states (Pr and Pfr). Thus, the first objective of this chapter is to describe a method that provides mathematical functions for the parent state spectra. The functions will then be used in a linear combination to simulate the absorption spectrum of the given phytochrome photostate by a fitting procedure that yields the relative amounts of the parent states. The description is given for a phytochrome with Pfr and Pr as its parent states but can be applied without major modification to any two-state system (i.e., *15Z-15E* states).

3.1.1 Create Fit Functions for an Arbitrary Mixture of Pr and Pfr

1. Completely enrich the Pfr and Pr state of your chosen phytochrome and record their absorption spectra, using the same sample (e.g., for *bathy* phytochromes, complete the dark reversion, record the Pfr spectrum, illuminate the same sample with far-red light until the Pfr-specific band at ca. 750 nm is negligible, and record the Pr spectrum).
2. Subtract background scattering in the Q-band region (e.g., for bacterial phytochromes, ca. 550–850 nm) as a linear function to yield the "pure form spectra" of both Pr and Pfr. Convert wavelength scale to wave number (energy) scale (*see* **Note 8**).
3. Determine the isosbestic point, that is, the intersection of the pure form spectra, and record both its wave number, $\tilde{\nu}_{\text{iso}}$, and absorbance, κ. They will be used later to determine the global fitting constant c (Eq. 1).
4. If you detect substantial contributions of, for example, Pfr in the "pure form" Pr spectrum, subtract the unwanted spectrum (here Pfr) accordingly. This ensures that the absorption bands are not attributed to the wrong parent state. Now, reestablish the isosbestic point at $\tilde{\nu}_{\text{iso}}$ (and its absorbance κ) by rescaling the adjusted parent state spectrum (here Pr).
5. To obtain a mathematical representation of both parent state spectra (*see* Fig. 2), fit a linear combination of Gaussian band profiles to each spectrum. It is advised to use the minimum number of Gaussians necessary to achieve an appropriate goodness of fit, but avoid redundancies. This will yield the parent state functions $P^0_{\text{fr}}(\tilde{\nu})$ and $P^0_{\text{r}}(\tilde{\nu})$ (*see* **Note 9**).
6. Define a fit function to simulate the absorption spectrum, $A(\tilde{\nu})$, of your given phytochrome photostate:

$$A(\tilde{\nu}) = c \cdot \left[\alpha \cdot P^0_{\text{fr}}(\tilde{\nu}) + (1-\alpha) \cdot P^0_{\text{r}}(\tilde{\nu})\right] \qquad (1)$$

The fit function is constructed as a linear combination of $P^0_{\text{fr}}(\tilde{\nu})$ and $P^0_{\text{r}}(\tilde{\nu})$, where the variable α describes the relative amount of Pfr ([Pfr]/([Pfr] + [Pr])). The global fitting constant c rescales the parent state functions to your experimental data.

3.1.2 Apply the Fit Function to Your Phytochrome Sample

To determine the amounts of the parent states in your mixture, the function developed in the previous chapter will be applied. A MATLAB 2022b source code file is published on GitHub (*see* **Note 6**). It includes the code necessary for several phytochrome systems.

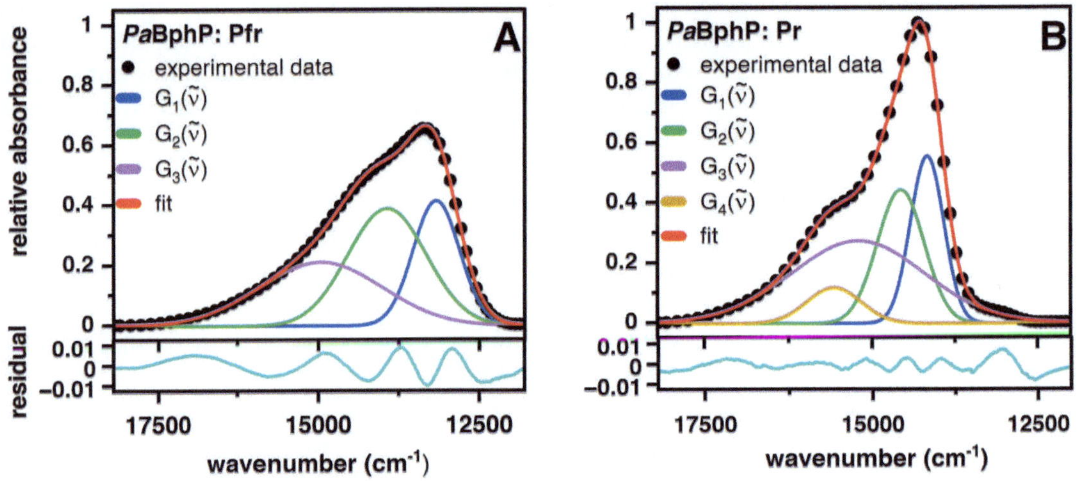

Fig. 2 Experimental pure form spectrum $P_{fr}^{0}(\tilde{\nu})$ (left) and $P_{r}^{0}(\tilde{\nu})$ (right) of bathy phytochrome PaBphP from *Pseudomonas aeruginosa*, simulated (OriginLab 2022b) by a sum of three and four Gaussian functions, respectively [8]

1. Illuminate phytochrome sample—while measuring the UV/Vis absorption spectrum after useful time intervals—until the photoequilibrated state has been established. Record the time necessary (*see* **Note 10**).
2. Subtract background scattering in the Q-band region (ca. 550–850 nm) as a linear function. Convert wavelength scale to wave number (energy) scale.
3. Determine the global scaling constant from the experimental data $c = A(\tilde{\nu}_{iso})/\kappa$ (*see* **Note 11**).
4. Now, simulate the resulting absorption spectrum via Eq. (1), yielding the relative amount of Pfr, α.

3.2 Autokinase Assays

It is of utmost importance to ensure that all steps are carried out in accordance with the highest standards of safety and in strict compliance with the respective laws and regulations. It is imperative that the operator utilizes personal protective equipment when handling radioactive material (*see* **Note 2**). It is recommended that each of the following steps be performed with minimal exposure to ambient light, particularly when the protein is not in the illumination setup. It is important to work quickly and meticulously to ensure that the protein is exposed to the least amount of light possible.

1. Determine the concentration of your protein sample and dilute it in 1x kinase buffer in a final volume of 10 μL (*see* **Note 12**). Choose a concentration that ensures the protein is clearly visible and stable on an SDS-PAGE gel (*see* **Note 13**). Store the protein mixture in a 1.5-mL reaction tube in the dark on ice and use it without delay.

2. Remove the lid of the reaction tube containing your protein sample permanently, put the tube in the illumination setup, and close the setup.
3. Illuminate your protein sample with the photodiode of chosen emission intensity and wavelength as described in the material section until the desired spectroscopic form is achieved (*see* **Note 14**).
4. Open the lid of the illumination kinase setup, remove 10 μL of your protein solution that has undergone irradiation but has not been incubated with radioactive-labeled ATP, and quickly close the lid. Mix it with 4 μL stop mix.
5. Open the lid of the illumination setup, initiate the kinase reaction with 2.5 μL of the 5× ATP mix, and close the lid quickly. Illuminate your protein sample at all times.
6. Incubate the reaction for a defined period of time. Alternatively, you can also conduct a time-course experiment and stop the reaction at varying time points (*see* **Note 15**).
7. Stop the reaction by opening the setup again, removing 10 μL of the protein solution, and mixing it with 4 μL stop mix.
8. If you chose the time-course experiment, prepare the desired number of reaction tubes with 4 μL 1× stop mix in advance. At any time, you want to test the kinase activity, remove 10 μL from your total protein sample, and put it in the tube with the stop mix to stop the reaction.
9. At this point, it is no longer necessary to work in the absence of light.
10. Use Microspin™ G-50 columns according to the manufacturer's instructions with the total sample volume to eliminate excess radiolabeled free ATP (*see* **Note 16**).
11. Start an SDS-PAGE gel immediately (*see* **Note 17**). Load 10 μL of the samples after centrifugation in the column into each well (*see* **Note 18**).
12. After migration, wash the gel with destaining solution and water. After that, carefully wrap it in cling film and transfer it to an imaging plate (e.g., Fujifilm) for 15 h (*see* **Note 19**).
13. Visualize the signals of the radioactively labeled proteins on the gel with a phosphorimager.
14. Stain the SDS-PAGE gel in Coomassie staining solution for 10 min and subsequently decolorize it with destaining solution or water under shaking conditions, allowing the visualization of the proteins in the gel (*see* **Note 20**).

4 Notes

1. Tris–HCl dissolves poorly at low temperatures. However, it is imperative to adjust the pH at 4 °C, as the pH of Tris–HCl solutions is temperature-dependent and a pH of 8.0 is required at 4 °C.
2. Working with radioactive material: Work behind lined absorbing material; keep as much distance between yourself and the radioactive source(s); wear a laboratory coat, gloves, if necessary, and safety goggles; and use a dosimeter. Dispose of your radioactive waste (including tips, containers, and liquids) and clean everything.
3. Always wear a mask when weighing SDS or acrylamide. The buffers with SDS should be stored at room temperature, as the SDS precipitates at 4 °C.
4. In case of doubts about the influence of the measuring light of the spectrometer on the Pr/Pfr ratio, control experiments should be carried out, for example, by varying the intensity of the measuring light (neutral density filter) and/or by varying the scanning speed (integration time at each wavelength and spectral resolution). These measures change the total dose of photons and can thus reveal (unwanted) photoconversion.
5. In both setups, that is, the UV/Vis spectrometer and the kinase assay, the same targeted photostationary state, whose kinase activity is to be measured, is generated by irradiating the sample with a photodiode. Keep in mind that both the ratio Pr/Pfr and the time constant for the formation of the photoequilibrated state are determined by the intensity and wavelength of the exciting light and by the rate of dark reversion (caution is required in case of systems with fast dark reversion). Thus, in both setups, protein concentration and the illumination procedure—such as duration, light source, and its geometry with respect to the sample cuvette—must be the same in order to obtain the same photostationary state. The given intensities (of the usually divergent diode emission) refer to the distance between diode and sample in the KIM. To ensure that (minor) differences in the illumination geometry between KIM and spectrometer have no influence on the Pr/Pfr ratio, check this by carrying out appropriate control experiments, for example, by varying the distance between the diode and the sample in the spectrometer.
6. Link to GitHub repository: https://github.com/MRStrack/Characterising-phytochrome-photostates-using-UV-Vis-absorption-spectra.

7. Good results were achieved using a sample volume of ca. 100 µL and diodes whose emission divergence matched the geometrical aperture of the reaction tube.
8. Spectral absorption bands of a molecule represent vibronic transitions due to the absorption of energy ($E = h\nu$; Planck constant h, and frequency $\nu = c/\lambda$) from photons. Thus, for their physically rigorous quantitative description, a conversion from wavelength (λ, which is inversely proportional to energy) to wave number ($\tilde{\nu} = 1/(100\lambda)$) is necessary.
9. For several *bathy* phytochromes, parameters are listed in [8]. The "Multiple Curves/Multiple Peaks" tool from *OriginLab 2022b* was used.
10. We recommend measuring the spectra, while the photodiodes are still active to prevent the state of the phytochrome from changing significantly as a result of the dark reversion. Artifacts caused by scattered diode light must be avoided.
11. It is recommended to use *c* as a reasonable starting parameter, while its exact value is determined during the fitting process.
12. Increase the final volume if you want to do a time-course experiment by 10 µL of your protein solution for each time point you want to test. Add another 10 µL for potential pipetting errors. However, we recommend that the final volume does not exceed 100 µL to ensure that the entire sample is illuminated uniformly.
13. We recommend a protein concentration of 5–10 µM. If necessary, initiate a test SDS-PAGE with your diluted protein solution mixed with SDS sample buffer to define the optimal protein concentration for the experiment.
14. As described above, the optimal time point was determined by UV-Vis experiments.
15. Use 2.5 µL of the 5× ATP mix for each 10 µL protein solution to initiate the reaction.
16. Use one column for every time point (10 µL protein solution + 4 µL stop mix) to prevent the residual-free [γ-^{32}P]-ATP from radiating excessively on the radiogram.
17. The percentage of acrylamide in the SDS-PAGE gel and the running time must be individually adapted to each protein.
18. To indicate the orientation of the SDS-PAGE gel, use a prestained marker.
19. In different experiments, try to keep the SDS-PAGE gel on the imaging plate for the same length of time. The longer the gel is applied on the imaging plate and the fresher the used [γ-^{32}P]-ATP is, the more intense the signals after visualization.

20. The SDS-PAGE gel must be stained with staining solution as a control. The colorized gel demonstrates whether the same amount of protein has been applied to each well. Unequal protein amounts generate signals with unequal intensities.

Acknowledgments

This work was in part funded by the Landesforschungsschwerpunkt Rheinland-Pfalz BioComp. We thank Philip Wolf for technical support and construction and realization of the Kinase Illumination Module (KIM).

References

1. Rockwell NC, Lagarias JC (2010) A brief history of phytochromes. Chemphyschem Eur J Chem Phys Phys Chem 11(6):1172–1180. https://doi.org/10.1002/cphc.200900894
2. Rockwell NC, Su YS, Lagarias JC (2006) Phytochrome structure and signaling mechanisms. Annu Rev Plant Biol 57:837–858. https://doi.org/10.1146/annurev.arplant.56.032604.144208
3. Giraud E, Zappa S, Vuillet L, Adriano JM, Hannibal L, Fardoux J, Berthomieu C, Bouyer P, Pignol D, Verméglio A (2005) A new type of bacteriophytochrome acts in tandem with a classical bacteriophytochrome to control the antennae synthesis in *Rhodopseudomonas palustris*. J Biol Chem 280(37):32389–32397. https://doi.org/10.1074/jbc.M506890200
4. Karniol B, Vierstra RD (2003) The pair of bacteriophytochromes from *Agrobacterium tumefaciens* are histidine kinases with opposing photobiological properties. Proc Natl Acad Sci USA 100(5):2807–2812. https://doi.org/10.1073/pnas.0437914100
5. Multamäki E, Nanekar R, Morozov D, Lievonen T, Golonka D, Wahlgren WY, Stucki-Buchli B, Rossi J, Hytönen VP, Westenhoff S, Ihalainen JA, Möglich A, Takala H (2021) Comparative analysis of two paradigm bacteriophytochromes reveals opposite functionalities in two-component signaling. Nat Commun 12(1):4394. https://doi.org/10.1038/s41467-021-24676-7
6. Scheerer P, Michael N, Park JH, Nagano S, Choe HW, Inomata K, Borucki B, Krauss N, Lamparter T (2010) Light-induced conformational changes of the chromophore and the protein in phytochromes: bacterial phytochromes as model systems. Chemphyschem Eur J Chem Phys Phys Chem 11(6):1090–1105. https://doi.org/10.1002/cphc.200900913
7. Xue P, El Kurdi A, Kohler A, Ma H, Kaeser G, Ali A, Fischer R, Krauß N, Lamparter T (2019) Evidence for weak interaction between phytochromes Agp1 and Agp2 from *Agrobacterium fabrum*. FEBS Lett 593(9):926–941. https://doi.org/10.1002/1873-3468.13376
8. Huber C, Strack M, Schultheiss I, Pielage J, Mechler X, Hornbogen J, Diller R, Frankenberg-Dinkel N (2024) Darkness inhibits autokinase activity of bacterial bathy phytochromes. J Biol Chem 300(4):107148. https://doi.org/10.1016/j.jbc.2024.107148
9. Mukherjee S, Jemielita M, Stergioula V, Tikhonov M, Bassler BL (2019) Photosensing and quorum sensing are integrated to control *Pseudomonas aeruginosa* collective behaviors. PLoS Biol 17(12):e3000579. https://doi.org/10.1371/journal.pbio.3000579
10. Tasler R, Moises T, Frankenberg-Dinkel N (2005) Biochemical and spectroscopic characterization of the bacterial phytochrome of *Pseudomonas aeruginosa*. FEBS J 272(8):1927–1936. https://doi.org/10.1111/j.1742-4658.2005.04623.x
11. Yang X, Ren Z, Kuk J, Moffat K (2011) Temperature-scan cryocrystallography reveals reaction intermediates in bacteriophytochrome. Nature 479(7373):428–432. https://doi.org/10.1038/nature10506
12. Lamparter T, Esteban B, Hughes J (2001) Phytochrome Cph1 from the cyanobacterium *Synechocystis* PCC6803. Purification, assembly, and quaternary structure. Eur J Biochem 268(17):4720–4730. https://doi.org/10.1046/j.1432-1327.2001.02395.x

13. Lamparter T, Michael N, Mittmann F, Esteban B (2002) Phytochrome from *Agrobacterium tumefaciens* has unusual spectral properties and reveals an N-terminal chromophore attachment site. Proc Natl Acad Sci USA 99(18):11628–11633. https://doi.org/10.1073/pnas.152263999
14. Butler WL, Hendricks SB, Siegelman HW (1964) Action spectra of phytochrome *in vitro*. Photochem Photobiol 3(4):521–528. https://doi.org/10.1111/j.1751-1097.1964.tb08171.x
15. Giraud E, Lavergne J, Vermeglio A (2010) Characterization of bacteriophytochromes from photosynthetic bacteria: histidine kinase signaling triggered by light and redox sensing. Methods Enzymol 471:135–159. https://doi.org/10.1016/S0076-6879(10)71009-0
16. Antelo GT, Sanchez-Lamas M, Goldbaum FA, Otero LH, Bonomi HR, Rinaldi J (2020) A spectroscopy-based methodology for rapid screening and characterization of phytochrome photochemistry in search of Pfr-favored variants. Photochem Photobiol 96(6):1221–1232. https://doi.org/10.1111/php.13313
17. Bai Y, Rottwinkel G, Feng J, Liu Y, Lamparter T (2016) Bacteriophytochromes control conjugation in Agrobacterium fabrum. J Photochem Photobiol B 161:192–199. https://doi.org/10.1016/j.jphotobiol.2016.05.014
18. Conforte V, Otero LH, Toum L, Sirigu S, Antelo GT, Rinaldi J, Foscaldi S, Klinke S, Chavas LMG, Vojnov AA, Goldbaum FA, Malamud F, Bonomi HR (2021) Pr-favoured variants of the bacteriophytochrome from the plant pathogen *Xanthomonas campestris* hint on light regulation of virulence-associated mechanisms. FEBS J 288(20):5986–6002. https://doi.org/10.1111/febs.15883
19. Inomata K, Hammam MA, Kinoshita H, Murata Y, Khawn H, Noack S, Michael N, Lamparter T (2005) Sterically locked synthetic bilin derivatives and phytochrome Agp1 from *Agrobacterium tumefaciens* form photoinsensitive Pr- and Pfr-like adducts. J Biol Chem 280(26):24491–24497. https://doi.org/10.1074/jbc.M504710200
20. Rottwinkel G, Oberpichler I, Lamparter T (2010) Bathy phytochromes in rhizobial soil bacteria. J Bacteriol 192(19):5124–5133. https://doi.org/10.1128/JB.00672-10
21. Smith RW, Helwig B, Westphal AH, Pel E, Borst JW, Fleck C (2017) Interactions between PhyB and PIF proteins alter thermal reversion reactions *in vitro*. Photochem Photobiol 93(6):1525–1531. https://doi.org/10.1111/php.12793
22. Takala H, Björling A, Berntsson O, Lehtivuori H, Niebling S, Hoernke M, Kosheleva I, Henning R, Menzel A, Ihalainen JA, Westenhoff S (2014) Signal amplification and transduction in phytochrome photosensors. Nature 509(7499):245–248. https://doi.org/10.1038/nature13310
23. Rockwell NC, Lagarias JC (2024) Cyanobacteriochromes: a rainbow of photoreceptors. Annu Rev Microbiol 78:61–81. https://doi.org/10.1146/annurev-micro-041522-094613

Chapter 18

Detection of Histidine Kinase Activities of Bacteriophytochromes with Phos-tag™ Acrylamide Gel Electrophoresis

Iida Tuure and Heikki Takala

Abstract

Phosphorylation is a key regulatory mechanism of cellular functions, and a majority of known bacteriophytochromes (BphPs) act as histidine kinases (HKs) in two-component signaling. These red/far-red light-sensing HKs play a role in diverse light responses in bacteria. In this chapter, we introduce protocols to easily track the net kinase and phosphatase activities of BphP HKs. The protocols provided here apply Phos-tag™ acrylamide gel electrophoresis for protein samples under controlled illumination conditions. They can be applied to BphP HKs and with a few modifications to photoreceptor HKs in general.

Key words Bacteriophytochrome, Sensory photoreceptor, Phos-tag, Kinase activity, Phosphatase activity, Two-component system, Histidine kinase, Response regulator

1 Introduction

Bacterial two-component systems (TCSs) are ubiquitous signaling systems that enable bacteria to adapt to diverse environmental changes. TCSs also regulate a wide range of cellular processes, from respiration to biofilm formation and virulence [1]. They consist of a sensor histidine kinase (SHK) protein that receives the environmental signal and a response regulator (RR) protein that usually functions as a transcription factor. SHKs are modular proteins that consist of a sensory module and a histidine kinase (HK) module, as well as additional transduction modules that relay and integrate the signal from the sensory module to the HK module [2]. The HK module is further divided into a dimerization histidine phosphotransfer (DHp) and a catalytic ATP-binding (CA) domain.

The majority of HKs are bifunctional, displaying both kinase and phosphatase activities [2, 3]. These two activities are in an equilibrium that is controlled by the N-terminal sensory module.

Andreas Winkler (ed.), *Phytochromes: Methods and Protocols*, Methods in Molecular Biology, vol. 2970,
https://doi.org/10.1007/978-1-0716-4791-2_18,

The enzymatic activities can be further divided into three catalytic reactions: autophosphorylation, phosphotransfer, and dephosphorylation. During autophosphorylation, the γ-phosphoryl group of the CA-bound ATP is transferred to a conserved histidine residue located at the DHp domain. In the phosphotransfer reaction, the phosphoryl group is transferred to a conserved aspartate at the cognate RR. In the dephosphorylation reaction, the phosphoryl group of the phosphorylated RR (p-RR) is hydrolyzed.

Bacteriophytochromes (BphPs) are photoreceptor proteins responsible for sensing red/far-red light wavelengths in bacteria [4]. They consist of a photosensory module (PSM) that senses the incident red/far-red light with its biliverdin chromophore and an output module that relays the signal to the cell. The majority of BphPs function as SHKs in TCS, that is, have an HK bidomain as their output module [5]. In response to incident red/far-red light, they switch between structurally and functionally distinct red light-absorbing (Pr) and far-red light-absorbing (Pfr) states. Depending on the BphP type, their dark-adapted resting state is either the Pr state (prototypical) or the Pfr state (bathy) [6].

Most studies to date indicate that BphP-HKs act as kinases in their resting Pr state and that this kinase activity is hindered in the Pfr state [6–9]. However, the majority of BphP studies have concentrated on kinase activity, whereas phosphatase activity of these bifunctional enzymes has been overlooked. This may be partially due to experimental challenges and limitations posed by the detection methods. Conventionally, the kinase activity of BphPs has been studied using radiolabeled [γ-^{32}P]ATP accompanied with sodium dodecyl sulfate–polyacrylamide gel electrophoresis (SDS-PAGE) [10]. This approach, however, requires cautious work and special equipment to handle radioactive reagents. Furthermore, assays that detect fluorescence signals caused, for example, by (dis)appearance of a reaction substrate, binding of a phosphospecific antibody, or FRET [11], may impose challenges when combined with strict illumination requirements of BphP samples.

In this chapter, we describe a label-free method to study BphP HK activities that applies Phos-tag™ acrylamide gel electrophoresis. This method is based on the instructions provided by FUJIFILM Wako Chemicals and modified to meet the special needs of BphP samples. Phos-tag™ is an alkoxide-bridged dinuclear metal complex molecule that is able to trap phosphorylated proteins [12]. Conjugated with acrylamide, Phos-tag™ allows separation of phosphorylated and non-phosphorylated proteins through SDS-PAGE and thus does not require special laboratory equipment. Phos-tag™ acrylamide gel electrophoresis also allows analysis of unstable histidine- and aspartate-phosphorylated proteins and thus is suitable for TCS studies [13]. Figure 1a describes the general overview of the Phos-tag™ detection of BphPs. BphPs are incubated in the presence of ATP and cognate RR, either in

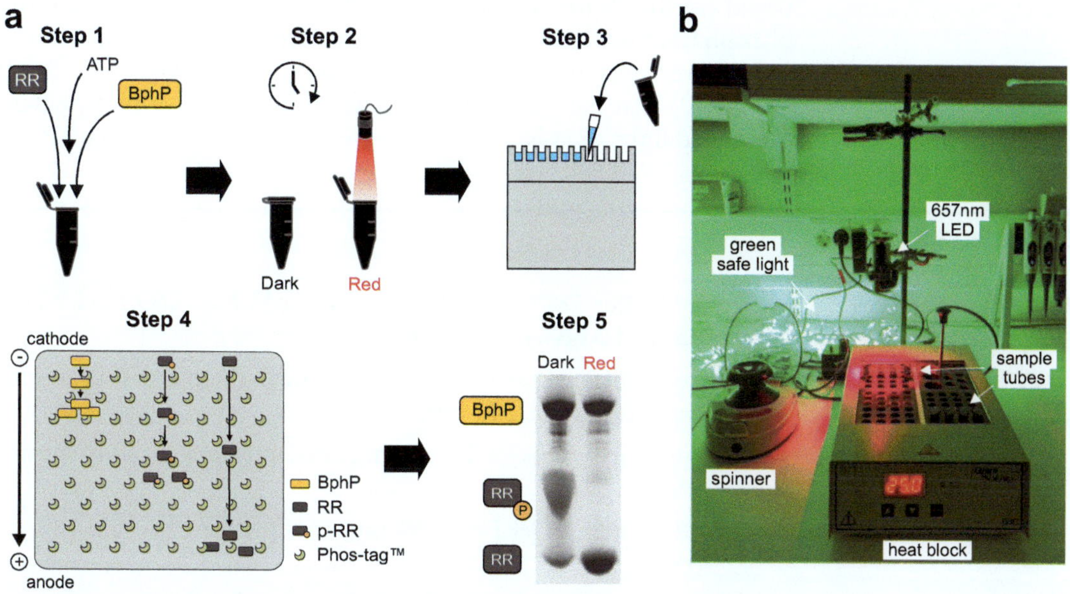

Fig. 1 Workflow and setting of Phos-tag™ activity assays. (**a**) General workflow for the detection of histidine kinase activities of phytochrome proteins using Phos-tag™ polyacrylamide gel electrophoresis. Step 1: Phytochrome (BphP) and its cognate response regulator protein (RR) are added into black microcentrifuge tubes, and the reactions are initiated with the addition of ATP. Step 2: Reactions are incubated for an appropriate time either under constant red light or in darkness. Step 3: Reactions are loaded on a Phos-tag™ polyacrylamide gel. Step 4: Electric current is applied to the gel, and the gel separates proteins based on their sizes. Small proteins move faster on the gel than larger proteins. Additionally, Phos-tag™ interacts with phosphorylated proteins and slows down their migration on the gel. Step 5: The Phos-tag™ gel has separated the phosphorylated and non-phosphorylated RRs from each other and from phytochromes. The gel is ready for visualization and analysis. (**b**) Experimental setting used in our laboratory

the dark or under red light, to support kinase or phosphatase reactions. Samples are then loaded on a Phos-tag™ acrylamide SDS-PAGE gel where they migrate under constant current until they are resolved based on their size and phosphorylation status.

In the following sections, we describe methods to study both kinase and phosphatase activities of BphPs. By applying a bifunctional BphP chimera *Dr*F1, which has a net kinase activity in the dark (Pr) and net phosphatase activity under red light (Pfr) [14, 15], we demonstrate how to detect both kinase activity (*see* Subheading 3.2) and phosphatase activity (*see* Subheading 3.3) using the Phos-tag™ assay. We additionally describe two methods for generating phosphorylated RR (p-RR) for phosphatase activity detection. The first one utilizes the kinase activity of BphP for enzymatic generation of p-RR (*see* Subheading 3.1.1), and the second one uses acetyl phosphate for chemical generation of p-RR (*see* Subheading 3.1.2). Some BphPs may not show kinase activity [16], whereas some RRs may not respond to the acetyl phosphate treatment; hence, the appropriate method can be chosen based on the BphP-RR pair under investigation. To showcase the phosphatase detection of BphPs with acetyl-phosphate-generated p-RR, we assay the activity of *Deinococcus radiodurans* BphP (*Dr*BphP) that has been shown to act as an exclusive phosphatase under red light, that is, in the Pfr state [16] (*see* Subheading 3.3.2).

2 Materials

2.1 Generation of Phosphorylated Response Regulator

2.1.1 Kinase Treatment

1. Assay buffer (2×): 40 mM Tris–HCl pH 8.0, 160 mM NaCl, and 5 mM $MgCl_2$ (*see* **Note 1**).
2. Protein buffer, that is, the buffer where the protein is stored: 30 mM Tris–HCl, pH 8.0.
3. RR protein (*Bj*FixJ): 10–30 mg/mL purified protein (purity >90%).
4. BphP protein (*Dr*F1): 5–30 mg/mL purified protein (purity >90%).
5. ATP: 20 mM ATP in ddH_2O. Use freshly dissolved. Can be stored in −80 °C for later use.

2.1.2 Acetyl Phosphate Treatment

1. Assay buffer (2×): 40 mM Tris–HCl pH 8, 160 mM NaCl, and 5 mM $MgCl_2$ (*see* **Note 1**).
2. Protein buffer, that is, the buffer where the protein is stored: 30 mM Tris–HCl, pH 8.0.
3. RR protein (*Dr*RR): 10–30 mg/mL purified protein (purity >90%).
4. Acetyl phosphate: 200 mM acetyl phosphate (purity ≥85%) in ddH_2O. Use freshly dissolved.
5. Heat block or water bath.
6. Tabletop microcentrifuge.
7. Spin columns (e.g., Amicon Ultra-0.5 Centrifugal Filters—10 kDa cutoff).
8. Microcentrifuge tubes, 1.5 mL.
9. Spectrophotometer (e.g., NanoDrop One, Thermo Scientific) for protein concentration determination.

2.2 Kinase/Phosphatase Reactions

For details of the experimental setting used in our laboratory, *see* Fig. 1b.

1. Assay buffer (2×): 40 mM Tris–HCl pH 8, 160 mM NaCl, and 5 mM $MgCl_2$ (*see* **Note 1**).
2. Protein buffer, that is, the buffer where the protein is stored: 30 mM Tris–HCl, pH 8.0.

3. RR protein (*Bj*FixJ or *Dr*RR): 10–30 mg/mL purified protein (purity >90%).
4. BphP protein (*Dr*F1 or *Dr*BphP): 5–30 mg/mL purified protein (purity >90%).
5. ATP: 20 mM ATP in ddH_2O. Use freshly dissolved. Can be stored in −80 °C for later use.
6. Black microcentrifuge tubes, 1.5 mL.
7. Tabletop centrifuge or spinner.
8. Heat block or water bath.
9. Stand.
10. Red light-emitting device (LED) source (e.g., Mightex SLS-0310-C, 657 nm).
11. Light power meter (e.g., Thorlabs PM100D).
12. Stop buffer supplemented with $ZnCl_2$: 1 mM $ZnCl_2$, 0.01% (w/v) bromophenol blue, 150 mM Tris–HCl pH 6.8, 62.5% (v/v) glycerol, 5% (w/v) SDS, and 5% (v/v) 2-mercaptoethanol (*see* **Note 2**).

2.3 Phos-tag™ Gel Electrophoresis

2.3.1 Casting Phos-tag™ Gels

1. 30% acrylamide/Bis solution.
2. 5 mM Phos-tag™ acrylamide (FUJIFILM Wako Chemicals, AAL-107) (*see* **Note 3**).
3. 1.4 M Bis-Tris–HCl pH 6.8. Store at +4 °C.
4. 10 mM $ZnCl_2$ (purity >98%).
5. 2 M sucrose.
6. ddH_2O.
7. 10% ammonium persulfate (APS).
8. Tetramethylethylenediamine (TEMED).
9. Gel casting system (e.g., Mini-PROTEAN© Tetra Cell Casting Module by Bio-Rad) (*see* **Note 4**).
10. 70% ethanol.

2.3.2 Gel Electrophoresis

1. Buffer A: 0.5 M Tris–HCl pH 7.8, 0.5 M MOPS, and 0.5% (w/v) SDS (*see* **Note 5**). Store at +4 °C in the dark.
2. Buffer B: 0.5 M $NaHSO_3$ (*see* **Note 5**). Store at +4 °C in the dark.
3. Gel running device (e.g., Mini-PROTEAN® Tetra Vertical Electrophoresis Cell and PowerPac Universal Power Supply by Bio-Rad).
4. Staining solution: 0.04% (w/v) SERVA7 Blue R, 25% (v/v) isopropanol, and 10% (v/v) acetic acid.
5. Destaining solution: 10% (v/v) acetic acid.

6. Gel imaging device (e.g., ChemiDoc MP Imaging System by Bio-Rad).

3 Methods

3.1 Generation of Phosphorylated Response Regulator

Subheading 3.1.1 describes the generation of p-RR by utilizing the kinase activity of BphP HK (*Dr*F1). Subheading 3.1.2 describes a chemical alternative for p-RR generation, which can be applied if the RR is susceptible to acetyl phosphate treatment (*see* **Note 6**).

3.1.1 Kinase Treatment

1. Dilute *Bj*FixJ stock to 100 μM with protein buffer (*see* **Note** 7).
2. Dilute *Dr*F1 stock to 10 μM with protein buffer (*see* **Notes** 7 and **8**).
3. Prepare and label a black microcentrifuge tube for the reaction.
4. Prepare a master mix: Add $n \times 5$ μL of assay buffer (2×) to the black microcentrifuge tube (n is the number of desired reactions plus 1 for pipetting margin).
5. Add $n \times 2$ μL of *Bj*FixJ (100 μM) to the tube and mix gently by pipetting.
6. Add $n \times 2$ μL of *Dr*F1 (10 μM) to the same tube and mix gently by pipetting.
7. Spin the sample tubes briefly with a spinner or a centrifuge.
8. Optional: Pre-illuminate the reaction with far-red light to assure the kinase-active Pr state of *Dr*F1 (*see* **Note 9**).
9. Add $n \times 1$ μL of 20 mM ATP, mix briefly by pipetting or tapping, close the tubes, and spin them with a spinner or a centrifuge.
10. Incubate the master mix in the dark, on a heat block set to 25 °C for 30 min (*see* **Note 10**).
11. Continue with the phosphatase reaction depicted in Subheading 3.3.1 (*see* **Note 11**).

3.1.2 Acetyl Phosphate Treatment

1. Prepare the reaction mix in a microcentrifuge tube by adding 100 μL of assay buffer (2×), 10 μL of *Dr*RR, and 90 μL of 200 mM acetyl phosphate (*see* **Note 12**).
2. Incubate at 37 °C for approximately 30 min.
3. Exchange buffer to protein buffer with four subsequent concentration cycles with spin columns, according to manufacturer's instructions. In each cycle, concentrate p-*Dr*RR by centrifugation (e.g., 12,000× *g*, 5 min, +4 °C) and dilute with protein buffer for the maximum supported column volume. For the sample elution, place the spin column upside

down into a new collection tube and centrifuge (e.g., 1000× *g*, 3 min).

4. Check the concentration, for example, with NanoDrop spectrophotometer, and dilute to 100 μM with protein buffer (*see* **Notes** 7 and **13**).

3.2 Kinase Reactions

This subheading describes how to study kinase activity of BphP HKs by applying *Dr*F1, which phosphorylates its cognate RR *Bj*FixJ in the dark Pr state [14, 15] (*see* Fig. 2a).

1. Dilute *Dr*F1 stock to 10 μM with protein buffer (*see* **Notes 7** and **8**).
2. Dilute *Bj*FixJ stock to 100 μM with protein buffer (*see* **Note** 7).
3. Mark a black microcentrifuge tube for each dark and illuminated reaction (*see* **Note 14**).
4. Add 5 μL of assay buffer (2×) to each tube.
5. Add 2 μL of *Bj*FixJ dilution (100 μM) to the tubes and mix by pipetting.
6. Add 2 μL of *Dr*F1 dilution (10 μM) to the tubes and mix by pipetting.
7. Mix briefly by pipetting and spin down with a spinner or centrifuge.
8. Optional: Pre-illuminate the dark reactions with far-red light and illuminated reactions with red light (*see* **Note 15**).
9. Add 1 μL of 20 mM ATP to the reactions, mix quickly, and immediately spin the tubes briefly with a spinner or centrifuge (*see* **Note 16**).
10. Incubate reactions at +25 °C in dark or under constant red light (*see* **Notes 17–19**) for 2–30 min (*see* **Note 20**).
11. Stop the reactions by adding 3 μL of stop buffer to each reaction (*see* **Note 21**).

3.3 Phosphatase Reactions

Subheading 3.3.1 describes a method how to study phosphatase activity of BphPs with both kinase and phosphatase activities. For this, we apply *Dr*F1 that functions as a phosphatase in the red-illuminated (Pfr) state [15] (*see* Fig. 2b). Subheading 3.3.2 describes how to assay phosphatase activity when p-RR is produced chemically (*see* Subheading 3.1.2). For this, *Dr*BphP which functions only as a phosphatase [16] is applied (*see* Fig. 3a, b). The details of the protocols should be modified based on the samples used.

3.3.1 DrF1

1. Mark a black microcentrifuge tube for each dark and illuminated reaction.

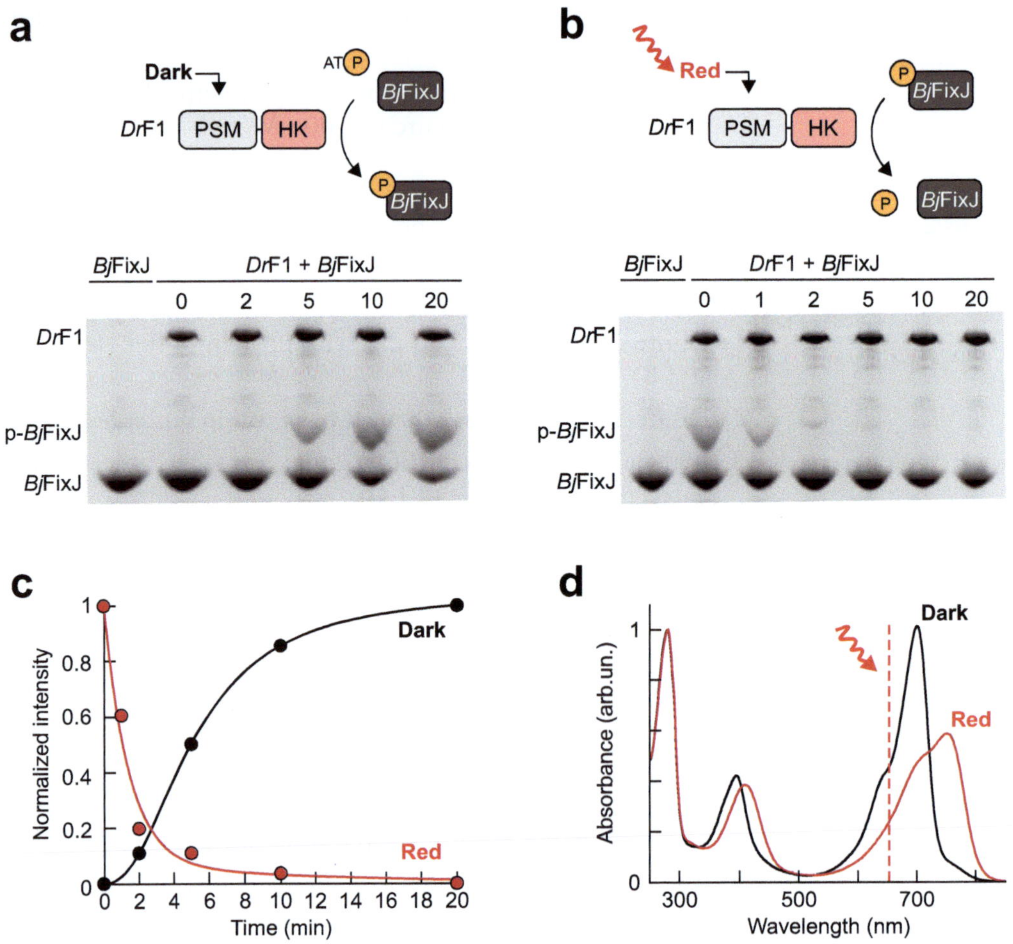

Fig. 2 Kinase and phosphatase activities of *Dr*F1. (**a**) Kinase activity of *Dr*F1 detected with Phos-tag™ assay. *Top*: Schematic presentation showing how *Dr*F1 phosphorylates (by using ATP) its cognate RR *Bj*FixJ in the dark. *Bottom*: Gel showing the appearance of p-*Bj*FixJ at various time points in the dark. (**b**) Phosphatase activity of *Dr*F1 detected with Phos-tag™ assay. *Top*: Schematic presentation showing how *Dr*F1 dephosphorylates its cognate RR *Bj*FixJ under red light. *Bottom*: Gel showing the disappearance of p-*Bj*FixJ at various time points under red light. (**c**) Densitometric analysis of the p-*Bj*FixJ amount in the gels of panel a (dark) and panel b (red), normalized to the maximal p-*Bj*FixJ value of each experiment. (**d**) UV-vis absorption spectra of *Dr*F1 in its dark (Pr) state and red-illuminated (Pfr) state. The spectral position of 657 nm light-emitting device (LED) used for red illumination is shown as a dotted vertical line

2. Divide 10 μL of the p-*Bj*FixJ-containing master mix (*see* Subheading 3.1.1) into the tubes. Spin briefly (*see* **Note 8**).
3. Incubate reactions at +25 °C under constant red light or in the dark (*see* **Notes 17–19**) for 2–30 min (*see* **Note 20**).
4. Stop the reactions by adding 3 μL of stop buffer (*see* **Note 21**).

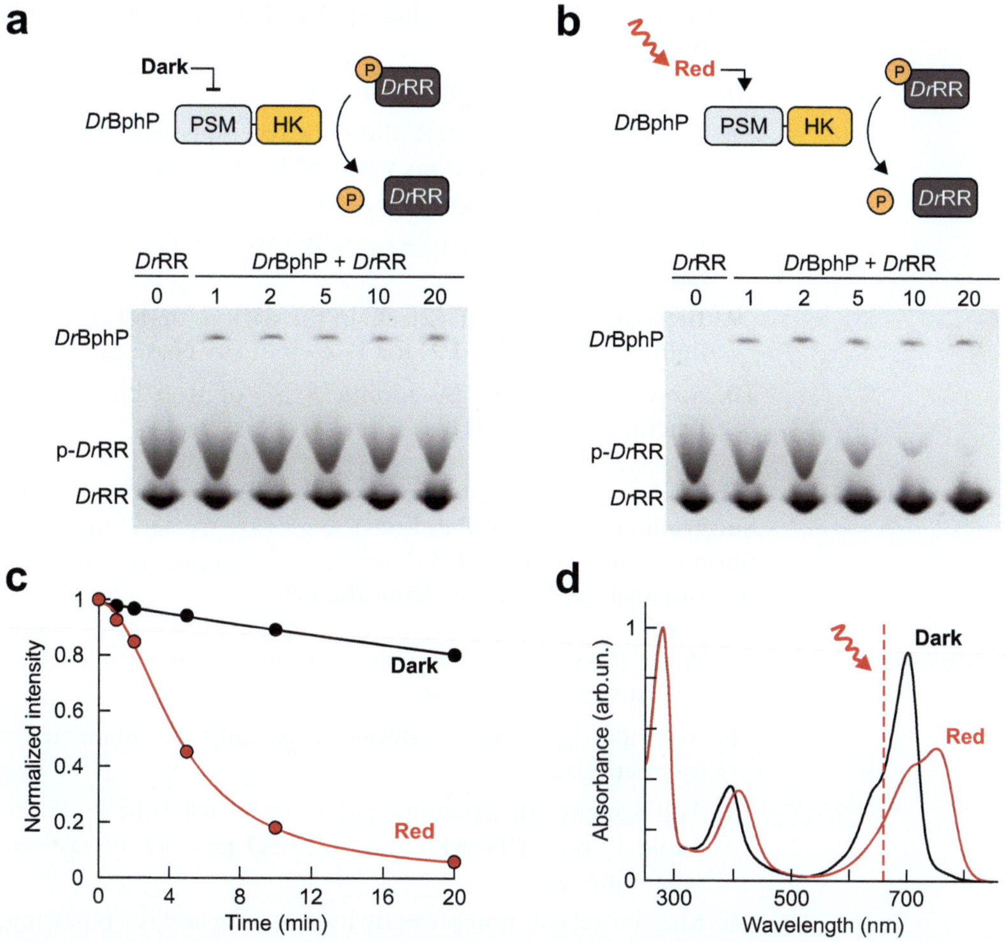

Fig. 3 Phosphatase activity of *Dr*BphP in dark and under red light. (**a**) Activity of *Dr*BphP in dark detected with Phos-tag™ assay. *Top*: Schematic presentation showing how darkness inhibits phosphatase activity of *Dr*BphP, that is, dephosphorylation of p-*Dr*RR. *Bottom*: Gel showing the change in the p-*Dr*RR amount at various time points in the dark. (**b**) Phosphatase activity of *Dr*BphP under red light detected with Phos-tag™ assay. *Top*: Schematic presentation showing how *Dr*BphP dephosphorylates its cognate RR *Dr*RR under red light. *Bottom*: Gel showing the disappearance of p-*Dr*RR at various time points. (**c**) Densitometric analysis of the p-*Dr*RR amount in the gels of panel a (dark) and panel b (red), normalized to the maximal p-*Dr*RR value of each experiment. (**d**) UV-vis absorption spectra of *Dr*BphP in its dark (Pr) state and red-illuminated (Pfr) state. The spectral position of 657 nm LED used for red illumination is shown as a dotted vertical line

3.3.2 DrBphP

1. Dilute *Dr*BphP stock to 1 μM with protein buffer (*see* **Notes 7** and **8**).
2. Mark a black microcentrifuge tube for each dark and illuminated reaction (*see* **Note 14**).
3. Add 5 μL of assay buffer (2×) to each tube.
4. Add 2 μL of p-*Dr*RR (100 μM; *see* Subheading 3.1.2) to the tubes and mix by pipetting.

5. Add 2 μL of *Dr*BphP dilution (1 μM) to the tubes and mix by pipetting.
6. Spin briefly with a spinner or centrifuge.
7. Optional: Pre-illuminate the dark reactions with far-red light and illuminated reactions with red light (*see* **Note 15**).
8. Add 1 μL of 20 mM ATP to the reactions, mix quickly, and immediately spin the tubes briefly with a spinner or centrifuge (*see* **Note 16**).
9. Incubate reactions at +25 °C in the dark or under constant red light (*see* **Notes 17–19**) for 1–20 min (*see* **Note 20**).
10. Stop the reactions by adding 3 μL of stop buffer to each reaction (*see* **Note 21**).

3.4 Phos-tag™ Gels

Subheading 3.4.1 describes casting of the Phos-tag™ gels, whereas Subheading 3.4.2 describes how the gels are treated in gel electrophoresis. Subheading 3.4.3 describes an example on how to quantify and analyze the results from the gel.

3.4.1 Casting Phos-tag™ Gels

1. Wash glass plates first with detergent and then rinse with ddH_2O and 70% ethanol.
2. Assemble gel casting device according to manufacturer's instructions.
3. Add reagents for separating gel into a 50-mL tube according to Table 1. Add 10% APS and TEMED just before casting the separating gel.
4. Mix the gel solution properly by inverting and/or pipetting, let it settle for 20–60 s, and transfer it with a Pasteur pipette between the glass plates.

Table 1
Recipe for Phos-tag™ polyacrylamide gels used in our laboratory. The amounts are for single 1.5-mm gel

Ingredients	Separating gel (9%)	Stacking gel (4%)
30% acrylamide	3 mL	650 μL
1.4 M Bis-Tris–HCl pH 6.8	2.5 mL	1.25 mL
5 mM Phos-tag™ (AAL-107)	25 μL	–
10 mM $ZnCl_2$	100 μL	–
2 M sucrose	1.25 mL	1.25 mL
ddH_2O	3 mL	1.8 mL
10% APS	50 μL	25 μL
TEMED	10 μL	5 μL

5. Add few drops of ddH_2O on top of the separating gel solution between the glasses. This makes the top of the separating gel even.
6. Let the separating gel solidify for >15 min. Solidifying process can be followed from the leftover solution in the 50-mL tube.
7. Remove the small film of water on the top of the separating gel (e.g., by pouring).
8. Add reagents for the stacking gel into a 50-mL tube according to Table 1. Add 10% APS and TEMED just before casting the stacking gel.
9. Mix the gel solution properly by inverting and/or pipetting, let it settle for a 20–60 s, and transfer it with a Pasteur pipette on top of solidified separating gel between the glass plates.
10. Place a comb on top of the glass plates.
11. Let the stacking gels solidify for >15 min (*see* **Note 22**).

3.4.2 Gel Electrophoresis

1. Prepare running buffer by mixing 100 mL of buffer A and 5 mL of buffer B. Fill the volume to 500 mL with ddH_2O (*see* **Note 23**).
2. Assemble gel running device and remove the comb from the gel.
3. Pour the running buffer into the chamber and rinse the wells with running buffer by gentle pipetting.
4. Load the reactions into the wells.
5. Run the gel at 40 mA at room temperature (RT) for 80 min or an appropriate time (*see* **Note 24**).
6. Stain the gel with staining solution and destain the gel with destaining solution (*see* **Note 25**).
7. Image the gel with an imaging device and export the gel images as high-resolution format for publication and analysis.

3.4.3 Analysis

The gels can be quantified for changes in the p-RR amount, which gives an indication of net kinase and phosphatase activity of BphP HK (*see* Figs. 2c and 3c). However, quantifying data from the gels like this should be done with extra care, as it requires high-quality gels and several (>3) experimental repeats. Here, we showcase the analysis process of our example gels.

1. Open the gel image in ImageJ software (https://imagej.net/ij/).
2. Subtract background by using a large rolling ball radius and sliding paraboloid option to ensure uniform background of the gel (*see* **Note 25**).

3. Define a single rectangular selection that encircles all single RR and p-RR bands to be quantified one at a time.
4. Move the selection on top of each protein band and measure their mean intensity.
5. Choose a representative background position (no protein bands or dirt) for the rectangular selection and measure mean intensity value for the selection.
6. Subtract the intensity value of the background selection from the protein band selection and plot the data with MATLAB (MathWorks Inc.) or similar.

4 Notes

1. We have also successfully used alternative buffer: 50 mM Tris–HCl pH 7.8, 10 mM $MgCl_2$, 8 mM 2-mercaptoethanol, and 10% ethylene glycol [16, 17].
2. Other kinds of SDS-PAGE sample loading buffers can be used. Added $ZnCl_2$ is optional but improves the quality of the gel bands.
3. Wako Chemicals offers two methods for gel preparation [12]. We apply here Zn^{2+}–Phos-tag™ SDS-PAGE using a neutral-pH buffer system, which in our hands has proved superior to Mn^{2+}–Phos-tag™ SDS-PAGE using Laemmli's buffer system.
4. Gels with 1.5 mm thickness are easier to handle than thinner (e.g., 0.75 mm) gels.
5. Corresponds to "Solution P" and "Solution O", respectively. However, manufacturer uses shorter versions in their instructions. It is recommended to degas buffer B Solution O (Sol. O) and seal it to minimize its contact with air. Oxidation of this buffer compromises the performance of the gel electrophoresis.
6. We have noted that some RRs (e.g., *Dr*RR) respond well to acetyl phosphatase treatment, but others (e.g., *At*RR) may not [16]. The chemical p-RR generation is more applicable than the enzymatic, as is does not involve other protein components (like BphP) that may complicate subsequent experiments.
7. The concentrations to be used depend on the BphPs under investigation and the study question. 10 × molar amount of RR compared to BphP is a good starting point.
8. As BphPs are sensitive to red/far-red light, it is preferable to work under green safe light.

9. We usually apply separate high-power laser diode modules for far-red (Roithner Lasertechnik GmbH RLDB780-30-3) or red (Roithner Lasertechnik GmbH RLDB650-24-3) pre-illumination.
10. The time required for saturating p-RR concentration depends on the reaction conditions and BphP used. Therefore, an experiment described in Subheading 3.2 may be required to determine the incubation time.
11. The reaction will continue spontaneously, eventually depleting ATP from the sample. Therefore, the phosphatase assay (*see* Subheading 3.3.1) should proceed shortly after the kinase treatment.
12. The amount of *Dr*RR depends on the need. Usually, we treat 10 μL of 30 mg/mL to ensure a sufficient amount of p-*Dr*RR for subsequent experiments. In the case of *Dr*RR, the p-*Dr*RR yield from this reaction is approximately 50% [17].
13. The resulting p-*Dr*RR can be used immediately, or flash-frozen and stored in −80 °C as aliquots. The phosphate moiety has a high tendency to hydrolyze [17], meaning that p-*Dr*RR should be kept on ice and used within a few hours after preparation. p-*Dr*RR is stable over extended time periods if flash-frozen and stored in −80 °C.
14. The reaction components and protein should be kept on ice. The assay reactions can be prepared either on ice or at room temperature.
15. This can be done with separate lasers (*see* **Note 9**). A pre-illumination step is applicable for samples that have slow dark reversion, that is, reversion from the illuminated state to the dark resting state.
16. Adding ATP starts the reaction, and therefore, the subsequent treatments should be done relatively quickly. Alternatively, ATP addition can be conducted on ice, which allows more time for this step by slowing the reaction.
17. Determine the light intensity needed for BphP photoconversion spectroscopically, that is, the required intensity to reach and maintain the desired Pr or Pfr population. Verify the light intensity used in the activity assays with a power meter. For example, we have used red light peak intensity of approximately 13 mW/cm^2 before [17].
18. You can also facilitate the saturation of the Pfr state by briefly pre-illuminating the samples with a red laser (*see* **Note 9**) and then place the samples under red LED illumination.
19. Illumination conditions (intensity and wavelength) depend on the BphP to be studied (*see* Figs. 2d and 3d) and on the

research question. Usually, the illumination condition should be saturating but not harmful to the sample.

20. The time points and overall reaction time depend on assay conditions and BphP used. Therefore, the best reaction schedule would be determined by trial and error.
21. The samples should be run on the gel soon after the reaction, due to potential hydrolysis of the phosphate in p-RR. Alternatively, samples can be frozen and run at a later time.
22. Although the gel performance may be maintained for several months according to manufacturer's instructions, we prepare the gels fresh to assure best performance.
23. The running buffer should be stored in +4 °C. It can be reused, but the best results have been attained when the buffer is fresh.
24. We usually include a pre-stained protein marker ladder (e.g., PageRuler™, Thermo Scientific) in the gel, which helps in tracking the sample migration and estimation of the unphosphorylated protein sizes.
25. The longer the gel is destained, the better its quality for post-analysis. If destaining is conducted properly, the background subtraction step in the gel image analysis (*see* Subheading 3.4.1) is not necessary.

Acknowledgments

Financial support was provided by the Research Council of Finland (grants 330678 and 358092). We acknowledge Elina Multamäki (University of Oulu) and Moona Kurttila (University of Jyväskylä) for preceding assay development.

References

1. Capra EJ, Laub MT (2012) Evolution of two-component signal transduction systems. Ann Rev Microbiol 66:325–347
2. Möglich A (2019) Signal transduction in photoreceptor histidine kinases. Protein Sci 28: 1923–1946
3. Buschiazzo A, Trajtenberg F (2019) Two-component sensing and regulation: how do histidine kinases talk with response regulators at the molecular level? Ann Rev Microbiol 73:507–528
4. Takala H, Edlund P, Ihalainen JA et al (2020) Tips and turns of bacteriophytochrome photoactivation. Photochem Photobiol Sci 19:1488–1510
5. Gourinchas G, Etzl S, Winkler A (2019) Bacteriophytochromes – from informative model systems of phytochrome function to powerful tools in cell biology. Curr Opin Struct Biol 57:72–83
6. Karniol B, Vierstra RD (2003) The pair of bacteriophytochromes from Agrobacterium tumefaciens are histidine kinases with opposing photobiological properties. Proc Natl Acad Sci USA 100:2807–2812
7. Huber C, Strack M, Schultheiß I et al (2024) Darkness inhibits autokinase activity of bacterial bathy phytochromes. J Biol Chem 300: 107148
8. Lamparter T, Michael N, Mittmann F et al (2002) Phytochrome from Agrobacterium tumefaciens has unusual spectral properties

and reveals an N-terminal chromophore attachment site. Proc Natl Acad Sci USA 99: 11628–11633

9. Giraud E, Zappa S, Vuillet L et al (2005) A new type of bacteriophytochrome acts in tandem with a classical bacteriophytochrome to control the antennae synthesis in Rhodopseudomonas palustris. J Biol Chem 280:32389–32397
10. Bhoo SH, Davis SJ, Walker J et al (2001) Bacteriophytochromes are photochromic histidine kinases using a biliverdin chromophore. Nature 414:776–779
11. Wang Y, Ma H (2015) Protein kinase profiling assays: a technology review. Drug Discov Today Technol 18:1–8
12. Kinoshita E, Kinoshita-Kikuta E, Takiyama K et al (2006) Phosphate-binding tag, a new tool to visualize phosphorylated proteins*. Mol Cell Proteomics 5:749–757
13. Kinoshita-Kikuta E, Kinoshita E, Eguchi Y et al (2015) Functional characterization of the receiver domain for phosphorelay control in hybrid sensor kinases. PLoS One 10:e0132598
14. Multamäki E, García de Fuentes A, Sieryi O et al (2022) Optogenetic control of bacterial expression by red light. ACS Synth Biol 11: 3354–3367
15. Meier SSM, Multamäki E, Ranzani AT et al (2024) Leveraging the histidine kinase-phosphatase duality to sculpt two-component signaling. Nat Commun 15:4876
16. Multamäki E, Nanekar R, Morozov D et al (2021) Comparative analysis of two paradigm bacteriophytochromes reveals opposite functionalities in two-component signaling. Nat Commun 12:4394
17. Kurttila M, Etzl S, Rumfeldt J et al (2022) The structural effect between the output module and chromophore-binding domain is a two-way street via the hairpin extension. Photochem Photobiol Sci 21:1881–1894

Chapter 19

Assessing the Role of PPKs in Phosphorylating Phytochrome A Using the Protoplast Transient Expression System

Ziyi Feng and Jigang Li

Abstract

Phytochrome A (phyA) is the far-red (FR) light photoreceptor in plants, and its phosphorylation status plays a critical role in modulating its protein activity and function. The protoplast transient expression system has become an increasingly efficient approach for analyzing promoter activities, protein subcellular localization, protein–protein interactions, etc. Recently, we employed the protoplast transient expression system to assess the role of the protein kinases PHOTOREGULATORY PROTEIN KINASES (PPKs) in phosphorylating phyA. TANDEM ZINC-FINGER/PLUS3 (TZP), a phyA-interacting protein previously shown to mediate phyA phosphorylation in vivo, was coexpressed with phyA and PPKs in *Arabidopsis* protoplasts. After the treatment with FR light, we observed that PPKs can directly phosphorylate phyA and that TZP enhances PPKs-mediated phosphorylation of phyA in *Arabidopsis* protoplasts. In this chapter, we will describe the detailed experimental procedures and the key points for implementing this system. This system may also be feasible for validating the relationship between other kinases and their potential substrates.

Key words phyA, PPKs, TZP, Protoplast, Phosphorylation

1 Introduction

Photoreceptors perceive light signals and then transmit them to downstream signaling components, triggering various biological responses in plants. Phytochromes primarily absorb and respond to red (R, 600–700 nm) and far-red (FR, 700–750 nm) light and function as the molecular light switches that regulate multiple aspects of plant growth and development [1–4]. Notably, phytochrome A (phyA) is the primary FR light photoreceptor and also contributes to R light signal transduction, playing a crucial role in regulating plant adaptations to the light environment [3, 4]. Phytochromes exist *in vivo* in two interconvertible forms, that is, R light-absorbing Pr form and FR light-absorbing Pfr form [5, 6]. It has

Andreas Winkler (ed.), *Phytochromes: Methods and Protocols*, Methods in Molecular Biology, vol. 2970,
https://doi.org/10.1007/978-1-0716-4791-2_19,

long been believed that the Pfr form is the biologically active form; however, the Pr form was recently suggested to have biological activity when localized in the nucleus [7].

As early as the 1980s, oat (*Avena sativa*) phyA was established as a model for studying phyA phosphorylation [8, 9]. It was demonstrated that oat phyA is phosphorylated *in vivo*, and several key residues, including Ser8 and Ser18 in the N-terminal domain and Ser599 in the hinge region of oat phyA, were shown to be phosphorylation sites [10–14]. The phosphorylation status of these residues influences phyA's activity, stability, subcellular localization, and interactions with downstream signaling components, thereby regulating plant responses to light [15–18]. However, the phosphorylation sites of *Arabidopsis* phyA have not yet been identified. A previous study examined the patterns of phyA proteins by immunoblotting and observed that *Arabidopsis* phyA proteins exhibited two distinct forms, corresponding to the phosphorylated (slower-migrating phyA band) and unphosphorylated phyA (faster-migrating phyA band) in FR light [19]. Notably, it was observed that the abundance of phosphorylated phyA correlated well with the extent of phyA activity, suggesting that phosphorylated phyA may represent a more active form [18]. Recently, PHOTOREGULATORY PROTEIN KINASES (PPKs), a family of four light-regulated kinases (PPK1–4) in *Arabidopsis*, were identified as the first class of protein kinases that directly phosphorylate phyA [20]. In addition, TANDEM ZINC-FINGER/PLUS3 (TZP), a recently characterized phyA-interacting protein, undergoes liquid–liquid phase separation upon FR light exposure and facilitates PPKs-mediated phyA phosphorylation by promoting their colocalization and interaction [20].

To date, two *in vitro* phosphorylation assay systems have been used to detect phytochrome phosphorylation. The most commonly used method involves the addition of radioactive γ-^{32}P-ATP to serve as a phosphate donor for substrate proteins, followed by the detection via autoradiography [20–22]. Another method utilizes ATP-γ-S, which undergoes alkylation to convert thiophosphorylation sites into thiophosphate esters, which can be recognized by thiophosphate esters-specific antibodies in immunoblotting [23, 24]. For both approaches, the respective proteins should be purified and then added into the *in vitro* kinase assay systems. Phytochrome proteins can be purified from various hosts, such as yeast strains [21], *Escherichia coli* (*E. coli*) [25], or insect cells [26]. However, efficient transient systems for detecting phytochrome phosphorylation *in vivo* remain limited. Currently, sodium dodecyl sulfate–polyacrylamide gel electrophoresis (SDS-PAGE) or Phos-tag gel systems are often employed to separate the total proteins from wild-type (or kinase mutant) seedlings grown under specific light conditions, thus assessing the phosphorylation status of phytochrome proteins *in vivo*.

The protoplast transient expression system allows the introduction of exogenous DNA, RNA, or proteins into protoplasts via polyethylene glycol (PEG)–calcium fusion, electroporation, or microinjection [27]. This system has been widely used in promoter activity, protein subcellular localization, and protein–protein interaction analyses [28, 29]. It was also employed to investigate plant physiological processes and signal transduction pathways, including light signaling, sugar signaling, stress responses, and auxin signaling [30–32]. In this chapter, we describe an experimental procedure using a protoplast-based system to investigate PPKs-mediated phosphorylation of *Arabidopsis* phyA. Specifically, we demonstrate that TZP promotes PPKs-mediated phyA phosphorylation in *Arabidopsis* protoplasts in FR light (Fig. 1). This system may also be used for validating the role of additional kinases involved in phytochrome phosphorylation and the relationship between other kinases and their potential substrates.

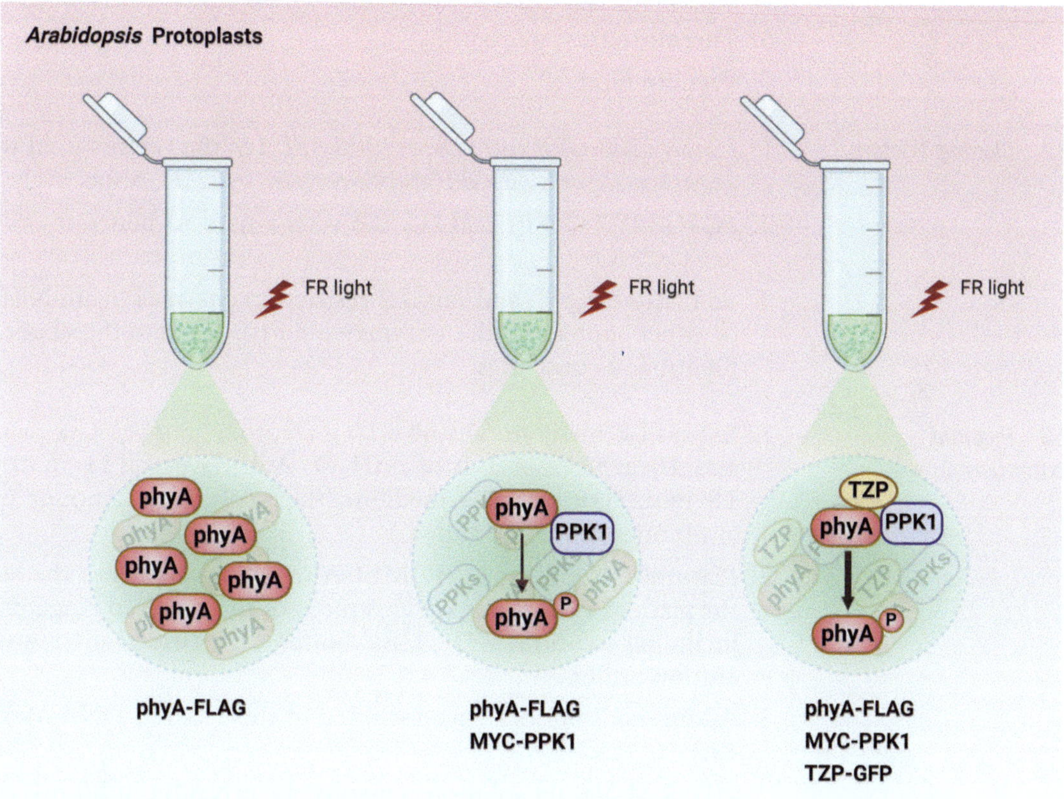

Fig. 1 Schematic illustration of the protoplast transient expression system to show PPK1/TZP-mediated phosphorylation of phyA in FR light. When phyA-FLAG is transiently coexpressed with MYC-PPK1 in *Arabidopsis* protoplasts and treated with FR light, phyA-FLAG is phosphorylated by MYC-PPK1. The presence of TZP-GFP further enhances PPK1-mediated phosphorylation of phyA

2 Materials

Prepare all solutions using ultrapure water (prepared by purifying deionized water to attain a sensitivity of 18 MΩ-cm at 25 °C) and analytical-grade reagents. Unless otherwise specified, prepare and store all reagents at room temperature. When handling β-mercaptoethanol, adhere to safety precautions and work within a fume hood. Do not mix waste solvents, and dispose of the waste solvents and β-mercaptoethanol-containing solutions in accordance with local regulations.

2.1 Accessories and Supplies

1. Laminar flow hood.
2. Incubator and shaker set to 37 °C.
3. Temperature-controlled dark room with a weak green safety light (*see* **Note 1**).
4. Light growth chamber (Percival Scientific) with a fluence rate of 50 μmol m^{-2} s^{-1} for continuous FR light.
5. Cryogenic high-speed centrifuge.
6. Thermostatic water bath.
7. Vacuum filter.

2.2 Cloning Vectors

1. *pSuper1300-GFP* and *pSuper1300-3FLAG*: the vectors used to generate *Super:TZP-GFP* and *Super:phyA-FLAG*, respectively.
2. *pCM1307-2×35SPTL-MYC*: the vector used to generate *35S:MYC-PPK1*.
3. *Sal*I, *Xba*I, *Kpn*I, and *BamH*I restriction enzymes (from NEB or other suppliers): the enzymes used to generate the above-mentioned constructs.

2.3 Plasmid Extraction

1. Liquid LB medium: Dissolve 10 g tryptone, 5 g yeast extract, and 10 g NaCl in 1 L of ddH_2O. Autoclave at 121 °C for 15 min. (For selective medium, the appropriate amount of antibiotics should be added.)
2. Plasmid DNA Maxi Kit (OMEGA or other suppliers): the kit for plasmid extraction contains most of the required reagents, including solutions I–III, HBC buffer, DNA wash buffer, and elution buffer.
3. Additional solutions (not included in the Plasmid DNA Maxi Kit):
   (i) 3 M NaOH solution: Dissolve 12 g NaOH in 80 mL of ddH_2O, add ddH_2O to a final volume of 100 mL, and mix thoroughly (*see* **Note 2**).

(ii) 5 M NaCl solution: Dissolve 29.2 g NaCl in 80 mL of ddH_2O, add ddH_2O to a final volume of 100 mL, and mix thoroughly.

(iii) 75% (v/v) ethanol: Measure 75 mL of 100% ethanol, add ddH_2O to a final volume of 100 mL, and mix thoroughly.

2.4 Protein Expression in Arabidopsis *Mesophyll Protoplasts*

1. 0.2 M KCl solution: Dissolve 1.491 g KCl in 80 mL of ddH_2O and add ddH_2O to a final volume of 100 mL.
2. 2 M KOH solution: Dissolve 11.222 g KOH in 80 mL of ddH_2O and add ddH_2O to a final volume of 100 mL (*see* **Note 2**).
3. 0.5 M 4-morpholineethanesulfonic acid (MES) (pH 5.7) solution: Dissolve 9.762 g MES in 80 mL of ddH_2O, adjust the pH to 5.7 with 2 M KOH, and add ddH_2O to a final volume of 100 mL.
4. 1 M $CaCl_2$ solution: Dissolve 14.701 g $CaCl_2{\cdot}2H_2O$ in 80 mL of ddH_2O and add ddH_2O to a final volume of 100 mL.
5. 1.54 M NaCl solution: Dissolve 9 g NaCl in 80 mL of ddH_2O and add ddH_2O to a final volume of 100 mL.
6. Enzyme solution: Mix 2.25 g cellulase R10 (YaKult Honsha), 0.45 g macerozyme R10 (YaKult Honsha), 1.09 g mannitol, 150 μL 0.2 M KCl, and 600 μL 0.5 M MES (pH 5.7) in 12.5 mL of ddH_2O. Heat in a 55 °C water bath for 10 min; after cooling, add 150 μL 1 M $CaCl_2$, 0.015 g bovine serum albumin (BSA), and 6 μL 14.3 M β-mercaptoethanol, and add ddH_2O to a final volume of 15 mL.
7. W5 solution: Mix 10 mL 1.54 M NaCl, 12.5 mL 1 M $CaCl_2$, 250 μL 0.2 M KCl, and 1 mL 0.5 M MES (pH 5.7); adjust the pH to 5.8 with 2 M KOH; and add ddH_2O to a final volume of 100 mL.
8. WI solution: Add 0.1085 g mannitol, 1 mL 0.2 M KCl, and 800 μL 0.5 M MES (pH 5.7) in 80 mL of ddH_2O. Adjust pH to 5.7 with 2 M KOH, and add ddH_2O to a final volume of 100 mL.
9. MMG solution [27]: Dissolve 7.3 g mannitol, 0.305 g $MgCl_2$, and 0.1 g MES in 80 mL ddH_2O. Adjust pH to 5.6 with 2 M KOH, and add ddH_2O to a final volume of 100 mL.
10. 40% (w/v) PEG4000 solution: Add 0.364 g mannitol, 4 g PEG4000, and 1 mL 1 M $CaCl_2$ in 4 mL of ddH_2O. After being fully dissolved, add ddH_2O to a final volume of 10 mL (*see* **Note 3**).

All the above solutions should be filtered and sterilized using 0.45-μm filter membranes and stored at 4 °C.

2.5 Protein Extraction and Immunoblotting

1. 1 M Tris-HCl (pH 6.8/ pH 7.5) solution: Dissolve 12.11 g Tris base in 80 mL of ddH_2O, adjust the pH to 6.8 or 7.5 with concentrated HCl, and add ddH_2O to a final volume of 100 mL (*see* **Note 4**).
2. 1 M $MgCl_2$ solution: Dissolve 20.33 g $MgCl_2$ in 80 mL of ddH_2O and add ddH_2O to a final volume of 100 mL.
3. 0.5 M EDTA (pH 8.0) solution: Dissolve 18.61 g EDTA in 80 mL of ddH_2O, adjust the pH to 8.0 with solid NaOH, and add ddH_2O to a final volume of 100 mL (*see* **Notes 2** and **5**).
4. Non-denatured protein extraction buffer: Mix 50 mL 1 M Tris-HCl (pH 7.5), 30 mL 5 M NaCl, 10 mL 1 M $MgCl_2$, 2 mL 0.5 M EDTA (pH 8.0), and 1 mL NP-40. Add ddH_2O to a final volume of 1 L and store at 4 °C.
5. 6× SDS loading buffer: Mix 37.5 mL 1 M Tris-HCl (pH 6.8), 12 g sodium dodecyl sulfate (SDS), 60 mL glycerol, 60 mg bromophenol blue (BPB), and 9.3 g dithiothreitol (DTT). Add ddH_2O to a final volume of 100 mL, aliquot into 1.5-mL centrifuge tubes, and store at −20 °C.
6. Alkaline phosphatase (AP) buffer: Dissolve 12.1 g Tris base, 5.84 g NaCl, and 14.4 g $MgCl_2 \cdot 6H_2O$ in 800 mL ddH_2O. Adjust the pH to 9.5 with HCl, and add ddH_2O to a final volume of 1 L. Filter-sterilize using 0.45-μm filter membranes and store at room temperature.
7. 100× NBT solution: Dissolve 0.1 g of nitro blue tetrazolium (NBT) in 100 mL 70% DMF. Mix thoroughly, aliquot into 1.5-mL centrifuge tubes, and store at −20 °C in the dark.
8. 100× BCIP solution: Dissolve 1 g of 5-bromo-4-chloro-3'-indolyl phosphate (BCIP) in 60 mL N,N-dimethyl formamide (DMF). Mix thoroughly, aliquot into 1.5-mL centrifuge tubes, and store at −20 °C in the dark.
9. Lambda protein phosphatase (λ-PPase) (from New England Biolabs (NEB) or other suppliers) is used for phosphatase treatment. The 10× NEB buffer: NEB buffer for protein metallophosphatases (PMP) and 10 mM $MnCl_2$ solutions are supplied with the product.
10. Primary antibodies include anti-phyA (1:1000 [v/v]) [33], anti-TZP (1:1000 [v/v]) [34], anti-MYC (1:2000 [v/v], catalog no. MF083; Mei5 Biotechnology), anti-HSP (1:2000 [v/v], catalog no. AbM51099-31-PU; Beijing Protein Innovation), anti-rabbit IgG (whole molecule), alkaline phosphatase (AP, 1:8000 [v/v], catalog no. A3687; Sigma-Aldrich), and goat anti-mouse IgG (whole molecule), horseradish peroxidase (HRP, 1:8000 [v/v], catalog no. A9044; Sigma-Aldrich) (*see* **Note 6**).

2.6 Primers

3 Methods

The overall workflow of this section includes the following: generate the required constructs, extract the amounts of plasmids needed for protoplast transformation, transform the plasmids into *Arabidopsis* mesophyll protoplasts, express the respective proteins transiently in *Arabidopsis* protoplasts (with appropriate light treatments if necessary), extract the proteins from *Arabidopsis* protoplasts (with phosphatase treatment if necessary), and perform immunoblotting to analyze the patterns of the respective proteins.

3.1 Cloning and Construct Preparation

1. Generate the *Super:phyA-FLAG* construct by amplifying the full-length phyA coding sequence via polymerase chain reaction (PCR) using the primers phyA-F and phyA-R (Table 1) and cloning into the *Sal*I site of the *pSuper1300-3FLAG* vector [35].
2. Generate the *Super:TZP-GFP* construct by amplifying the full-length TZP coding sequence via PCR using the primers TZP-F and TZP-R and cloning into the *Xba*I-*Kpn*I sites of the *pSuper1300-GFP* vector [35].
3. Generate the *35S:MYC-PPK1* construct by amplifying the full-length PPK1 coding sequence via PCR using the primers PPK1-F and PPK1-R and cloning into the *BamH*I-*Sal*I sites of the *pCM1307-2×35SPTL-MYC* vector [35].

3.2 Plasmid Extraction

1. Inoculate 5-mL LB medium containing the appropriate antibiotics with several colonies. This step should be performed in a laminar flow hood to prevent contamination by other microorganisms (*see* **Note** 7).
2. Shake the bacterial culture at 200 rpm at 37 °C for 8–10 h.

Table 1
Primers used to generate the constructs

Construct	Primer name	5′–3′
Super:phyA-FLAG	phyA-F	ACT<u>GTCGAC</u>ATGTCAGGCTCTAGGCCGAC
	phyA-R	ATC<u>GTCGAC</u>CCTTGTTTGCTGCAGCGAGTTC
Super:TZP-GFP	TZP-F	ACTG<u>TCTAGA</u>ATGGGAGATGGAGATGAGCAA
	TZP-R	AGTC<u>GGTACC</u>AAAGCCTAACATTTTTCTCTGC
35S:MYC-PPK1	PPK1-F	TGACAG<u>GGATCC</u>ATGCCGGAGCTTCGCCGT
	PPK1-R	TGACAG<u>GTCGAC</u>TCAAGATACAGTTCGGCCATA

3. Transfer 2.5 mL of the bacterial culture into 250 mL of LB medium containing the corresponding antibiotic. Incubate overnight with shaking at 200 rpm at 37 °C.
4. Measure the OD_{600} of the overnight culture (the OD_{600} should be up to ~1.5). Transfer 250 mL of the overnight culture into an appropriate centrifuge bottle and centrifuge at 3500× *g* for 20 min at room temperature.
5. Decant and discard the supernatant.
6. Add 12 mL Solution I containing RNase A. Vortex briefly and mix thoroughly to fully resuspend the cells (*see* **Note 8**). Transfer the cell suspension to a 50-mL centrifuge tube capable of withstanding 10,000× *g*.
7. Add 12 mL Solution II. Invert and rotate the tube gently 10-12 times until a cleared lysate is obtained. If necessary, incubate at room temperature for 2–3 min (*see* **Note 9**).
8. Add 16 mL Solution III. Invert and rotate the tube gently until white flocculent precipitates form. If necessary, incubate at room temperature for 2–3 min (*see* **Note 10**).
9. Centrifuge at 10,000× *g* for 10 min at room temperature (preferably at 4 °C). Transfer the suspension to a fresh 50-mL centrifuge tube.
10. Add 3 mL 3 M NaOH to the HiBind DNA Maxi Column. Let sit at room temperature for 4 min and then centrifuge at 1800× *g* for 3 min at room temperature. Discard the filtrate.
11. Carefully transfer the cleared supernatant from **step 9** into the HiBind DNA Maxi Column. Centrifuge at 1800× *g* for 5 min at room temperature and discard the filtrate.
12. Repeat **step 11** until all cleared supernatant has passed through the HiBind DNA Maxi Column.
13. Add 10 mL HBC buffer to the HiBind DNA Maxi Column. Centrifuge at 1800× *g* for 3 min at room temperature and discard the filtrate (*see* **Note 11**).
14. Add 15 mL DNA wash buffer to the HiBind DNA Maxi Column. Centrifuge at 1800× *g* for 3 min at room temperature and discard the filtrate (*see* **Note 12**). Repeat this step for a second wash with DNA wash buffer.
15. Centrifuge at 1800× *g* for 10 min at room temperature to completely dry the HiBind column matrix.
16. Transfer the HiBind DNA Maxi Column to a nuclease-free 50-mL centrifuge tube. Add 6 mL elution buffer to the center of the column matrix and let sit at room temperature for 5 min.
17. Centrifuge at 1800× *g* for 5 min at room temperature.

18. Add the 260 μL 5 M NaCl and 4.4 mL isopropanol to the collected eluate. Vortex briefly, then evenly distribute into 6× 2-mL centrifuge tubes.
19. Centrifuge at 10,000× *g* for 30 min at 4 °C. Discard the supernatant.
20. Add 1 mL prechilled 75% (v/v) ethanol to wash the DNA pellet by gently inverting the tube.
21. Centrifuge at 10,000× *g* for 10 min at 4 °C and discard the supernatant. Repeat **steps 20** and **21** for a second wash with ethanol (*see* **Note 13**).
22. Allow the DNA pellet to air-dry or dry in a laminar flow hood.
23. Add 50 μL ddH_2O to each tube to fully dissolve the DNA pellet and then combine into a single tube. Determine the DNA concentration and store at −80 °C.

3.3 Protein Expression in Arabidopsis *Mesophyll* Protoplasts

1. Preheat a thermostatic water bath to 55 °C. Prepare 15 mL of protoplast transient enzyme solution as described above, filter-sterilize using a 0.45-μM filter and store in a sterile Petri dish (*see* **Note 14**).
2. Harvest healthy *Arabidopsis* plants grown under a 16-h/8-h (light/dark) photoperiod. Select the fifth to eighth leaves, cut them into fine strips using a sterile blade, and immerse them in the enzyme solution (*see* **Note 15**).
3. Wrap the Petri dish with aluminum foil and subject it to a vacuum infiltration for 30 min. Incubate the leaf strips in enzymatic solution at room temperature in the dark for 3 h.
4. Gently shake the Petri dish to facilitate the release of protoplasts from the leaf tissue. Continue shaking until the enzymatic hydrolysate turns green. Filter the protoplasts through a 100-μm nylon mesh into a 50-mL round-bottom centrifuge tube (*see* **Note 16**).
5. Centrifuge horizontally at 100× *g* for 2 min at room temperature. Discard the supernatant and keep the pellet on ice (*see* **Note 17**).
6. Resuspend the protoplasts pellet in 10 mL of prechilled W5 solution and shake gently. Incubate on ice for 30 min (*see* **Note 18**).
7. Mix equal amounts of the respective plasmids (e.g., 10 μg per plasmid) in 2-mL round-bottom centrifuge tubes (*see* **Note 19**).
8. Repeat **step 5** to centrifuge the protoplasts again. Discard the supernatant and keep the pellet on ice.
9. Add an appropriate volume of MMG solution according to the amount of settled protoplasts and mix gently.

10. Pipette a small volume of protoplast suspension onto a slide. Observe the protoplasts under a microscope to assess their status and density (*see* **Note 20**).
11. Slowly add 200 μL of protoplast suspension to the mixed plasmids from **step** 7. Mix gently and then add 210 μL of 40% (w/v) PEG4000 solution. Invert the tubes gently from side to side to ensure thorough mixing. Incubate at room temperature for 10 min to allow transformation (*see* **Note 21**).
12. After transformation, add 820 μL of prechilled W5 solution. Invert the tubes slowly several times to stop the reaction.
13. Centrifuge horizontally at 100× *g* for 2 min at room temperature. Carefully aspirate and discard the supernatant.
14. Slowly add 1 mL of WI solution [27] along the tube wall. Mix gently by inverting the tubes several times. Lay the tubes flat in a light-protected box at 21 °C and incubate overnight for 16 h.
15. On the morning of the next day, mix the protoplasts gently by inverting the tubes. Transfer the tubes to a FR light chamber for specified light treatment (*see* **Note 22**).
16. After FR light treatment, gently mix the protoplasts again by inversion. Place the tubes in a suitable centrifuge under dark condition. Centrifuge horizontally at 100× *g* for 2 min at room temperature. Carefully remove the supernatant in the dark, immediately freeze the pellet in liquid nitrogen, and store at −80 °C for later use (*see* **Note 23**).

3.4 Protein Extraction and Immunoblotting

The following steps should be performed on ice to prevent protein degradation.

1. Add 100 μL of non-denatured protein extraction buffer to the protoplasts collected in **step 16** of the above section. Vortex to dissolve the pellet thoroughly.
2. Centrifuge at 10,000× *g* for 5 min at 4 °C and transfer the supernatant to a fresh 1.5-mL tube.
3. Repeat **step 2** to ensure complete protein extraction.
4. Lambda protein phosphatase (λ-PPase) treatment (optional):
   (i) Transfer 40 μL of the protein solution to a fresh 1.5-mL tube.
   (ii) Add 5 μL of 10× NEB buffer and 5 μL of 10 mM $MnCl_2$ to make a total reaction volume of 50 μL.
   (iii) Add 1 μL of active or inactive (boiled) λ-PPase (*see* **Note 24**) (Fig. 2).
   (iv) Incubate at 30 °C for 30 min.
5. Mix the supernatant with 6× SDS loading buffer and vortex thoroughly. Boil the samples at 100 °C for 10 min.

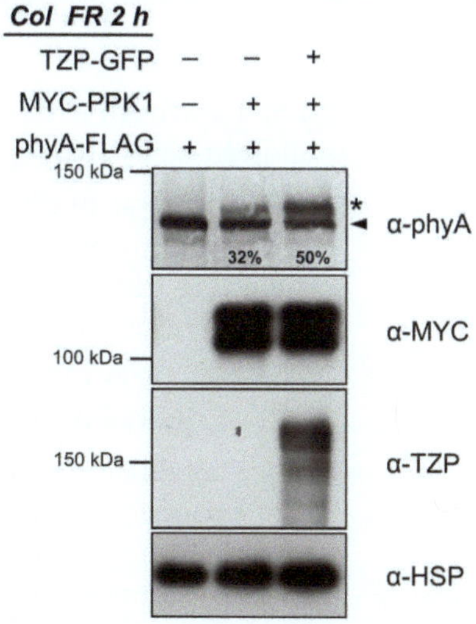

Fig. 2 PPK1 phosphorylates phyA in *Arabidopsis* protoplasts. *Super:phyA-FLAG* and *35S:MYC-PPK1* plasmids were transiently cotransfected into Col protoplasts. The total proteins were extracted and treated with active (+) or inactive (boiled) or without (–) λ-PPase for 30 min and then analyzed by immunoblots. The asterisk and arrowhead indicate the phosphorylated and unphosphorylated phyA forms, respectively

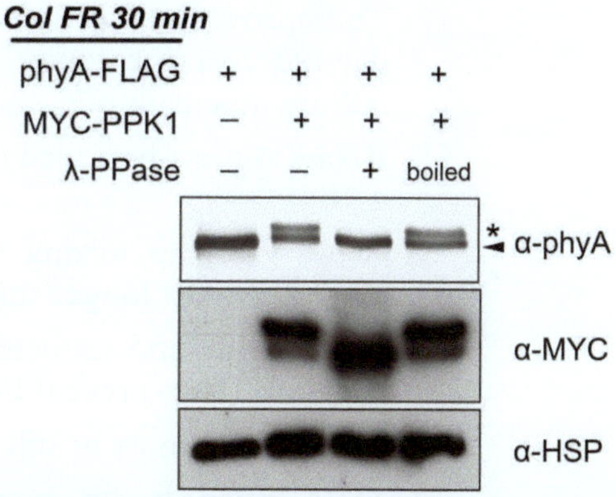

Fig. 3 TZP promotes PPK1-mediated phosphorylation of phyA in *Arabidopsis* protoplasts. *Super:phyA-FLAG*, *Super:TZP-GFP*, and *35S:MYC-PPK1* plasmids were transiently cotransfected into Col protoplasts. The percentages of phosphorylated phyA in the immunoblots were calculated by dividing the intensities of phosphorylated phyA bands by the total phyA band intensities. The asterisk and arrowhead indicate the phosphorylated and unphosphorylated phyA forms, respectively

6. Perform SDS-PAGE and immunoblotting to analyze the protein patterns of phyA-FLAG, TZP-GFP, and MYC-PPK1 in protoplasts using antibodies against phyA, TZP, MYC, and HSP, respectively (*see* **Note 25**) (Figs. 2 and 3).

4 Notes

1. The darkroom with dim green light should be strictly used for all light-sensitive steps to prevent unintended light exposure.
2. Due to the strongly alkaline and corrosive nature of NaOH/KOH, appropriate personal protective equipment (PPE), including gloves, safety goggles, and masks, must be worn to prevent potential injury.
3. The 40% (w/v) PEG4000 solution should be freshly prepared and equilibrated to room temperature prior to use.
4. Since an alkaline reagent cannot be used for pH adjustment in this process, concentrated HCl must be added carefully to prevent the pH from becoming too acidic. In addition, due to the strong acidity of concentrated HCl, appropriate PPE, including gloves, safety goggles, and masks, must be worn to prevent potential injury.
5. Since HCl cannot be used for pH adjustment in this process, solid NaOH must be added carefully to prevent the pH from exceeding 8.0.
6. Commercial antibodies from other suppliers may also be applicable.
7. To improve the activity of bacterial culture, fresh one- or two-day-old *E. coli* colonies will yield the best results for liquid culture inoculation.
8. RNase A must be added to Solution I before use and stored at 4 °C.
9. Avoid vigorous mixing of the lysate, and the lysis reaction should not last longer than 5 min.
10. Thoroughly and immediately mix the solution after adding Solution III to prevent localized precipitation.
11. HBC buffer must be diluted with isopropanol before use.
12. DNA wash buffer must be diluted with 100% ethanol before use.
13. Carefully discard the supernatant to prevent the loss of the DNA pellet.
14. Enhance the dissolution of enzymes in a 55 °C water bath for 10 min.
15. To achieve better enzymatic hydrolysis, apply the enzyme solution to both sides of the leaves.
16. Handle the process carefully and slowly to prevent rupture of the protoplasts.
17. The slow acceleration and deceleration rates should be used to prevent fragmentation of the protoplasts. Additionally, carefully aspirate and discard the supernatant.

18. Gently and slowly invert the tube to prevent breakage of the protoplasts.
19. Calculate the required volumes of plasmids based on their concentrations. Adjust the volumes with ddH_2O to ensure the same total volume of plasmids in each tube. The total volume of plasmids added to protoplasts should be less than 10 μL to achieve a high transformation efficiency.
20. The protoplasts are in good condition if they are full, spherical, and unbroken and at a uniform density under the microscope.
21. Each transformation reaction should last for the same time to ensure consistent transformation efficiency.
22. Protoplasts can be subjected to different light treatments according to the experimental design.
23. Healthy protoplasts will form a circular pellet at the bottom of the tube after centrifugation. If the pellet fails to settle at the bottom of tube or exhibits in an irregular shape, the protoplasts may not be in good condition.
24. The presence of inactivated (boiled) λ-PPase and the absence of active λ-PPase (-) both serve as negative controls in the experiment.
25. The alkaline phosphatase (AP)-conjugated secondary antibody and detection system were employed to detect the phosphorylated and unphosphorylated phyA bands.

Acknowledgments

This work was supported by grants from the National Key R&D Program of China (2024YFA1306700), National Natural Science Foundation of China (32225006 to J.L.), and the Beijing Natural Science Foundation (5232011 to J.L. and 5254033 to Z.F.).

References

1. Quail PH, Boylan MT, Parks BM, Short TW, Xu Y, Wagner D (1995) Phytochromes: photosensory perception and signal transduction. Science 268:675–680
2. Fankhauser C (2000) Phytochromes as light-modulated protein kinases. Semin Cell Dev Biol. 11:467–473
3. Li J, Li G, Wang H, Deng XW (2011) Phytochrome signaling mechanisms. Arabidopsis Book 9:e0148
4. Cheng MC, Kathare PK, Paik I, Huq E (2021) Phytochrome signaling networks. Annu Rev Plant Biol 72:217–244
5. Franklin KA, Quail PH (2010) Phytochrome functions in *Arabidopsis* development. J Exp Bot 61:11–24
6. Rausenberger J, Tscheuschler A, Nordmeier W et al (2011) Photoconversion and nuclear trafficking cycles determine phytochrome's response profile to far-red light. Cell 146: 813–825
7. Li J, Hiltbrunner A (2021) Is the Pr form of phytochrome biologically active in the nucleus? Mol Plant 14:535–537
8. Quail PH, Briggs WR, Pratt LH (1978) In vivo phosphorylation of phytochrome. Carnegie

Institution Annual Report. Department of Plant Biology, Carnegie Institution of Washington, Stanford, pp 342–344
9. Hunt RE, Pratt LH (1980) Partial characterization of undegraded oat phytochrome. Biochemistry 19:390–394
10. Han YJ, Kim HS, Kim YM et al (2010) Functional characterization of phytochrome autophosphorylation in plant light signaling. Plant Cell Physiol 51:596–609
11. McMichael RW Jr, Lagarias JC (1990) Phosphopeptide mapping of Avena phytochrome phosphorylated by protein kinase in vitro. Biochemistry 29:3872–3878
12. Lapko VN, Jiang XY, Smith DL, Song PS (1996) Protein kinase A-catalyzed phosphorylation and conformational changes in phytochrome A. Biochemistry 35:6585–6594
13. Lapko VN, Jiang XY, Smith DL, Song PS (1997) Post-translational modification of oat phytochrome A: phosphorylation of a specific serine in a multiple serine cluster. Biochemistry 36:10595–10599
14. Lapko VN, Jiang XY, Smith DL, Song PS (1999) Mass spectrometric characterization of oat phytochrome A: isoforms and posttranslational modifications. Protein Sci 8:1032–1044
15. Kim JI, Shen Y, Han YJ et al (2004) Phytochrome phosphorylation modulates light signaling by influencing the protein-protein interaction. Plant Cell. 16:2629–2640
16. Jordan ET, Cherry JR, Walker JM, Vierstra RD (1996) The amino-terminus of phytochrome A contains two distinct functional domains. Plant J. 9:243–257
17. Trupkin SA, Debrieux D, Hiltbrunner A, Fankhauser C, Casal JJ (2007) The serine-rich N-terminal region of *Arabidopsis* phytochrome A is required for protein stability. Plant Mol Biol 63:669–678
18. Zhou Y, Yang L, Duan J et al (2018) Hinge region of *Arabidopsis* phyA plays an important role in regulating phyA function. Proc Natl Acad Sci U S A 115:E11864–E11873
19. Saijo Y, Zhu D, Li J et al (2008) *Arabidopsis* COP1/SPA1 complex and FHY1/FHY3 associate with distinct phosphorylated forms of phytochrome A in balancing light signaling. Mol Cell 31:607–613
20. Feng Z, Wang M, Liu Y et al (2024) Liquid-liquid phase separation of TZP promotes PPK-mediated phosphorylation of the phytochrome A photoreceptor. Nat Plants 10:798–814
21. Shin AY, Han YJ, Baek A et al (2016) Evidence that phytochrome functions as a protein kinase in plant light signalling. Nat Commun 7: 11545
22. Hoang QTN, Cho JY, Choi DM et al (2021) Protein kinase activity of phytochrome A positively correlates with photoresponses in *Arabidopsis*. Front Plant Sci 12:706316
23. Liu X, Jiang W, Li Y et al (2023) FERONIA coordinates plant growth and salt tolerance via the phosphorylation of phyB. Nat Plants 9: 645–660
24. Zhao Y, Shi H, Pan Y et al (2023) Sensory circuitry controls cytosolic calcium-mediated phytochrome B phototransduction. Cell 186: 1230–1243.e14
25. Burgie ES, Li H, Gannam ZTK, McLoughlin KE, Vierstra RD, Li H (2023) The structure of *Arabidopsis* phytochrome A reveals topological and functional diversification among the plant photoreceptor isoforms. Nat Plants 9:1116–1129
26. Wang Z, Wang W, Zhao D et al (2024) Light-induced remodeling of phytochrome B enables signal transduction by phytochrome-interacting factor. Cell 187:6235–6250.e19
27. Yoo SD, Cho YH, Sheen J (2007) *Arabidopsis* mesophyll protoplasts: a versatile cell system for transient gene expression analysis. Nat Protoc 2:1565–1572
28. Sheen J (2001) Signal transduction in maize and *Arabidopsis* mesophyll protoplasts. Plant Physiol 127:1466–1475
29. Chen S, Tao L, Zeng L, Vega-Sanchez ME, Umemura K, Wang GL (2006) A highly efficient transient protoplast system for analyzing defence gene expression and protein-protein interactions in rice. Mol Plant Pathol 7:417–427
30. Sheen J, Zhou L, Jang JC (1999) Sugars as signaling molecules. Curr Opin Plant Biol 2: 410–418
31. Tena G, Asai T, Chiu WL, Sheen J (2001) Plant mitogen-activated protein kinase signaling cascades. Curr Opin Plant Biol 4:392–400
32. Fujii H, Chinnusamy V, Rodrigues A et al (2009) In vitro reconstitution of an abscisic acid signalling pathway. Nature 462:660–664
33. Zhang S, Li C, Zhou Y et al (2018) TANDEM ZINC-FINGER/PLUS3 is a key component of phytochrome A signaling. Plant Cell 30: 835–852
34. Li C, Qi L, Zhang S et al (2022) Mutual upregulation of HY5 and TZP in mediating phytochrome A signaling. Plant Cell 34:633–654
35. Qi L, Liu S, Li C et al (2020) PHYTOCHROME-INTERACTING FACTORS interact with the ABA receptors PYL8 and PYL9 to orchestrate ABA signaling in darkness. Mol Plant 13:414–430

Chapter 20

A Fast and Reliable Nuclei Extraction Method for Phytochrome Functional Analysis

Ilse Villmann, Akira Nagatani, and Mathias Zeidler

Abstract

Phytochromes are plant photoreceptors that sense red and far-red light, regulating numerous developmental processes during de-etiolation and subsequent growth. A critical aspect of phytochrome signaling involves their light-dependent translocation from the cytoplasm to the nucleus, where they mediate diverse physiological responses. Upon activation, both phytochrome A (phyA) and phytochrome B (phyB) accumulate in the nucleus, where they interact with transcription factors and modulate the localization and/or function of other regulatory proteins, such as E3 ubiquitin ligases. While the nuclear functions of phytochromes are well-documented, emerging evidence suggests they may also play signaling roles in the cytoplasm. To further elucidate the distinct roles of phytochromes in the cytoplasm and nucleus and to identify their potential interaction partners in each compartment, a protocol has been developed for isolating intact nuclei and a separate cytoplasmic fraction from *Arabidopsis thaliana* seedlings. This method enables the identification of interaction partners through downstream analyses, such as cross-linking combined with mass spectrometry or other protein–protein interaction techniques.

The protocol involves shearing Arabidopsis seedlings followed by purification of nuclei using a Percoll gradient. This approach is relatively quick and straightforward, making it suitable for etiolated or light-exposed seedlings.

Key words Phytochrome, Nuclear import, Nuclear, cytoplasmic partitioning, Protein extraction

1 Introduction

Plants perceive light quality and quantity in the red (R) to far-red (FR) region of the daylight spectrum through a specialized family of photoreceptors called phytochromes [1]. Phytochromes assemble with an open tetrapyrrole chromophore, phytochromobilin, to form a dimer of approximately 250 kDa [2, 3]. In *Arabidopsis thaliana*, the phytochrome family comprises five members, of which phytochrome A (phyA) and phytochrome B (phyB) are the most functionally significant [4, 5].

Phytochromes exist in two spectrally and functionally distinct states: the red-light-absorbing Pr form and the far-red-light-

Andreas Winkler (ed.), *Phytochromes: Methods and Protocols*, Methods in Molecular Biology, vol. 2970,
https://doi.org/10.1007/978-1-0716-4791-2_20,

absorbing Pfr form. The ratio of these forms (Pr/Pfr) reflects the light environment and governs phytochrome activity. PhyB predominantly regulates plant responses in natural light environments, including neighbor detection and daylength measurement as well as temperature sensing [6], while phyA plays the major role during germination and early seedling development [7]. The transition from etiolated growth beneath the soil to photomorphogenesis upon soil emergence is one of the most dramatic developmental changes mediated by phyA [8].

Upon conversion to the active Pfr form, both phyA and phyB accumulate in the nucleus [9–12]. PhyA nuclear translocation requires the assistance of small transport facilitators, FAR-RED ELONGATED HYPOCOTYL1 (FHY1) and its homolog FAR-RED ELONGATED HYPOCOTYL1-LIKE (FHL) [13–15]. By contrast, phyB nuclear import appears to be independent of such facilitators and may involve interactions with transcription factors in the cytoplasm which then facilitate cotransport [16].

In the nucleus, phytochromes regulate transcription through various mechanisms, including direct interaction with transcription factors, initiation of transcription factor cascades [17, 18], and modulation of transcription factor abundance via the E3-ubiquitin ligase–proteasome pathway [19, 20]. While phytochrome functions are predominantly nuclear, some signaling activities occur in the cytoplasm. For example, phyA has been shown to regulate processes such as the modulation of gravitropism and the enhancement of phototropism, highlighting its cytoplasmic roles [14, 21].

To investigate phytochrome functions, their interactions with partner proteins, and secondary modifications during phytochrome-regulated developmental responses in the nucleus and cytoplasm, a robust method for isolating *Arabidopsis thaliana* nuclei and cytoplasmic fractions has been developed. This protocol facilitates the compartment-specific analysis of proteins associated with phytochromes, enabling the identification of post-translational modifications, interaction partners, and signaling components localized within distinct cellular environments.

By separating nuclear and cytoplasmic fractions, it becomes possible to study the local state or partner context of phytochrome signaling. While the nuclear fraction allows detailed examination of transcriptional regulation and the interaction of phytochromes with transcription factors or chromatin-associated proteins [22], the cytoplasmic fraction provides access to signaling components and processes outside the nucleus, such as those regulating gravitropism, phototropism, or cytoplasmic signal transduction cascades.

The protocol is optimized for Arabidopsis seedlings, utilizing mechanical disruption followed by a Percoll gradient centrifugation step to isolate intact nuclei [23]. This method ensures high purity and yields of nuclear and cytoplasmic fractions, making it suitable for various downstream analyses, including mass spectrometry, immunoprecipitation, or protein cross-linking studies. The approach is versatile and can be applied to etiolated seedlings or plants grown under light conditions to study phytochrome function across developmental stages and environmental contexts.

2 Materials

1. Nuclei extraction buffer (NIB): 10 mM Mes-KOH, pH 6.0, 20 mM NaCl, 10 mM EDTA, 0.3 mM Spermidine, 1 mM Spermidine, 1.2% Triton X-100 (v/v), 0.5 M saccharose (4 °C).
2. Cytoplasmic extraction buffer 2x (CIB): 100 mM Tris-HCl, pH 7.8 at (4 °C), 200 mM NaCl, 30 mM EDTA, pH 8.0, 10 mM DTT (DL-Dithiothreitol), 2 mM PMSF (Phenylmethansulfonylfluorid) (4 °C).
3. 4,6-Diamidino-2-phenylindole dihydrochloride (DAPI) solution: 5 mg/ml 4,6-diamidino-2-phenylindole dihydrochloride (DAPI) in H_2O.
4. H_2O: pure H_2O (Milli-Q or similar).
5. Percoll.
6. Liquid nitrogen.
7. Gauze/mull (e.g., Miracloth, Millipore).
8. Nylon mesh (20 μm pore width).
9. Cooled centrifuge with swing out rotor for 1500× *g*, cooled Eppendorf centrifuge.
10. Rotary tool, drilling machine like Dremel or similar.
11. Razor blades (new).
12. Mortar and pistil.
13. Fluorescence microscope for control and DAPI stain visualization (optional): Here, a Leica DM6000B with filter cube A4 (Ex: 360 nm BP 400 nm Em: 460 nm) was used.
14. Sodium dodecyl sulfate–polyacrylamide gel electrophoresis (SDS-PAGE) equipment, Western blot equipment, polyvinylidene fluoride (PVDF) membrane, anti-histone H3 antibody (Agrisera, AS10 710), anti-phosphoenolpyruvate carboxylase (PEPC) antibody (Agrisera, AS09 458) (optional).

3 Methods

3.1 Sampling

1. To obtain cytoplasmic and nuclear fraction, 6 g of seedlings are used for each treatment (dark, light, etc.). One gram is used for

the cytoplasmic protein extraction and 5 g for the nuclei isolation.

2. Freeze 1 g seedlings in liquid nitrogen directly after light treatment for cytoplasmic proteins (*see* Subheading 3.3) and keep frozen.
3. Use 5 g of fresh seedlings for nuclei isolation (*see* Subheading 3.2).

3.2 Nuclei Isolation

1. Cut 5 g seedlings in 20–30 mL 1x NIB with a razor blade attached to a Dremel or similar instrument (*see* Fig. 1) in a 250-mL beaker for ca. 25 s on ice (*see* **Note 1**).
2. Rinse eight layers of gauze/mull first with water, then with NIB (*see* **Note 2**).
3. Filter the extract trough these eight layers of gauze/mull.
4. Rinse the 20 μm nylon mesh with NIB.
5. Filter the extract and then through the nylon mesh.
6. Centrifuge the filtered extract at 1500× *g* for 10 min at 4 °C, no brakes.
7. Discard the supernatant and resuspend the pellet in 2 mL 1 × NIB (*see* **Note 3**).
8. Layer the resuspended pellet on a two-layer Percoll gradient (lower layer 2 mL 50% Percoll in NIB, upper layer 2 mL 25% Percoll in NIB).
9. Centrifuge at 7800× *g* 30 min 4 °C in a swing-out centrifuge, no brakes (*see* **Note 4**).

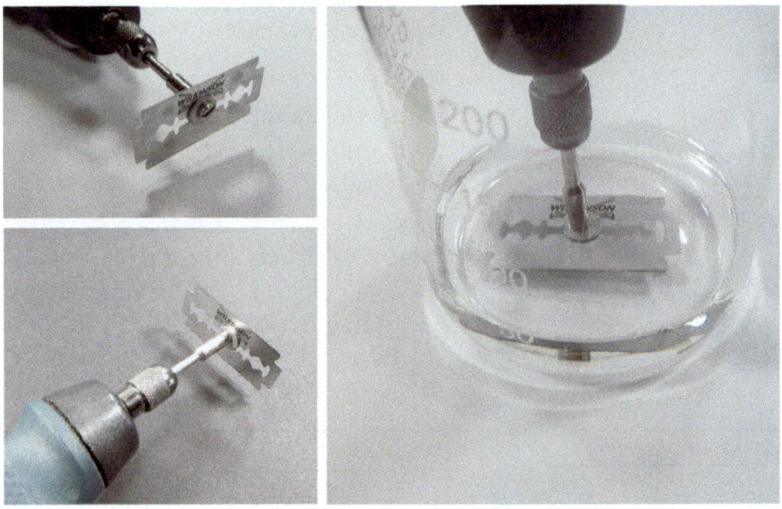

Fig. 1 Example of razor blade attached to a rotary tool/drilling machine. The small multi-tool driller usually contains a holder for thin abrasive disks. This should be used to fix a razor blade onto the tool as shown in the picture

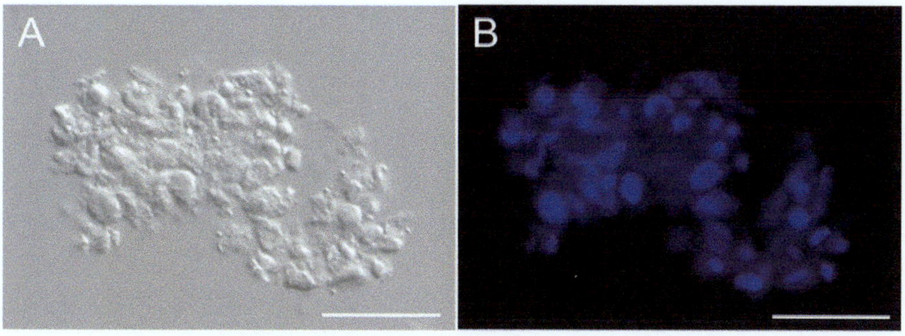

Fig. 2 Microscopic analysis of purified DAPI-stained nuclei from 3-day-old dark grown Arabidopsis seedlings. (**a**) Differential interference contrast (DIC) microscopy and (**b**) same nuclei preparation as in A, viewed with fluorescence microscopy. Scale bar represents 25 μm

10. There should be a white flaky interphase between the layers.
11. Take this interphase, double its volume with 1x NIB and repeat gradient once more.
12. Take the interphase and double its volume with 1x NIB and centrifuge in an Eppendorf tube at 500× *g* for 10 min at 4 °C with no brake.
13. Discard the supernatant, resuspend the pellet in 1 x NIB, and pool eventually.
14. Centrifuge like in **step 12**.
15. The final pellet should be the purified nuclei. The purity of the nuclei can be verified by staining an aliquot with the same volume DAPI solution and observation with fluorescence microscopy (Fig. 2).

3.3 Cytoplasmic Protein Extraction

1. To obtain the cytoplasmic fraction, use the frozen seedlings from **step 3.1**, and grind in a mortar with liquid nitrogen until a fine powder is visible (*see* **Note 5**).
2. Weight still frozen powder and add the same volume of 2x CIB. Thaw quickly by shaking.
3. Centrifuge at 20,000× *g* for 10 min at 4 °C.
4. Take the supernatant to a new tube; this contains the cytoplasmic proteins.
5. The success of the fractionation can be verified by Western blotting with specific antibodies to nuclear and cytoplasmic proteins (Fig. 3).

4 Notes

1. The beaker should be fixed in a box filled with ice, so that it is cooled but cannot rotate during the cutting process. Always

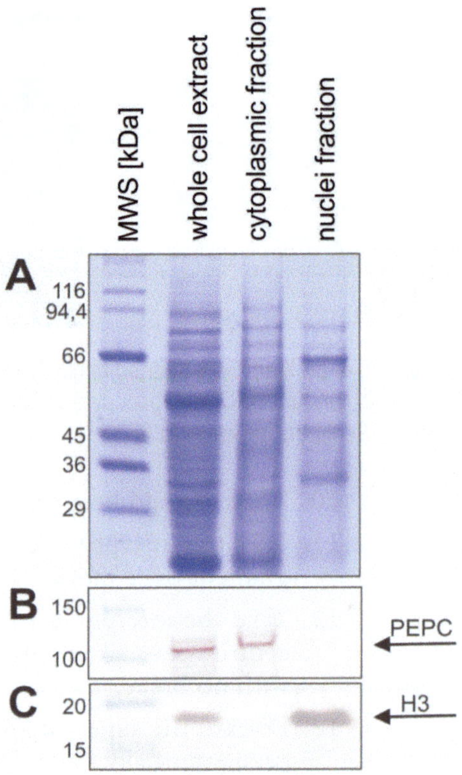

Fig. 3 Analysis of total cell extract and fractions by SDS-PAGE and Western blot. (**a**) Coomassie-stained SDS gel with protein molecular weight marker in the first lane, total cell extract, and cytosolic and nuclei fraction. (**b** and **c**) Same gel as in (**a**) blotted on PVDF membrane and probed with (**b**) anti-phosphoenolpyruvate carboxylase (PEPC) as cytosolic marker protein and (**c**) anti-histone H3 as nuclear marker protein

use fresh razor blades. Be aware that foam develops; let foam settle before proceeding to the next step. The time and speed of the cutting process are critical for the success and should be optimized for your instrument. Hypocotyl pieces should still be visible after cutting and do not over-homogenize the sample. Check a test sample under a microscope at different times. Usually 20–30 s minimal speed with a standard Dremel (3000–5000 rpm) is sufficient.

Please be advised that the proposed rotating razor blade is a serious safety risk: Use cut safe gloves and be super careful with the instrument! Only switch it on if the blade is inside the beaker as shown in Fig. 1. Never switch it on if the blade is outside of the beaker. If available, a stainless steel beaker should be used.

2. Minimize the amount of gauze/mull; you will need to avoid losing too much extract in this step. A small funnel lined with

the gauze/mull to the brink will do the trick. The rinse and equilibration with NIB are essential.

3. The supernatant can be used as cytoplasmic fraction; however, we found it to still contain nuclear proteins, and the overall protein concentration is very low. To obtain the cytoplasmic fraction, follow the steps from outlined in Subheading 3.1.
4. All the steps up to the phase separation can be done under very low fluence rate green "safety" light or in darkness, for experienced users.
5. The mortar and pistil should be precooled with liquid nitrogen, so that the nitrogen does not boil in the mortar anymore. If the nitrogen is evaporated, some can be added to grind for a longer time. The progress can be monitored by taking a small sample and thaw it on a glass slide, cover it with a cover slip, and view it under the microscope. Only small cell debris should be visible.

Acknowledgment

This work was supported by DFG (Deutsche Forschungsgemeinschaft) grants ZE485/2-2 and ZE485/3-1 to MZ and JSPS (Japan Society for the Promotion of Science) grant to AN and IV.

References

1. Quail PH, Boylan MT, Parks BM et al (1995) Phytochromes: photosensory perception and signal transduction. Science 268(5211): 675–680. https://doi.org/10.1126/science.7732376
2. Lagarias JC, Rapoport H (1980) Chromopeptides from phytochrome. The structure and linkage of the Pr form of the phytochrome chromophore. J Am Chem Soc 102:4821–4828. https://doi.org/10.1021/ja00534a042
3. Rockwell NC, Su YS, Lagarias JC (2006) Phytochrome structure and signaling mechanisms. Annu Rev Plant Biol 57:837–858. https://doi.org/10.1146/annurev.arplant.56.032604.144208
4. Sharrock RA, Quail PH (1989) Novel phytochrome sequences in *Arabidopsis thaliana*: structure, evolution, and differential expression of a plant regulatory photoreceptor family. Genes Dev 3(11):1745–1757. https://doi.org/10.1101/gad.3.11.1745
5. Clack T, Mathews S, Sharrock RA (1994) The phytochrome apoprotein family in Arabidopsis is encoded by five genes: the sequences and expression of *PHYD* and *PHYE*. Plant Mol Biol 25(3):413–427. https://doi.org/10.1007/BF00043870
6. Quint M, Delker C, Franklin KA et al (2016) Molecular and genetic control of plant thermomorphogenesis. Nat Plants 2:15190. https://doi.org/10.1038/nplants.2015.190
7. Casal JJ, Candia AN, Sellaro R (2014) Light perception and signalling by phytochrome A. J Exp Bot 65(11):2835–2845. https://doi.org/10.1093/jxb/ert379
8. Montgomery BL (2016) Spatiotemporal phytochrome signaling during photomorphogenesis: from physiology to molecular mechanisms and back. Front Plant Sci 7:480. https://doi.org/10.3389/fpls.2016.00480
9. Sakamoto K, Nagatani A (1996) Nuclear localization activity of phytochrome B. Plant J 10(5):859–868. https://doi.org/10.1046/j.1365-313x.1996.10050859.x
10. Yamaguchi R, Nakamura M, Mochizuki N et al (1999) Light-dependent translocation of a phytochrome B-GFP fusion protein to the nucleus in transgenic Arabidopsis. J Cell Biol

145(3):437–445. https://doi.org/10.1083/jcb.145.3.437

11. Hisada A, Hanzawa H, Weller JL et al (2000) Light-induced nuclear translocation of endogenous pea phytochrome A visualized by immunocytochemical procedures. Plant Cell 12(7): 1063–1078
12. Kircher S, Kozma-Bognar L, Kim L et al (1999) Light quality dependent nuclear import of the plant photoreceptors phytochrome A and B. Plant Cell 11(8):1445–1456. https://doi.org/10.1105/tpc.12.7.1063
13. Hiltbrunner A, Viczian A, Bury E et al (2005) Nuclear accumulation of the phytochrome A photoreceptor requires FHY1. Curr Biol 15(23):2125–2130. https://doi.org/10.1016/j.cub.2005.10.042
14. Rösler J, Klein I, Zeidler M (2007) Arabidopsis *fhl/fhy1* double mutant reveals a distinct cytoplasmic action of phytochrome A. Proc Natl Acad Sci U S A 104(25):10737–10742. https://doi.org/10.1073/pnas.0703855104
15. Helizon H, Rösler-Dalton J, Gasch P et al (2018) Arabidopsis phytochrome A nuclear translocation is mediated by a far-red elongated hypocotyl 1–importin complex. Plant J 96(6): 1255–1268. https://doi.org/10.1111/tpj.14107
16. Pfeiffer A, Nagel MK, Popp C et al (2012) Interaction with plant transcription factors can mediate nuclear import of phytochrome B. Proc Natl Acad Sci U S A 109(15): 5892–5897. https://doi.org/10.1073/pnas.1120764109
17. Tepperman JM, Zhu T, Chang HS et al (2001) Multiple transcription-factor genes are early targets of phytochrome A signaling. Proc Natl Acad Sci U S A 98(16):9437–9442. https://doi.org/10.1073/pnas.161300998
18. Leivar P, Quail PH (2011) PIFs: pivotal components in a cellular signaling hub. Trends Plant Sci 16(1):19–28. https://doi.org/10.1016/j.tplants.2010.08.003
19. Hoecker U (2017) The activities of the E3 ubiquitin ligase COP1/SPA, a key repressor in light signaling. Curr Opin Plant Biol 37: 63–69. https://doi.org/10.1016/j.pbi.2017.03.015
20. Zeidler M (2020) Fine-tuning by the touch of big dudes: SUMOs role in light signaling. Mol Plant 13(7):943–945. https://doi.org/10.1016/j.molp.2020.06.001
21. Rösler J, Jaedicke K, Zeidler M (2010) Cytoplasmic phytochrome action. Plant Cell Physiol 51(8):1248–1254. https://doi.org/10.1093/pcp/pcq091
22. Kim C, Kwon Y, Jeong J et al (2023) Phytochrome B photobodies are comprised of phytochrome B and its primary and secondary interacting proteins. Nat Commun 14(1): 1708. https://doi.org/10.1038/s41467-023-37421-z
23. Nagatani A, Jenkins GI, Furuya M (1988) Non-specific association of phytochrome to nuclei during isolation from dark-grown pea (*Pisum sativum* cv. Alaska) plumules. Plant Cell Physiol 29(7):1141–1145. https://doi.org/10.1093/oxfordjournals.pcp.a077614

Chapter 21

Opto-APCs: Displaying Phytochrome B on the Surface of Living Cells for Optogenetic Studies

Anna K. Ehret, Marissa Russ, O. Sascha Yousefi, and Wolfgang W. Schamel

Abstract

Interactions between proteins on the surface of two cells are pivotal in determining cell fate. These might be receptor–ligand pairs or adhesion molecules. Mechanistic and functional details of these interactions can be studied by optogenetics using parts of the photoreceptor phytochrome B (PhyB) of *Arabidopsis thaliana*. In these cases, PhyB serves as a membrane-bound ligand for engineered receptors or adhesion molecules. Here, we provide a protocol to generate cells that display PhyB as a membrane-bound protein on their surface to serve as the ligand. This is done by expressing a transmembrane SpyCatcher protein on the cells and covalently coupling purified recombinant PhyB-SpyTag to these cells. Coming from immunology, we call the PhyB-coupled cells opto-antigen-presenting cells (opto-APCs).

Key words Phytochrome B, Ligand, Receptor, Optogenetics, SpyCatcher, SpyTag

1 Introduction

Cells rely on precise responses to external stimuli, often mediated by ligand–receptor or adhesion molecule interactions that govern cellular behavior and fate. Understanding these interaction dynamics requires tools to manipulate ligand concentration, binding affinity, interaction half-life, and temporal binding patterns. Likewise, spatial control enables precise activation or inhibition of specific targets using focused light stimulation, allowing for localized modulation of cellular functions, reduced off-target effects, and improved experimental precision. Optogenetics, a revolutionary approach, enables precise control over these parameters using light-sensitive proteins [1, 2].

Non-opsin-based optogenetic systems utilize photoreceptors from plants or cyanobacteria that change conformation upon light absorption. For instance, the photoreceptor phytochrome B (PhyB) from *Arabidopsis thaliana* interacts with its binding

Andreas Winkler (ed.), *Phytochromes: Methods and Protocols*, Methods in Molecular Biology, vol. 2970,
https://doi.org/10.1007/978-1-0716-4791-2_21,

partner, phytochrome-interacting factor (PIF), in response to red light, while far-red light disrupts the interaction [2, 3]. This reversible binding mechanism enables spatiotemporal regulation of molecular interactions with high precision.

We previously engineered an optogenetic system using the first 651 amino acids of PhyB as a recombinant protein (PhyB1-651, from now on called PhyB). This truncated PhyB incorporates the light-responsive chromophore phycocyanobilin and serves as the ligand for engineered receptors or adhesion molecules that contain the first 100 amino acids of PIF [4–6]. Additionally, we modified PIF to enable its transport through the secretory pathway as part of engineered receptors and adhesion molecules [4–6]. This system allowed us to control ligand binding to the PIF-fused T-cell receptor (TCR) or the adhesion molecule αVβ3 integrin by light. We showed that we can modulate TCR activation by controlling the half-life of the PhyB:PIF-TCR interaction by light. Prolonged ligand–receptor binding induced T-cell activation, whereas shorter interactions acted as antagonists, demonstrating how optogenetics can elucidate the kinetics of signaling processes [7].

In native physiological contexts, TCR ligands are transmembrane proteins on antigen-presenting cells (APCs), enabling two-dimensional interactions during cell–cell contact [8]. Therefore, we developed a system with which we can display PhyB as a ligand on the surface of engineered cells (the opto-APCs) [9]. In plants, PhyB is a cytosolic protein and cannot be sent through the secretory pathway of other cells to the plasma membrane in an active form; thus, we engineered the system to recombinantly express PhyB in bacteria and use the purified protein to bind to the opto-APCs [9]. These cells can be any cell line that expresses a transmembrane SpyCatcher, since PhyB is eventually covalently coupled to it by being fused to the SpyTag (Fig. 1). We used the third generation of the SpyCatcher–SpyTag system [10]. This innovation facilitates optogenetic control of two-dimensional ligand-receptor dynamics, bridging the gap between soluble systems and physiological cell–cell interactions.

In this article, we provide a protocol for the covalent coupling of recombinant PhyB-mCherry-SpyTag to the SpyCatcher-expressing cells (Fig. 1, 3.1), called "loading" in this publication. We use PhyB coupled to mCherry-SpyTag for easier determination of the loading efficiency, but one can also use only PhyB-SpyTag if the mCherry interferes with the readouts. Protocols to produce and purify recombinant PhyB have been published, and herewith, we would like refer to them [11–13]. In this article, we also provide a protocol to test for the loading efficiency (Fig. 1, 3.2) and another one to validate that the loaded PhyB is functional (Fig. 1, 3.3).

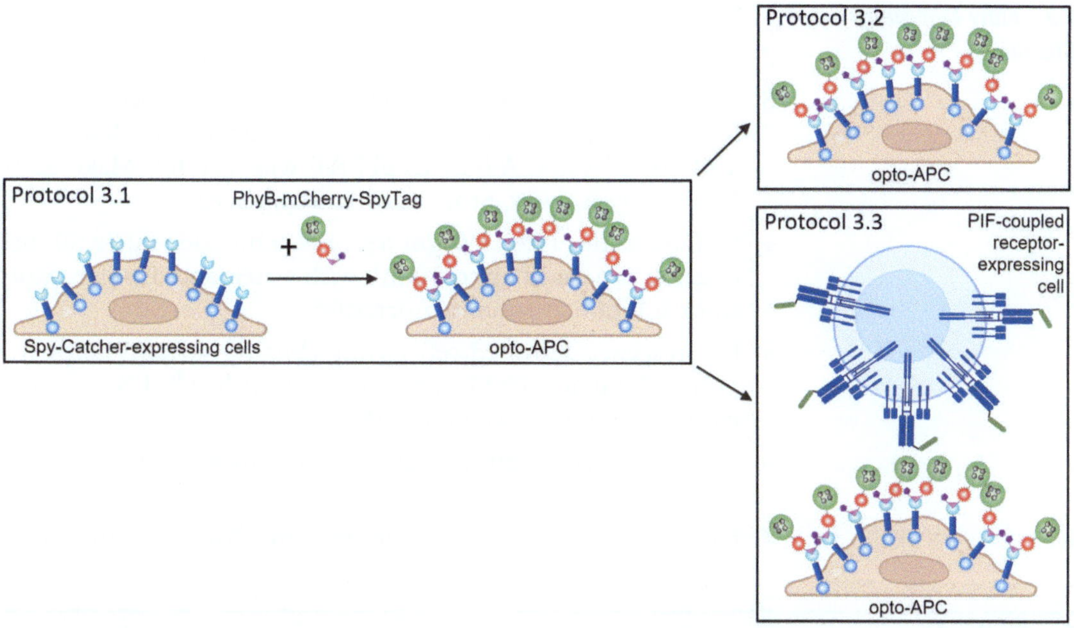

Fig. 1 Schematic depiction of steps described in this protocol. (Figure generated with BioRender®)

2 Materials

2.1 Loading of SpyCatcher-Expressing Cells with PhyB-mCherry-SpyTag

1. Roswell Park Memorial Institute (RPMI) medium: Add desired amount of fetal bovine serum (FBS) to 500 mL RPMI-1640 (5 mL for 1% and 50 mL for 10%). Supplement medium with 0.5% penicillin and streptomycin as well as 100 mM 4-(2-hydroxyethyl)-1-piperazineethanesulfonic acid (HEPES) (*see* **Note 1**).
2. SpyCatcher cell line: Cell line stably expressing the SpyCatcher construct (*see* **Note 2**).
3. 96F-bottom black-walled-clear-bottom plate for cell culture.
4. PhyB-mCherry-SpyTag protein (*see* **Note 3**).
5. Tris(2-carboxyethyl)phosphine hydrochloride (TCEP; 0.5 M): Weight 1.43 g of TCEP into graduated cylinder and dissolve in 6 mL of water. Adjust to pH 7.0 with NaOH. Make up to 10 mL with water. Aliquot into 500 μL and store stock solution at −20 °C.
6. PIF-TCR-expressing Jurkat T cells (*see* **Note 4**).
7. Cell culture sterile bench.
8. Centrifuge: Suitable for 1.5-mL reaction tubes.
9. Opto-Plate-96 illuminator [14] and the software optoConfig-96 [15].
10. Cell culture incubator: With cooling function.

2.2 Flow Cytometry Staining

1. 96V-bottom plate.
2. Centrifuge: Suitable for 96-well plate centrifugation.
3. Ethylenediaminetetraacetic acid (EDTA; 0.5 M): Weight 5.85 g of EDTA into graduated cylinder and dissolve in 400 mL of water. Adjust to pH 8.0 with NaOH. Make up to 500 mL with water. Store at room temperature.
4. Sodium azide (10%): Weight 5 g of sodium azide into a 50 -mL reaction tube, add up to 50 mL with water, and dissolve. Store stock solution at room temperature.
5. Fluorescence-activated cell sorting (FACS) buffer: Phosphate-buffered saline (PBS) supplemented with 1% FBS, 1 mM EDTA, and 0.02% sodium azide.
6. Activation marker antibodies: Various antibodies to stain activation markers on T cells.
7. Flow cytometer device and the software FlowJo for analysis.

3 Methods

This protocol describes the step-by-step procedures to load SpyCatcher-expressing cells with PhyB-mCherry-SpyTag (*see* Subheading 3.1), validate the loading (*see* Subheading 3.2), and assess the functional activity of the loaded PhyB (*see* Subheading 3.3). Each subheading outlines specific experimental steps necessary to ensure reproducibility and accuracy.

3.1 Loading SpyCatcher-Expressing Cells with PhyB-mCherry-SpyTag

This subheading describes the process of loading PhyB-mCherry-SpyTag to SpyCatcher-expressing cells under controlled conditions to maintain protein functionality and ensure a sterile work environment. All steps in which the recombinant protein PhyB is present should be carried out under green light to avoid photon absorption by PhyB. To ensure a sterile work environment, cells and reagents should be handled in a cell culture sterile bench.

1. Seed 50,000 SpyCatcher-expressing cells in 100 μL 1% FBS RPMI medium per well in a 96F-bottom black-walled-clear-bottom plate (*see* **Note 5**). Incubate overnight at 37 °C in a 5% CO_2 humidified atmosphere (*see* **Note 6**).
2. On the next day, the medium has to be removed carefully.
3. Prepare the PhyB-mCherry-SpyTag protein solution by mixing the desired amount with 10% FBS RPMI medium and adding 0.8 mM TCEP (*see* **Note** 7).
4. Add 50 μL per well of the PhyB-mCherry-SpyTag solution and incubate for the desired loading time at 37 °C in a 5% CO_2 humidified atmosphere (*see* **Note** 7).
5. Carefully remove the excess of the PhyB-mCherry-SpyTag-containing medium.

3.2 Validation of the PhyB-mCherry-SpyTag Loading onto the SpyCatcher-Expressing Cells

Here, we describe how to confirm successful loading of PhyB-mCherry-SpyTag onto SpyCatcher-expressing cells using fluorescence-activated cell sorting (FACS) analysis. This subheading describes the necessary steps for sample preparation.

From now on, all steps are carried out at 4 °C, if not otherwise specified.

1. Add 180 μL FACS buffer to each well and incubate one minute on ice.
2. Detach cells by pipetting up and down, and transfer cell solution into 96V-bottom plate.
3. Pellet the cells at 300× *g* for 4 min and wash one time with FACS buffer.
4. Resuspend cell pellet in 180 μL FACS buffer before analyzing the mCherry signal in the flow cytometer (Fig. 2) (*see* **Note 8**).

3.3 Validation That the Loaded PhyB on the Opto-APCs Is Functional

In this subheading, we outline the experimental steps required to evaluate the functional activity of the loaded PhyB on the opto-APCs. In our case, we used optogenetic stimulation of SNAP-PIF-TCR-expressing T cells by the opto-APCs. All steps in which PhyB is present should to be carried out under green light to avoid photon absorption by PhyB. To ensure a sterile work environment, cells and reagents should be handled in a cell culture sterile bench.

1. During **step 4** of Subheading 3.1, the SNAP-PIF-TCR-expressing Jurkat T cells can be prepared. Therefore, pellet the cells for 10 min at 300× *g* and adjust to a cell concentration of 2.5×10^5 cells/mL in 10% FBS RPMI medium containing 0.8 mM TCEP.

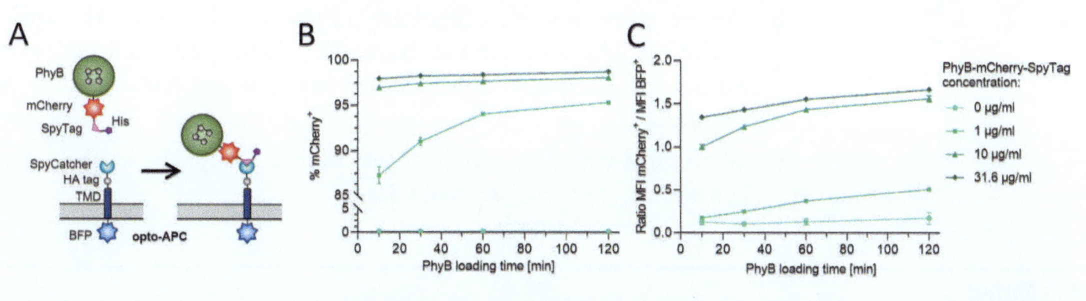

Fig. 2 Determination of the optimal PhyB-mCherry-SpyTag loading time and concentration. (**A**) Scheme of the loading of PhyB-mCherry-SpyTag to the membrane-bound SpyCatcher on the opto-APCs. PhyB-mCherry-SpyTag contains a His6 tag for protein purification, but for the sake of simplicity, we have not included it in the name. (**B** and **C**) Testing the loading of PhyB-mCherry-SpyTag to the opto-APCs (in this case, transmembrane SpyCatcher-BFP-expressing HEK 293T cells). The HEK 293T cells were incubated with different concentrations of the purified PhyB-mCherry-SpyTag protein for 10, 30, 60, or 120 min at 37 °C before analyzing by flow cytometry. The experiment was performed in triplicates, and the standard error of the mean (SEM) is shown. (**B**) The percentage of PhyB-mCherry-SpyTag positive cells and (**C**) the ratio of the median fluorescence intensity (MFI) of BFP to the MFI of mCherry are shown

2. After completing the incubation time in **step 4** of Subheading 3.1, illuminate the opto-APCs with far-red light to switch all PhyB molecules into a state in which they cannot bind to PIF, the so-called 'OFF or $PhyB_{Pr}$ state'.
3. After medium removal (**step 5**, Subheading 3.1), add 200 μL per well of the PIF-expressing Jurkat T cell suspension.
4. For light stimulation, close the plate with an opaque lid and place it onto the opto-Plate-96 illuminator (for set up details, refer to Chap. 22 of this edition) (*see* **Note 9**). After starting the light program, place the set up into a 37 °C cell culture incubator with cooling function and with a 5% CO_2 humidified atmosphere (*see* **Note 10**).

   For the experiment described in this chapter, an illumination pattern of pulsed 10 s, 550 μmol m^{-2} s^{-1} red light or 10 s, 1200 μmol m^{-2} s^{-1} far-red light followed by 15 min darkness for a duration of 6 h was applied.

From now on, all steps are carried out at 4 °C if not otherwise specified.

5. After the desired duration of light stimulation, resuspend the co-cultured cells and transfer them to a 96V-bottom plate.
6. Add 180 μL FACS buffer to the remaining cells in the 96F-bottom black-walled-clear-bottom plate and incubate on ice while pelleting the cells in the 96V-bottom plate with 300× *g* for 4 min; discard supernatant.
7. Resuspend the remaining cells in the 96F-bottom black-walled-clear-bottom plate and transfer them to the corresponding wells of the 96V-bottom plate. Pellet them as described before and remove supernatant.
8. Before staining surface markers on the T cells, dilute the antibodies in FACS buffer (*see* **Note 11**). Resuspend cell pellet in 30 μL of antibody dilution and stain them for 30 min in the dark.
9. Wash cells twice in FACS buffer before analyzing in the flow cytometer (Fig. 3) (*see* **Note 12**).

4 Notes

1. The used medium and supplementation has to be adjusted to the cell types used for the experiments.
2. We use lentiviral transduction to introduce our construct of choice into the cell line. Any cell line can be used to express the SpyCatcher construct. However, the basal activity of T cells (the cells being stimulated) may be affected by other proteins

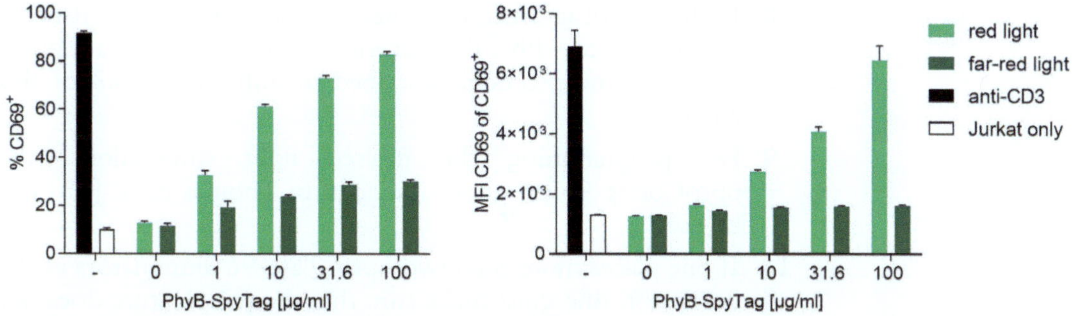

Fig. 3 Stimulation of SNAP-PIF-TCR-expressing Jurkat cells with opto-APCs. SNAP-PIF-TCR-expressing Jurkat cells were co-cultured with opto-APCs for 6 h while illuminating every 15 min with red or far-red light for 10 s. Jurkat cells not co-cultured with the opto-APCs were used as a negative control and stimulation with anti-CD3 antibodies were used as positive control. After the 6 h, cells were stained with a fluorescently labeled anti-CD69 antibody and analyzed by flow cytometry. Percentage of CD69-positive cells and the CD69 median fluorescent intensity (MFI) of the CD69-positive cells are displayed. The experiment was performed in triplicates, and data are shown with the standard error of the mean (SEM)

expressed on the SpyCatcher-expressing cells when co-cultured. This potential influence needs to be assessed for each specific experimental question.

3. For long time storage, the purified PhyB-mCherry-SpyTag protein is kept at −80 °C. Once thawed, it can be used for up to two weeks when stored at 4 °C. In our previous protocol [11], the production and purification of recombinant biotinylated PhyB is described, and the production and purification of the PhyB-mCherry-SpyTag protein is done in exactly the same manner. Note that PhyB-mCherry-SpyTag contains a His6 tag for purification, but for the sake of simplicity, we have not included it in the name.
4. Any cell type expressing PIF as part of the receptor or adhesion molecule on the surface can be used for the studies. Furthermore, newly developed tools such as the optogenetic receptor activation (opto-REACT) [16, 17] or the light-inducible T-cell engager (LiTE) [18] enable the stimulation of non-engineered T cells via the PhyB-PIF system.
5. This step is only necessary for adherend SpyCatcher-expressing cells. For detaching those cells PBS-EDTA needs to be used, and not trypsin, as this could diminish the SpyCatcher expression. Suspension cells can be seeded on the day of the stimulation.
6. This step should be executed in the afternoon, to make sure that the cells will not proliferate too much until the next day.
7. For our experiments, we used a concentration of 10 µg/mL of PhyB-mCherry-SpyTag and an incubation time of 30 min, before removing excess PhyB-mCherry-SpyTag.

8. If PhyB-SpyTag without mCherry is used for experiments, cells can be stained with a fluorescently labeled anti-His6 antibody, following the protocol described in Subheading 3.3 of this chapter.
9. For programming the desired light stimulations, the optoConfig-96 program described in Thomas et al. [15] can be used.
10. If one places more than two opto-Plate-96 illuminators in the incubator, one must make sure that the temperature does not rise too high and cool the incubator if necessary.
11. The antibody dilution can vary between different products. The right dilution has to be determined beforehand.
12. Due to the expression of blue fluorescent protein (BFP) by the opto-APCs, the APCs and T cells can be easily distinguished in the analysis.

Acknowledgment

We thank Kerstin Fehrenbach for excellent cell culture work; the technical workshop of the Faculty of Biology, University Freiburg, for the design and construction of illumination devices; and the BIOSS Signalling Factory for technical help. This work was supported by the German Research Foundation (DFG) under Germany's Excellence Strategy—EXC-2189—Project ID: 390939984 and under the Excellence Initiative of the German Federal and State Governments—EXC-294—and in part by the Ministry for Science, Research and Arts of the State of Baden-Württemberg. Further support was given by the DFG under FOR2799 (SCHA976/8-2 to WWS) and SFB1381 (Project ID: 403222702—A9 to WWS).

References

1. Fischer AAM, Kramer MM, Radziwill G, Weber W (2022) Shedding light on current trends in molecular optogenetics. Curr Opin Chem Biol 70:102196. https://doi.org/10.1016/j.cbpa.2022.102196
2. Toettcher JE, Weiner OD, Lim WA (2013) Using optogenetics to interrogate the dynamic control of signal transmission by the Ras/Erk module. Cell 155:1422–1434. https://doi.org/10.1016/j.cell.2013.11.004
3. Levskaya A, Weiner OD, Lim WA, Voigt CA (2009) Spatiotemporal control of cell signalling using a light-switchable protein interaction. Nature 461:997–1001. https://doi.org/10.1038/nature08446
4. Yousefi OS, Günther M, Hörner M et al (2019) Optogenetic control shows that kinetic proofreading regulates the activity of the T cell receptor. Elife 8:e42475. https://doi.org/10.7554/eLife.42475
5. Idstein V, Ehret AK, Yousefi OS, Schamel WW (2023) Engineering of an optogenetic T cell receptor compatible with fluorescence-based readouts. ACS Synth Biol 12:2857–2864. https://doi.org/10.1021/acssynbio.3c00429
6. Baaske J, Mühlhäuser WWD, Yousefi OS et al (2019) Optogenetic control of integrin-matrix interaction. Commun Biol 2:15. https://doi.org/10.1038/s42003-018-0264-7
7. Yousefi OS, Ruggieri M, Idstein V et al (2021) Cross-TCR antagonism revealed by

optogenetically tuning the half-life of the TCR ligand binding. Int J Mol Sci 22:4920. https://doi.org/10.3390/ijms22094920

8. Schamel WW, Alarcon B, Minguet S (2019) The TCR is an allosterically regulated macromolecular machinery changing its conformation while working. Immunol Rev 291:8–25. https://doi.org/10.1111/imr.12788
9. Russ M, Ehret AK, Hörner M et al (2023) Opto-APC: engineering of cells that display phytochrome B on their surface for optogenetic studies of cell-cell interactions. Front Mol Biosci 10:1143274. https://doi.org/10.3389/fmolb.2023.1143274
10. Keeble AH, Turkki P, Stokes S et al (2019) Approaching infinite affinity through engineering of peptide–protein interaction. Proc Natl Acad Sci 116:26523–26533. https://doi.org/10.1073/pnas.1909653116
11. Hörner M, Yousefi OS, Schamel W, Weber W (2020) Production, purification and characterization of recombinant biotinylated phytochrome B for extracellular optogenetics. Bio Protoc 10:e3541. https://doi.org/10.21769/BioProtoc.3541
12. Hörner M, Gerhardt K, Salavei P et al (2019) Production of phytochromes by high-cell-density E. coli fermentation. ACS Synth Biol 8:2442–2450. https://doi.org/10.1021/acssynbio.9b00267
13. Yousefi OS, Hörner M, Wess M et al (2020) Optogenetic tuning of ligand binding to the human T cell receptor using the opto-ligand-TCR system. Bio Protoc 10:e3540. https://doi.org/10.21769/BioProtoc.3540
14. Bugaj LJ, Lim WA (2019) High-throughput multicolor optogenetics in microwell plates. Nat Protoc 14:2205–2228. https://doi.org/10.1038/s41596-019-0178-y
15. Thomas OS, Hörner M, Weber W (2020) A graphical user interface to design high-throughput optogenetic experiments with the optoPlate-96. Nat Protoc 15:2785–2787. https://doi.org/10.1038/s41596-020-0349-x
16. Armbruster A, Ehret AK, Russ M et al (2024) OptoREACT: optogenetic receptor activation on nonengineered human T cells. ACS Synth Biol 13:752–762. https://doi.org/10.1021/acssynbio.3c00518
17. Armbruster A, Hörner M, Weber W (2025) Optogenetic clustering and stimulation of the T cell receptor in nongenetically modified human T cells. Optogenet Methods Mol Biol:149–162
18. Jaeger M, Anastasio A, Chamy L et al (2023) Light-inducible T cell engagers trigger, tune, and shape the activation of primary T cells. Proce Natl Acad Sci 120:e2302500120. https://doi.org/10.1073/pnas.2302500120

Chapter 22

Coupling Recombinant PhyB to Cell Culture Plates and Beads

Anna K. Ehret, Lisa Entringer, Anne Smedegaard Frederiksen, Vincent Idstein, Sara Hartmann, and Wolfgang W. Schamel

Abstract

Optogenetics enables precise and reversible control of receptor activation using light-responsive proteins. In one approach, the first 651 amino acids of the photoreceptor phytochrome B (PhyB) of *Arabidopsis thaliana* are used as a light-responsive ligand for engineered receptors. These receptors are fused to the PhyB-interacting factor 6 (PIF) and bind to PhyB under red, but not under far-red light. The recombinant biotinylated PhyB ligand can be conjugated either to streptavidin-coated plates or to beads, enabling PIF-fused receptor activation with high valency. This method provides high spatiotemporal precision, offering potential for studying receptor signaling in two-dimensional cell-free interactions. Here, we provide protocols to couple recombinant PhyB to plastic plates or polystyrene beads and validate the coupling using light-dependent stimulation of green fluorescent protein (GFP)-PIF-fused T-cell receptors (TCRs) in primary human T cells.

Key words Phytochrome B, Extracellular optogenetics, Primary T cells, Receptor activation

1 Introduction

Optogenetics has emerged as a powerful tool to control receptor activation with high spatiotemporal precision. By utilizing light-responsive proteins, this approach enables reversible, dose-dependent control of cellular receptors using cell-compatible visible and infrared light.

In the context of immunology, optogenetics has facilitated the activation of immune receptors, including T-cell receptors (TCR) [1, 2] and chimeric antigen receptors (CAR) [3–5] in T cells. One such optogenetic system leverages the red light-responsive interaction between the photoreceptor phytochrome B (PhyB) and the PhyB interacting factor (PIF) 6, derived from *Arabidopsis thaliana*. In this system, the first 100 amino acids of PIF6 are bound to the TCR via a single-chain variable fragment binding to the cluster of

Andreas Winkler (ed.), *Phytochromes: Methods and Protocols*, Methods in Molecular Biology, vol. 2970,
https://doi.org/10.1007/978-1-0716-4791-2_22,

differentiation (CD) 3 subunit of the TCR [6]. The light-responsive part of PhyB consists of the first 651 amino acids folded together with the chromophore phycocyanobilin, referred to as PhyB from now on. PhyB is purified as a recombinant biotinylated protein expressed in phycocyanobilin-producing bacteria, tetramerized with streptavidin, and thereby serves as the ligand to the PIF-bound TCR [6]. Under red light illumination, PhyB and PIF interact with each other [7], inducing receptor clustering and subsequent TCR activation. Indeed, TCR clustering is required for conformational changes at the TCR and TCR signaling [8, 9]. With far-red light, the interaction is terminated, stopping TCR signaling [1].

Here, we provide a detailed protocol of how to conjugate biotinylated PhyB to streptavidin-coated plates or polystyrene beads. Protocols on how to produce and purify recombinant biotinylated PhyB have been published [10–12]. We also provide a protocol to assess the functionality of the PhyB coated on a plastic plate or bead using primary human T cells as an example. Alternatively, recombinant PhyB can also be coupled to the surface of cells and thus serve as a membrane-bound ligand [6, 13]. A detailed protocol is available in Chap. 21.

In this context, the use of the PhyB-PIF system not only enhances our ability to study T-cell activation with fine spatiotemporal control but also holds potential for optimizing therapeutic strategies that require precise modulation of immune responses.

2 Materials

2.1 Coating Biotinylated Recombinant PhyB onto Streptavidin-Coated Surfaces and Beads

1. Streptavidin (1 mg/mL): Store stock at −20 °C.
2. 96F-bottom black-walled-clear-bottom plate for cell culture.
3. Biotinylated PhyB protein (*see* **Note 1**).
4. Tris(2-carboxyethyl)phosphine hydrochloride (TCEP; 0.5 M): Weight 1.43 g of TCEP into graduated cylinder and dissolve in 6 mL water. Adjust to pH 7.0 with NaOH. Make up to 10 mL with water. Aliquot into 500 μL and store stock solution at −20 °C.
5. Coating buffer: Phosphate-buffered saline (PBS) supplemented with 1.5 mM TCEP (*see* **Note 2**).
6. Optogenetic receptor activation (opto-REACT) protein (*see* **Note 1**).
7. Isolated human peripheral blood mononuclear cells (PBMCs) (*see* **Note 3**).
8. Biotin-free Roswell Park Memorial Institute (RPMI) medium: Preparation of the RPMI-1640 medium according to the manufacturer's protocol. Add 50 mL of biotin-free fetal bovine

serum (FBS) to 500 mL RPMI (*see* **Note 4**). Supplement medium with antibiotics and 100 mM 4-(2-hydroxyethyl)-1-piperazineethanesulfonic acid (HEPES).

9. Cell culture sterile bench.
10. Centrifuge: Suitable for 1.5-mL reaction tubes.
11. Opto-Plate-96 illuminator [14] and the software optoConfig-96 [15].
12. Cell culture incubator: With cooling function.
13. Spherotech streptavidin polystyrene beads: Store at 4 °C.

2.2 Staining of the Cells for Flow Cytometry

1. 96V-bottom plate.
2. Centrifuge: Suitable for 96-well plate centrifugation.
3. Sodium azide (10%): Weight 5 g of sodium azide into a 50-mL reaction tube, add up to 50 mL with water, and dissolve. Store stock solution at room temperature.
4. Fluorescence-activated cell sorting (FACS) buffer: PBS supplemented with 1% FBS and 0.02% sodium azide.
5. Activation marker antibodies: Various antibodies to stain activation markers on T cells.
6. Flow cytometer device and the software FlowJo for analysis.

3 Methods

This protocol describes the experimental procedures for coating plastic surfaces (*see* Subheading 3.1) and polystyrene beads (*see* Subheading 3.2) with biotinylated PhyB (Fig. 1), as well as the flow cytometry analysis of optogenetically stimulated T cells (*see* Subheading 3.3) as a proof of functional activity of the PhyB. Each subheading provides detailed steps to ensure reproducibility and accuracy in experimental execution.

3.1 Coating of Plastic Surfaces with Biotinylated PhyB

This subheading outlines the procedure for immobilizing biotinylated PhyB onto plastic surfaces using a streptavidin coating. Proper light conditions and sterile techniques are essential to maintain protein functionality and ensure experimental reliability. All steps in which PhyB is present have to be carried out under green light conditions to prevent that PhyB absorbs photons. To ensure a sterile work environment, the cells and reagents are handled in a cell culture sterile bench.

1. Coat wells of a 96F-bottom black-walled-clear-bottom plate with 10 μg/mL streptavidin in 50 μL PBS. Incubate at 37 °C in a 5% CO_2 humidified atmosphere for 4 h (*see* **Note 5**).

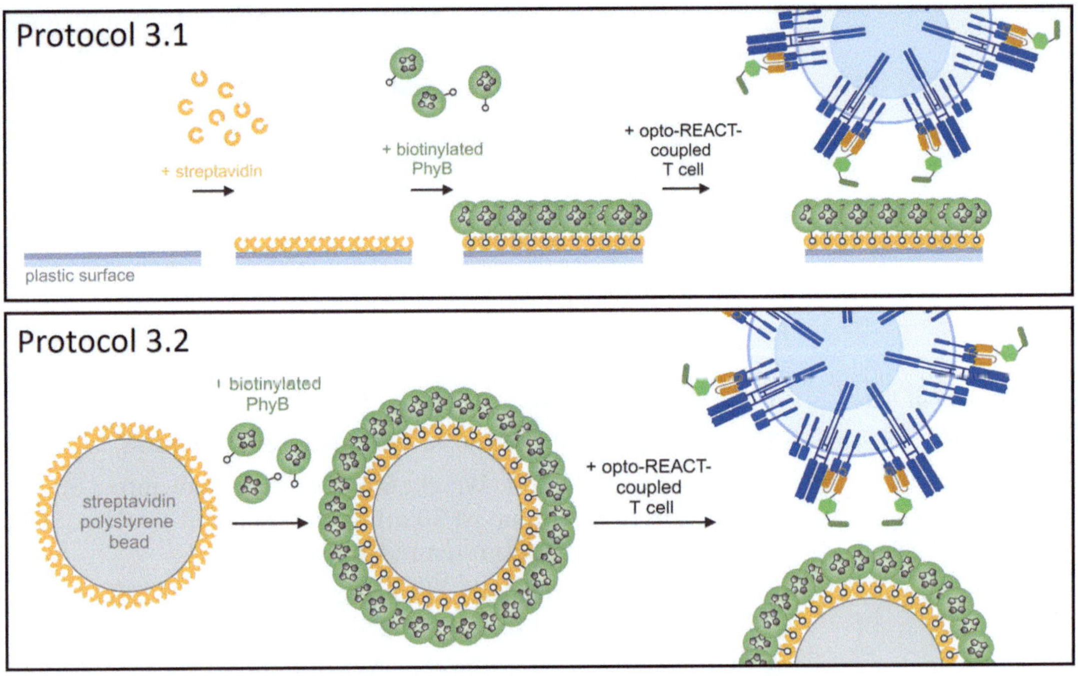

Fig. 1 Schematic depiction of steps described in this protocol. (Figure generated with BioRender®)

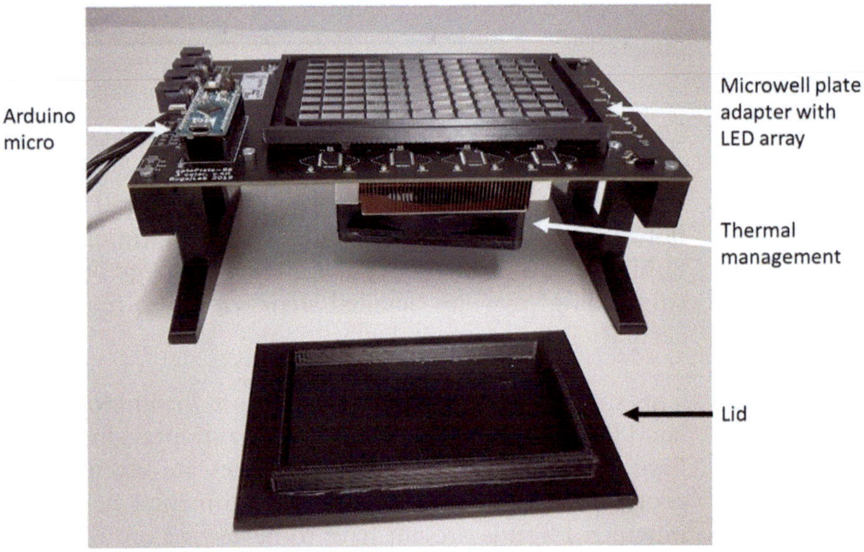

Fig. 2 The illumination device used. (Photograph of an optoPlate-96 illuminator)

2. Carefully remove the supernatant and coat the wells with 100 μg/mL recombinant biotinylated PhyB in 50 μL coating buffer. Incubate over night at 4 °C in darkness.

3. On the next day, the plate is illuminated with far-red light to switch all PhyB molecules into a state where they cannot bind to PIF6, the so-called 'OFF or $PhyB_{Pr}$ state'.
4. Carefully remove the supernatant and add 200 μL biotin-free RPMI medium containing 1×10^5 peripheral blood mononuclear cells (PBMCs, which include human T cells) bound to opto-REACT [13] per well (*see* **Note 6**).
5. For light stimulation, close the plate with an opaque lid and place it onto the opto-Plate-96 illuminator (Fig. 2) (*see* **Note 7**). After starting the light program, place the set up in a 37 °C cell culture incubator with 5% CO_2 humidified atmosphere (*see* **Note 8**).
6. For the experiment described in this protocol, an illumination pattern of pulsed 30 s, 550 μmol m^{-2} s^{-1} red light or 30 s, 1200 μmol m^{-2} s^{-1} far-red light followed by 14.5 min darkness for a duration of 24 h was applied.

3.2 Coating Polystyrene Streptavidin Beads with PhyB

This subheading describes the process of coating polystyrene streptavidin-coated beads with biotinylated PhyB. The PhyB-coated beads are then used for optogenetic stimulation of PBMCs. All steps in which PhyB is present should be carried out under green light. To ensure a sterile work environment, cells and reagents have to be handled in a cell culture sterile bench.

1. Take 1×10^5 polystyrene streptavidin-coated beads per well into a 1.5-mL reaction tube and dilute 1:1 with coating buffer. Centrifuge for 10 min at 15,000× *g* and 4 °C.
2. Remove the supernatant and wash once with the initial volume of the coating buffer.
3. Set the concentration to 1×10^6 beads in 50 μL coating buffer and add the desired concentration of biotinylated recombinant PhyB (*see* **Note 9**). Incubate for 1 h on ice in darkness.
4. In the meantime, prepare human PBMCs for stimulation by binding opto-REACT to the TCR (*see* **Note 6**). Therefore, resuspend PBMCs in biotin-free RPMI at a concentration of 1×10^6 PBMCs/mL and add 5 nM of opto-REACT (*see* **Note 6**). Incubate for 30 min at 37 °C in a 5% CO_2 humidified atmosphere. After the incubation, add 150 μL cell suspension per well to the 96F-bottom black-walled-clear-bottom plate.
5. Centrifuge beads for 10 min at 15,000× *g* and 4 °C and wash once with initial volume of coating buffer.
6. Resuspend beads at a concentration of 1.5×10^6 beads/mL in biotin-free RPMI.

7. Illuminate beads with far-red light to switch all PhyB molecules into a state in which they cannot bind to PIF6, the so-called 'OFF or $PhyB_{Pr}$' state.
8. Add 100 μL of bead suspension to each well.
9. For light stimulation, place the plate onto the opto-Plate-96 illuminator (Fig. 2) (*see* **Note 7**). After starting the light program, place the set up in a 37 °C cell culture incubator with 5% CO_2 humidified atmosphere (*see* **Note 8**).
10. For the experiment described in this protocol, an illumination pattern of pulsed 30 s, 550 μmol m^{-2} s^{-1} red light or 30 s, 1200 μmol m^{-2} s^{-1} far-red light followed by 14.5 min darkness for a duration of 72 h was applied.

3.3 Flow Cytometry Analysis of Optogenetically Stimulated T Cells

This subheading elucidates the steps required to prepare and analyze PBMCs after optogenetic stimulation using flow cytometry. Proper sample preparation and handling at low temperatures are essential for accurate results. All steps should be carried out at 4 °C if not otherwise specified.

1. After the desired duration of light stimulation, resuspend co-cultured PBMCs and transfer them to a 96V-bottom plate.
2. Pellet the PBMCs at 300× *g* for 4 min and wash with 180 μL FACS buffer twice.
3. Before staining the surface markers on the T cells, dilute the antibodies in FACS buffer (*see* **Note 10**).
4. Resuspend the cell pellet in 30 μL antibody solution and incubate the cells for 30 min in the dark.
5. Wash cells twice with 180 μL FACS buffer.
6. Analyze the cells by flow cytometry (Fig. 3).

4 Notes

1. For longtime storage, recombinant proteins are kept at −80 °C. Once thawed, they can be used for up to two weeks when stored at 4 °C. In our previous protocol [10], the production and purification of recombinant biotinylated PhyB is described. The production and purification of the opto-REACT are described in the following references [6, 13].
2. Prepare coating buffer always fresh, as TCEP will lose its reducing capability within 72 h when supplemented to PBS.
3. After isolation from human blood, PBMCs can be stored at −80 °C or in liquid nitrogen in freezing buffer (10% dimethyl sulfoxide, 90% FBS).

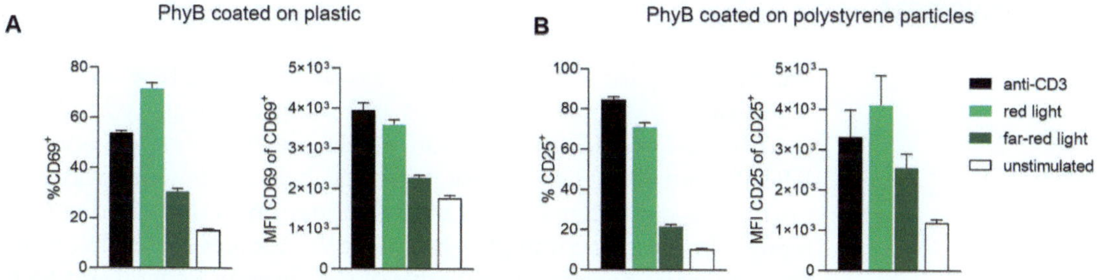

Fig. 3 Optogenetic activation of primary human T cells with PhyB-coated surfaces. (**A**) Opto-REACT-bound PBMCs were cultured in a streptavidin-PhyB-coated 96-well plate for 24 h while illuminating every 14.5 min with red or far-red light for 30 s. Non-opto-REACT-bound PBMCs were used as a negative control, and stimulation with anti-CD3 antibodies was used as positive control. Subsequently, cells were stained with a fluorescently labeled anti-CD69 antibody and analyzed by flow cytometry. Percentage of CD69-positive cells and the CD69 medain fluorescent intensity (MFI) of the CD69-positive cells are displayed. The experiment was performed with two healthy donors each in triplicates, and data are shown including the standard error of the mean (SEM). (**B**) Opto-REACT-bound PBMCs were co-cultured with PhyB-coated beads for 72 h while illuminating every 14.5 min with red or far-red light for 30 s. Controls used as described above. Cells were stained with a fluorescently labeled anti-CD25 antibody and analyzed by flow cytometry. Percentage of CD25-positive cells and the CD25 MFI of the CD25-positive cells are displayed. One exemplary experiment with one healthy donor performed in duplicate is shown, and data are shown including SEM

4. Biotin present in FBS may disrupt the binding of biotinylated PhyB to streptavidin. To prevent this, FBS can be pretreated with streptavidin beads to capture the free biotin, followed by bead removal through centrifugation and sterile filtration of the FBS.
5. Binding of streptavidin to plastic surfaces can also be performed over night at 4 °C. Moreover, streptavidin-coated cell culture plates are also commercially available.
6. Alternatively, Jurkat T cells or any other cells of your choice with a PIF-coupled receptor can be used. A detailed protocol on how to purify and bind opto-REACT to T cells can be found in Armbruster et al. [13].
7. For programming the desired light stimulations, the optoConfig-96 program described in Thomas et al. [15] can be used.
8. If more than two opto-Plate-96 illuminators are placed in the incubator, it is essential to ensure that the temperature does not rise too high and cool the incubator if necessary.
9. For our experiments, we used a concentration of 60 μg/mL of PhyB, but this has to be determined individually for each experimental setup.
10. The antibody dilution can vary between different products. The right dilution has to be determined beforehand.

Acknowledgment

We thank Kerstin Fehrenbach for excellent cell culture work, the technical workshop of the Faculty of Biology, University Freiburg, for the design and construction of illumination devices, and the BIOSS Signalling Factory for technical help. This work was supported by the German Research Foundation (DFG) under Germany's Excellence Strategy—EXC-2189—Project ID: 390939984 and under the Excellence Initiative of the German Federal and State Governments—EXC-294—and in part by the Ministry for Science, Research and Arts of the State of Baden-Württemberg. Further support was given by the DFG under FOR2799 (SCHA976/8-2 to WWS) and SFB1381 (Project ID: 403222702—A9 to WWS).

References

1. Yousefi OS, Günther M, Hörner M et al (2019) Optogenetic control shows that kinetic proofreading regulates the activity of the T cell receptor. Elife 8:e42475. https://doi.org/10.7554/eLife.42475
2. Yousefi OS, Ruggieri M, Idstein V et al (2021) Cross-TCR antagonism revealed by optogenetically tuning the half-life of the TCR ligand binding. Int J Mol Sci 22:4920. https://doi.org/10.3390/ijms22094920
3. Tischer DK, Weiner OD (2019) Light-based tuning of ligand half-life supports kinetic proofreading model of T cell signaling. Elife 8:e42498. https://doi.org/10.7554/eLife.42498
4. O'Donoghue GP, Bugaj LJ, Anderson W et al (2021) T cells selectively filter oscillatory signals on the minutes timescale. Proc Natl Acad Sci 118:e2019285118. https://doi.org/10.1073/pnas.2019285118
5. Harris MJ, Fuyal M, James JR (2021) Quantifying persistence in the T-cell signaling network using an optically controllable antigen receptor. Mol Syst Biol 17:e10091. https://doi.org/10.15252/msb.202010091
6. Armbruster A, Ehret AK, Russ M et al (2024) OptoREACT: optogenetic receptor activation on nonengineered human T cells. ACS Synth Biol 13:752–762. https://doi.org/10.1021/acssynbio.3c00518
7. Levskaya A, Weiner OD, Lim WA, Voigt CA (2009) Spatiotemporal control of cell signalling using a light-switchable protein interaction. Nature 461:997–1001. https://doi.org/10.1038/nature08446
8. Alarcon B, Schamel WW (2025) Allosteric changes underlie the outside-in transmission of activatory signals in the TCR. Immunol Rev 329:e13438. https://doi.org/10.1111/imr.13438
9. Minguet S, Swamy M, Alarcón B et al (2007) Full activation of the T cell receptor requires both clustering and conformational changes at CD3. Immunity 26:43–54. https://doi.org/10.1016/j.immuni.2006.10.019
10. Hörner M, Yousefi OS, Schamel W, Weber W (2020) Production, purification and characterization of recombinant biotinylated phytochrome B for extracellular optogenetics. Bio Protoc 10:e3541. https://doi.org/10.21769/BioProtoc.3541
11. Hörner M, Gerhardt K, Salavei P et al (2019) Production of phytochromes by high-cell-density E. coli fermentation. ACS Synth Biol 8:2442–2450. https://doi.org/10.1021/acssynbio.9b00267
12. Yousefi OS, Hörner M, Wess M et al (2020) Optogenetic tuning of ligand binding to the human T cell receptor using the opto-ligand-TCR system. Bio Protoc 10:e3540. https://doi.org/10.21769/BioProtoc.3540
13. Armbruster A, Hörner M, Weber W (2025) Optogenetic clustering and stimulation of the T cell receptor in nongenetically modified human T cells. Optogenet Methods Mol Biol:149–162
14. Bugaj LJ, Lim WA (2019) High-throughput multicolor optogenetics in microwell plates. Nat Protoc 14:2205–2228. https://doi.org/10.1038/s41596-019-0178-y
15. Thomas OS, Hörner M, Weber W (2020) A graphical user interface to design high-throughput optogenetic experiments with the optoPlate-96. Nat Protoc 15:2785–2787. https://doi.org/10.1038/s41596-020-0349-x

INDEX

Andreas Winkler (ed.), *Phytochromes: Methods and Protocols*, Methods in Molecular Biology, vol. 2970, https://doi.org/10.1007/978-1-0716-4791-2,

P

Q

R

S

T

V

X

If you have any concerns about our products,

Springer Nature Customer Service Center GmbH
Europaplatz 3, 69115 Heidelberg, Germany

in Hamburg, Germany